Ion Channels and Neurological Disease

Ion Channels and Neurological Disease

Editor

Carlo Musio

Basel • Beijing • Wuhan • Barcelona • Belgrade • Novi Sad • Cluj • Manchester

Editor
Carlo Musio
Institute of Biophysics (IBF)
National Research Council (CNR)
Trento
Italy

Editorial Office
MDPI AG
Grosspeteranlage 5
4052 Basel, Switzerland

This is a reprint of articles from the Special Issue published online in the open access journal *Life* (ISSN 2075-1729) (available at: https://www.mdpi.com/journal/life/special_issues/38L6X14SR5).

For citation purposes, cite each article independently as indicated on the article page online and as indicated below:

Lastname, A.A.; Lastname, B.B. Article Title. *Journal Name* **Year**, *Volume Number*, Page Range.

ISBN 978-3-7258-1929-4 (Hbk)
ISBN 978-3-7258-1930-0 (PDF)
doi.org/10.3390/books978-3-7258-1930-0

Contents

About the Editor

Carlo Musio

Carlo Musio holds a CNR Research Scientist (PI) permanent position at the Institute of Biophysics (IBF), Trento Unit, of the National Research Council CNR), Italy. He received his laureate degree in Biological Sciences (neurophysiological path) from the Dept. of Animal and Human Behavioral Sciences (now Dept. of Biology) at the University of Pisa, working on the neuronal basis of *Aplysia* behavior. Having been awarded a CNR Research Fellowship in Biophysics (equating a PhD), in 1989, he moved to the CNR Institute of Cybernetics "Eduardo Caianiello" (ICIB-CNR, now ISASI), Pozzuoli (NA), to carry out research on *Hydra* photoreception. From 1994 to 2011, at ICIB-CNR, he covered tenured positions until receiving a permanent one. At ICIB-CNR, he also covered positions of responsibility, e.g., PI, Head of "Commessa". In 2011, he moved to the CNR Institute of Biophysics, Trento (IBF-CNR), to establish and lead the "Neurosystems and photosensory Biophysics" lab, a research lab of the electrophysiology of natural and hybrid neurosystems in physiological and pathological conditions. From 2012 to 2016, he was President of the SIBPA—Italian Society of Pure and Applied Biophysics. Today, he still covers the following positions: Associate Editor of *Photochemistry and Photobiology* (Wiley-Blackwell), the official journal of the American Society for Photobiology (since 2011); *Biomolecular Concepts* (De Gruyter) (since 2023); and Appointed Member of the Scientific Advisory Board (SAB) of the European Society for Photobiology (ESP) (since 2010). In 2019, he was elected as the National Representative of the Research Staff in the Scientific Council of the CNR Dept. of Physical Sciences and Technologies of Matter (DSFTM) (term ended in 2023). He is a member of the Commission of Experts for Bilateral Agreements, CNR and the Commission of Experts for Joint International Laboratory Projects, CNR. He serves as a grant project reviewer for the following: NSF, USA; Villum Foundation, DK; Conacyt, MX; ARSLA, FR; MIUR-PRIN, Italian and foreign universities.

life

Editorial

Ion Channels and Neurological Disease

Carlo Musio

Institute of Biophysics—IBF, National Research Council—CNR, Via Sommarive 18, 38123 Trento, Italy; carlo.musio@cnr.it

1. Introduction

Ion channels are key elements in the control of membrane physiology and neurotransmission because ionic fluxes assure neuronal signal propagation across and between neurons through synaptic transmission [1,2]. The pathophysiology of ion channels may originate from either mutations of gene encoding components of the channel structure (channelopathy) or secondary dysfunctions; both conditions affect the intrinsic excitability of the cell and synaptic functions, leading to the pathophysiological signs of disease [3]. Accordingly, ion channels are considered suitable drug targets within modern pharmacology [4].

Neurological disorders represent pathological conditions that directly affect the nervous system or brain functions that produce single or composed neurodevelopmental, motor, sensory, or cognitive organic impairments whose etiology can be genotypical or idiopathic [5]. Most currently known neurological diseases, mainly neurodegenerative diseases (NDDs), exhibit alteration of neuronal excitability due to the dysfunction of the molecular and/or functional features of ion channels. In the majority of NDDs, the pathogenic role of ion channels has been widely demonstrated either for channelopathies or secondary dysfunction [6]. Nevertheless, the link between ion channel alterations that induce neuronal excitability and the onset of disease has been neglected in some disorders, while for others, this area of study is growing rapidly.

The aim of this Special Issue is to provide updates on the state of the art and new achievements in research on altered structure–function relationships in the ion channels that affect the pathophysiology of neurological disease. We are particularly interested in methods of drug screening and targeting that will allow for the development of novel therapeutic avenues for treating and alleviating these mostly incurable diseases. Multi- and inter-disciplinary research contributions—often combining structural, functional, and/or pharmacological approaches with different methods/techniques—are presented.

Citation: Musio, C. Ion Channels and Neurological Disease. *Life* **2024**, *14*, 758. https://doi.org/10.3390/life14060758

Received: 29 May 2024
Accepted: 12 June 2024
Published: 13 June 2024

2. The Special Issue

This Special Issue, entitled "Ion Channels and Neurological Disease" [7], belongs to *Life*'s "Pharmaceutical Science" section [8] and is a collection of ten peer-reviewed articles (seven reviews and three original articles) covering dysfunctions of the main typologies (sodium, potassium, calcium, chloride) of neuronal ion channels. Furthermore, new data on GABA receptors, nucleoporin, and the ion channels of neuroglia are reported. The main related neurological disorders include Alzheimer's disease, glioblastoma, Parkinson's disease, cerebellar ataxias. The Special Issue's contributions are grouped below ordered by functional classification, while they are listed at the end of the editorial along with the web site order.

2.1. Sodium (Na^+) Channels

A review on the impairments of voltage-gated sodium channels in neurological disorders opens the issue. Starting from the channels' conserved genes, *SCN1A*, etc. to the encoded proteins $Na_V1.1$, etc., a comprehensive survey outlines the structure, the function and the role of the Na_V channel mutations, emphasizing pharmacological therapeutic approaches in several disorders, like migraine, autism, and Alzheimer's (Contribution 9).

Accordingly, a review on the same channel types focuses their role in the pathophysiology of Alzheimer's disease. In particular, their role in mediating and tuning functional and altered neuronal excitability and the attenuation of hippocampal hyperactivity (which seems to ameliorate cognitive deficits) are considered promising elements to be addressed by potential therapeutic interventions (Contribution 6).

2.2. *Potassium (K^+) Channels*

Contribution 7 synthesizes all the recent literature from the Shakkottai lab, emphasizing the role of ion channels (and potassium ones in particular) in the dysregulation of neuronal excitability in several spinocerebellar ataxias (SCAs), a specific subset of polyglutaminic neurodegenerative diseases. Therefore, through the targeting of those channels, treatments that restore the altered intrinsic membrane excitability of cerebellar Purkinje neurons emerge as an important pharmacological and neurotherapeutic avenue.

2.3. *Calcium (Ca^{2+}) Channels*

Another original article and a review address the role that dysregulation of neuronal calcium channels plays in the pathogenesis of Alzheimer's disease (AD).

Said original research article (Contribution 1) reports the functional characterization of calcium and glutamate phenotypes involved in the pathogenesis of *PSEN1*-mutated familial Alzheimer's disease (FAD). They measured Ca^{2+} response dynamics in induced pluripotent stem cell (iPSC)-derived neurons carrying *PSEN1* mutations to glutamate, NMDA, AMPA and kainate, showing that alterations in Ca^{2+} and glutamate signaling can be considered an early functional FAD phenotype.

The review (Contribution 10) discusses a new challenging pathogenic element of AD which could reside in the functional feedback between Ca2+ signaling and lysosomal/autophagic dysfunctionality. Targeting and fine-tuning this functional link with drugs could represent a novel route to countering AD and (possibly) NDDs.

2.4. *Chloride (Cl^-) Channels*

Contribution 8 outlines the biophysical properties and the functional role of five types of transporters belonging to the voltage-gated ChLoride channels (CLC) family and located in the membranes of endosomes and lysosomes, which are intracellular organelles regulating the homeostatic and autophagic processes of the cell. These channels are crucial for anion/proton exchange and pH regulation: their mutations have been identified within the pathogenesis of several diseases, including neurodegenerative and neurodevelopmental disorders.

2.5. *GABA Receptors and Currents*

Two more contributions deal with the activation of GABA receptors and the biophysical properties of GABA currents.

The former (Contribution 3) group carried out an intriguing form of neural tissue preparation (i.e., slices of turtle spinal cord), exploring the activation of $GABA_B$ receptors in the terminals and axons of dorsolateral funiculus (DLF) eliciting functionally antagonist post-synaptic potential in motor neurons. The paper reports indications that $GABA_B$ receptors are activated by environmental GABA, the concentration of which is regulated by their release from interneurons and astrocytes.

The latter (Contribution 4) focuses on the biophysical alterations in $GABA_A$ reversal potential, showing them to be crucial features in several perturbed conditions, as a consequence of disequilibrium in the NKCC1-KCC2 (chloride–cation transporters) activity ratio. As $GABA_A$ current alteration has been identified in several neurological diseases, the paper reports imbalance in chloride–cation transporters to be a therapeutic target.

2.6. *Neuroglia Channels and Channel-Modulating Nucleoporins*

Two articles conclude this roundup on ion channels and neurological disease.

Contribution 5 provides a very thorough and up-to-date review of the functional roles of ion channels and ionotropic receptors in functional and pathological astrocytes, focusing mainly Alzheimer's disease (AD) and glioblastoma brain tumor (GBM). The authors underline that better collective understanding of ion channels and ionotropic receptors in astrocytes, which are involved in the above diseases, is required to develop novel therapeutic interventions and new strategies for treating brain disorders.

In Contribution 2, the tuning of neuronal excitability by nucleoporin NUP358 in mouse primary cortical neurons is reported; this regulating activity is exerted via a voltage-gated sodium channel. The authors stress the role of altered neuronal excitability as a potential key player in the pathogenesis of neurological disorders and present a window into ion channel pathophysiology in neurodegenerative diseases.

3. Conclusions

This Special Issue, "Ion Channels and Neurological Disease", presents several intriguing and up-to-date aspects of ion channels' role in the pathogenesis—and/or the pathological phenotype—of neurological disorders. Nowadays, there is a general consensus on the role exerted by ion channels in determining alterations in neuronal excitability. Thus, the altered structure/function relationship of ion channels can be considered a common feature of several neurological disorders; we may use this common feature to develop new therapeutic tools and avenues.

Finally, this Special Issue has received good attention and visibility in terms of total and single article views. Accordingly, a second edition of the Special Issue and a printed book of the first edition are in preparation. The second edition, co-edited by Carlo Musio (IBF-CNR Trento, Italy) and Marzia Martina (NRC Ottawa, Canada), is already open and in progress; further information and details are available on the official website [9].

Acknowledgments: My deep gratitude goes to all authors for their valuable contributions and all reviewers for their fair and constructive comments. I am very grateful to Michael Whalen (IBF-CNR, Trento, I) for accurate revision of the English and useful tips on the final manuscript. I wish to dedicate this effort to the beloved memory of my father, who perhaps knows why.

Conflicts of Interest: The author declares no conflict of interest.

List of Contributions

1. Targa Dias Anastacio, H.; Matosin, N.; Ooi, L. Familial Alzheimer's Disease Neurons Bearing Mutations in PSEN1 Display Increased Calcium Responses to AMPA as an Early Calcium Dysregulation Phenotype. *Life* **2024**, *14*, 625. https://doi.org/10.3390/life14050625.
2. Martínez-Rojas, V.A.; Pischedda, F.; Romero-Maldonado, I.; Khalaf, B.; Piccoli, G.; Macchi, P.; Musio, C. Nucleoporin Nup358 Downregulation Tunes the Neuronal Excitability in Mouse Cortical Neurons. *Life* **2023**, *13*, 1791. https://doi.org/10.3390/life13091791.
3. Delgado-Ramírez, X.; Alvarado-Cervantes, N.S.; Jiménez-Barrios, N.; Raya-Tafolla, G.; Felix, R.; Martínez-Rojas, V.A.; Delgado-Lezama, R. $GABA_B$ Receptors Tonically Inhibit Motoneurons and Neurotransmitter Release from Descending and Primary Afferent Fibers. *Life* **2023**, *13*, 1776. https://doi.org/10.3390/life13081776.
4. Kim, H.R.; Martina, M. Bidirectional Regulation of $GABA_A$ Reversal Potential in the Adult Brain: Physiological and Pathological Implications. *Life* **2024**, *14*, 143. https://doi.org/10.3390/life14010143.
5. Lia, A.; Di Spiezio, A.; Vitalini, L.; Tore, M.; Puja, G.; Losi, G. Ion Channels and Ionotropic Receptors in Astrocytes: Physiological Functions and Alterations in Alzheimer's Disease and Glioblastoma. *Life* **2023**, *13*, 2038. https://doi.org/10.3390/life13102038.
6. Baumgartner, T.J.; Haghighijoo, Z.; Goode, N.A.; Dvorak, N.M.; Arman, P.; Laezza, F. Voltage-Gated Na^+ Channels in Alzheimer's Disease: Physiological Roles and Therapeutic Potential. *Life* **2023**, *13*, 1655. https://doi.org/10.3390/life13081655.
7. Huang, H.; Shakkottai, V.G. Targeting Ion Channels and Purkinje Neuron Intrinsic Membrane Excitability as a Therapeutic Strategy for Cerebellar Ataxia. *Life* **2023**, *13*, 1350. https://doi.org/10.3390/life13061350.

8. Coppola, M.A.; Tettey-Matey, A.; Imbrici, P.; Gavazzo, P.; Liantonio, A.; Pusch, M. Biophysical Aspects of Neurodegenerative and Neurodevelopmental Disorders Involving Endo-/Lysosomal CLC Cl^-/H^+ Antiporters. *Life* **2023**, *13*, 1317. https://doi.org/10.3390/life13061317.
9. Barbieri, R.; Nizzari, M.; Zanardi, I.; Pusch, M.; Gavazzo, P. Voltage-Gated Sodium Channel Dysfunctions in Neurological Disorders. *Life* **2023**, *13*, 1191. https://doi.org/10.3390/life13051191.
10. Zhang, H.; Bezprozvanny, I. "Dirty Dancing" of Calcium and Autophagy in Alzheimer's Disease. *Life* **2023**, *13*, 1187. https://doi.org/10.3390/life13051187.

References

1. Hille, B. *Ion Channels of Excitable Membranes*, 3rd ed.; Sinauer Associates: Sunderland, MA, USA, 2001; p. 814.
2. Zheng, J.; Trudeau, M. (Eds.) *Handbook of Ion Channels*; CRC Press: Boca Raton, FL, USA, 2015; 691p.
3. Aschcroft, F. *Ion Channel and Disease*; Academic Press: Cambridge, MA, USA, 2000; p. 481.
4. Alexander, S.P.H.; Mathie, A.A.; Peters, J.A.; Veale, E.L.; Striessnig, J.; Kelly, E.; Armstrong, J.F.; Faccenda, E.; Harding, S.D.; Davies, J.A.; et al. The Concise Guide to Pharmacology 2023/24: Ion channels. *Br. J. Pharmacol.* **2023**, *180*, S145. [CrossRef] [PubMed]
5. Zigmond, M.J.; Rowland, P.L.; Coyle, J.T. *Neurobiology of Brain Disorders*, 2nd ed.; Elsevier: Amsterdam, NL, USA, 2022; p. 1134.
6. Pitt, G.S. *Ion Channels in Health and Disease*; Academic Press: Cambridge, MA, USA, 2016; p. 392.
7. Available online: https://www.mdpi.com/si/137860 (accessed on 11 June 2024).
8. Available online: https://www.mdpi.com/journal/life/sections/life-Pharma (accessed on 11 June 2024).
9. Available online: https://www.mdpi.com/si/203859 (accessed on 11 June 2024).

life

MDPI

Review

Voltage-Gated Sodium Channel Dysfunctions in Neurological Disorders

Raffaella Barbieri, Mario Nizzari, Ilaria Zanardi, Michael Pusch and Paola Gavazzo *

Institute of Biophysics, Via de Marini 6, 16149 Genova, Italy; raffaella.barbieri@ibf.cnr.it (R.B.); mario.nizzari@ibf.cnr.it (M.N.); ilaria.zanardi@ibf.cnr.it (I.Z.); michael.pusch@ibf.cnr.it (M.P.)
* Correspondence: paola.gavazzo@ibf.cnr.it

Abstract: The pore-forming subunits (α subunits) of voltage-gated sodium channels (VGSC) are encoded in humans by a family of nine highly conserved genes. Among them, *SCN1A*, *SCN2A*, *SCN3A*, and *SCN8A* are primarily expressed in the central nervous system. The encoded proteins Nav1.1, Nav1.2, Nav1.3, and Nav1.6, respectively, are important players in the initiation and propagation of action potentials and in turn of the neural network activity. In the context of neurological diseases, mutations in the genes encoding Nav1.1, 1.2, 1.3 and 1.6 are responsible for many forms of genetic epilepsy and for Nav1.1 also of hemiplegic migraine. Several pharmacological therapeutic approaches targeting these channels are used or are under study. Mutations of genes encoding VGSCs are also involved in autism and in different types of even severe intellectual disability (ID). It is conceivable that in these conditions their dysfunction could indirectly cause a certain level of neurodegenerative processes; however, so far, these mechanisms have not been deeply investigated. Conversely, VGSCs seem to have a modulatory role in the most common neurodegenerative diseases such as Alzheimer's, where SCN8A expression has been shown to be negatively correlated with disease severity.

Keywords: Nav channel blockers; epilepsy; FHM3; intellectual disability; neurodegeneration; Alzheimer's disease; Parkinson's disease; amyotrophic lateral sclerosis

Citation: Barbieri, R.; Nizzari, M.; Zanardi, I.; Pusch, M.; Gavazzo, P. Voltage-Gated Sodium Channel Dysfunctions in Neurological Disorders. *Life* **2023**, *13*, 1191. https://doi.org/10.3390/life13051191

Academic Editor: Carlo Musio

Received: 26 April 2023
Revised: 12 May 2023
Accepted: 14 May 2023
Published: 16 May 2023

1. Introduction

One in six people worldwide present a neurological condition. Age remains the biggest risk factor for developing a neural disease; thus this frequency is going to increase since the life spans in many countries continue to extend.

The topic of this review concerns the involvement of Nav channel isoforms in some inherited, genetic or non-genetic neurological or neurodegenerative pathologies affecting the Central Nervous System (CNS). A description of the molecular mechanisms underlying some common neural diseases is outlined and contextualized in light of impairments of the Nav channels.

The research activity on neural diseases, such as it is currently intended, began several decades ago, and for this reason, a large amount of information has been accumulated. However, over time widely divergent data have been published on specific points, probably due to the level of complexity of the topic. Throughout the paper, we will focus on some of these puzzles trying to provide the univocal and shared explanation that finally emerged from the scientific community itself.

VGSC Molecular Architecture, Function and Expression Patterns

The family of voltage-gated sodium channels (VGSC) includes nine isoforms (Nav1.1–Nav1.9) encoded in humans from nine different genes (*SCNA1-5*, *SCNA7-10*). They are composed of the principal alpha-subunit, which is essential and sufficient for channel functioning and can be eventually associated with the accessories ß-subunits (ß1–ß4) which regulate gating, kinetics and channel surface density.

The nine subtypes of the main α-subunit share a high degree of homology with more than 50% conserved sequences in the transmembrane and extracellular domain [1]. α-subunits are about 2000 amino acid residues long, have a molecular weight of 260 kDa and are organized in one long polypeptide chain folded in four linked internally repeated homologous domains (DI-IV), each containing six transmembrane segments (S1–S6), a pore region formed by the loop between S5 and S6 helices and a voltage sensor located at the level of the S4 segment containing several positively charged residues (see Figure 1).

Accessories ß subunits are 33–36 kDa transmembrane proteins harboring a single transmembrane region. They are associated with the main subunit by non-covalent interactions or disulfide bonds [2] and interact with cell adhesion molecules or intracellular matrix proteins such as ankyrin [3,4] thus contributing to cell migration and adhesion [5].

Among the nine VGSC human isoforms, Nav1.1, 1.2, 1.3 and 1.6 are the primary Nav channels expressed in the CNS, whereas Nav1.7, 1.8 and 1.9 are mostly restricted to the peripheral nervous system (PNS); Nav1.4 is expressed in skeletal muscles and Nav1.5 in the cardiac muscle [1,6]. Among the prevalent channels in central neurons, Nav1.1 and Nav1.3, encoded by the *SCNA1* and *SCNA3* genes respectively, are preferentially expressed in the cell body [7] while Nav1.2 is more concentrated in unmyelinated axons and dendrites [7] and Nav1.6 in myelinated axons and dendrites [8].

Besides the unique distribution along the different neuronal compartments, CNS Nav also exhibits a well-defined expression pattern in a variety of cortical neuron subtypes with specific roles in excitability. Even if broadly detectable in CNS, Nav1.1 is predominant in inhibitory Gamma Aminobutyric Acid-ergic (GABAergic) interneurons [9,10], while Nav1.2 prevails in excitatory neurons [11] and Glutamatergic neurons [12]. Similarly, Nav1.6 is expressed in neocortical excitatory neurons [13]. Nav1.3 is expressed in CNS only during embryonic and neonatal life, while during infancy its expression declines and is progressively replaced by other isoforms, especially Nav1.1 and 1.2 [14].

Voltage-gated sodium channels are determinants for the initiation, propagation and regulation of action potentials in neuronal circuits. Their conductance is strictly influenced by the transmembrane potential to the change of which they respond extremely quickly in the order of a few milliseconds. At resting potential, in the closed non-conductive state, their ion flux is prevented by the intracellular activation gate, located where the four S6 helices meet. Conversely, when transmembrane potential shifts to less negative values, an outward dislocation of the S4 helix occurs and, being this domain physically coupled to the pore region through the S4–S5 linkers, triggers pore opening and initiation of ion flux [15] (see Figure 1). S4 movement also initiates the process of fast inactivation, provoking the dislocation of the intracellular loop linking DIII and DIV, resulting in the occlusion of the pore. Nav channels also undergo another type of inactivation, named slow inactivation, provoked by prolonged or repetitive stimulation and connected with a movement of S6 in DIII with S1 in the DIV domain which induces pore collapse [16].

The channel activity is enhanced and a gain of function (GoF) phenotype is observed when, due to a mutation in the gene sequence or to the administration of a compound:

- the voltage of current activation shifts toward more negative values;
- inactivation shifts toward more positive values;
- the current persists longer;
- the recovery from inactivation is faster;
- the current density, i.e., the number of functioning channels expressed per membrane unit area increases

On the contrary, a loss of function phenotype (LoF) develops when current activation occurs at less negative voltages and inactivation at less positive, or when the current density decreases or recovery from inactivation is slower.

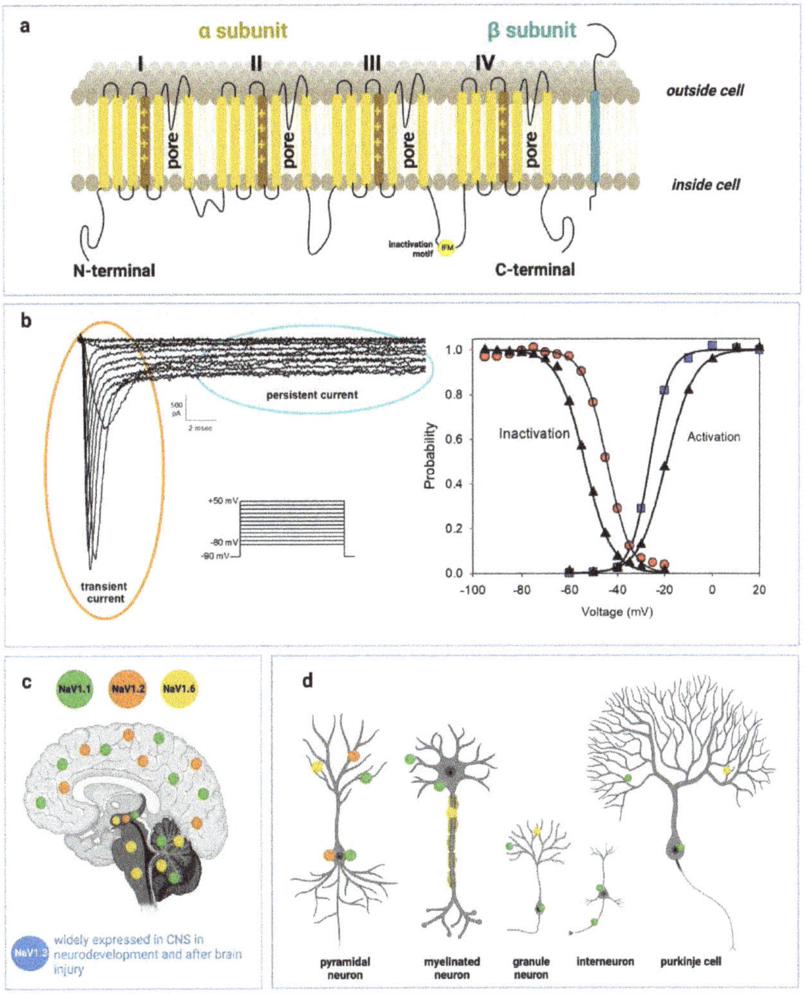

Figure 1. Voltage-gated sodium channel structure, function and distribution in CNS: (**a**) Schematic representations of α-subunit and auxiliary β-subunit of Nav channels. The α-subunit of the channel consists of four homologous domains (DI–DIV) made of six transmembrane helices (S1–S6). The voltage-sensor is localized in the fourth helix (S4) in each domain. The loops between S5 and S6 in each domain form the pore. (**b**) Left side: representative current traces recorded from a HEK293 cell transiently transfected with a cDNA encoding Q1489H/Nav1.1 mutant channel. Current traces were evoked from a series of 20 ms depolarizing pulses from −80 to +50 mV in 10 mV increments, starting from a holding potential of −90 mV. Q1489H mutation exhibits a large persistent current. Right side: curves of voltage dependence of steady-state activation (blu squares) or inactivation (red circles) are represented. The same curves are represented also for a *WT-SCN1A* transfected cell (black triangles) not reported in the figure. Lines stand for the Boltzmann function fits (**c**) Distribution of Nav1.1, Nav1.2, Nav1.3, and Nav1.6 in human brain regions. (**d**) Cellular and subcellular distribution of Nav1.1, Nav1.2, Nav1.3, and Nav1.6 in human brain.

Since the studies on the squid giant axon in the fifties [17] VGSCs have historically been among the first ion channels to have been hypothesized and then identified. However, a number of decades before, at the beginning of the twentieth century, the history of chemistry-driven ion channel drug discovery had its beginning with the identification of VGSC modulators such as lidocaine used as local anesthetics or anticonvulsants such as phenytoin [18] which are still in clinical use. Over time, VGSCs have been discovered to be targeted by many natural toxins and therapeutic drugs. This pharmacological interest has positively stimulated the search for structure determination at the atomic level of Nav channels or portions of them. This has been fundamental in making therapeutic drugs readily available to challenge the multiple diseases in which these channels are involved. Currently a well-documented classification of seven different drug toxin binding sites (site 1 to 7) and a local anesthetic binding site have been defined [19,20]. The drugs in use for the treatment of neurological and neurodegenerative diseases in which VGSCs are differently implicated will be described in the next sections of this review.

2. Methods

The information reported in this review was collected by interrogating Public Databases, mainly PUBMED and Scopus, combining the specific keywords of the review, as well as Nav channel, epilepsy, migraine, intellectual disability, and neurodegenerative disease.

Among the large amounts of recovered results, we chose to highlight the information reported in seminal papers and the most relevant recent findings which have the advantage of leveraging all prior knowledge.

3. VGSC Associated Neurological Disorders: Channelopathies

Mutations in VGSC isoforms cause diseases called channelopathies. Among the others, VGSCs have primary importance in genetic neurological disorders such as epilepsy and migraine. In the context of this topic, their functional impairments are widely investigated and pharmacological treatments have been defined or are under study. Epilepsy is one of the most common neurological disorders characterized by recurrent seizures that can be also associated with cognitive, psychological and social problems [21]. In the wide spectrum of the hundreds of epileptogenic genes, *SCN1A* and *2A* are the most relevant, but also *SCN3A* and *8A* have been found to be correlated with forms of epilepsy. However, even if the knowledge of the roles that Nav1.1, Nav1.2, and Nav1.6 channels play in epilepsy has increased greatly in the past decade, the prediction of the clinical outcome of a variant in any of these channels remains first unknown [22].

3.1. Nav1.1 Associated Epilepsies: Dravet Syndrome and GEFS+

Nav1.1 is widely expressed in the CNS and is by far the most frequent target of epileptogenic mutations which lead to several syndromes exhibiting a wide range of severity. At the moment about 1500 pathogenic mutations of *SCN1A* have been described [23] and the majority of them, over 900, are associated with the Severe Myoclonic Epilepsy of Infancy (SMEI), a rare and grave form of epilepsy described for the first time by Charlotte Dravet in 1978 renamed Dravet syndrome in 1989 [24]. This disease, an autosomal dominant disorder, displays its symptoms already in the first year of life with seizures often associated with high body temperatures, and worsens during the second year, when failure of motor coordination and cognitive impairments emerge. The syndrome in the majority of cases is associated with de novo mutations determining frame shifts or premature termination sequences in one copy of *SCN1A*, resulting in non-functional Nav1.1 channels and pathogenic haploinsufficiency [25]. As a consequence, Dravet patients fail to produce a sufficient level of functional channel and undergo a number of impairments actually correlated with a Nav1.1 LoF effect; but how is this decrease of Nav1.1 activity, which would predict reduced excitability in the brain, correlated with the occurrence of epileptic seizures, thus with an increase in excitability? An explanation of this apparent paradox was first proposed by Catterall's group [9]. A SMEI mouse model was generated through the ablation of Scn1a gene, and the heterozygous phenotype *Scn1a+/−* was assumed to mimic human SMEI. Currents recorded from hippocampal neurons showed a remarkable reduction of the Nav current in inhibitory GABAergic neurons of heterozygous and null homozygous animals with respect to wild type. However, the same reduction was not observed in glutamatergic excitatory pyramidal cells, thus indicating that Nav1.1 is predominant in GABAergic neurons. Accordingly loss-of function mutations of Nav1.1 results in a reduction of the brain inhibitory excitability determining an imbalance of brain excitation over inhibition which is at the basis of the epileptic seizure. This study besides being crucial for the advancement in the knowledge of the basic mechanisms underlying Dravet syndrome and many other forms of epilepsy as well, raises in the meanwhile the question of the need to design different therapies in relation to the Nav1.1 epileptogenic affected gene.

The above results have been overall confirmed in numerous subsequent studies performed on other mouse models and also on patients. However, a complex scenario is emerging in which, besides the damage induced from the presence of the mutation and of the subsequent seizures, compensatory remodeling mechanisms may take place depending on neuron type, genetic background and other factors, thus adding new variables to the genotype-phenotype correlation and to the clinical outcome of epilepsy and probably of other neurological diseases [26]. Additionally, generalized epilepsy with febrile seizure plus (GEFS+) syndrome is related to about 50 mutations of the Nav1.1 channel. GEFS+ is a milder form of epilepsy not associated with cognitive impairments and with symptoms usually well controlled by antiepileptic drugs. The confusing picture that emerged from initial studies of functional effects of GEFS mutations in transfected cells or *Xenopus laevis* oocytes [27,28] was later unambiguously clarified from in vivo study on a transgenic mouse model expressing the Scn1a GEFS+ mutation R1648H, consistent with the idea that R1648Q mutation led to a reduction in interneuron excitability [29] not associated with cognitive impairments. Several other milder forms of febrile seizures mostly in children have been also associated with Nav1.1 mutations even though it has been supposed that further precipitating factors, such as single nucleotide polymorphisms may contribute to the severity of the disease [30].

3.2. *Nav1.1 Associated Migraine: Familial Hemiplegic Migraine Type 3 (FHM3)*

Genetic analyses of a group of dominant monogenetic diseases called familiar hemiplegic migraine (FHM), led to the identification of specific migraine genes [31]. Three genes have been recognized as causative agents for FHM1, 2, and 3 respectively, all of which are involved in membrane ion transport. FHM3 is caused by mutations in *SCN1A* [32,33] and constitutes a severe subtype of migraine with aura, characterized by some degrees of hemiparesis, sometimes associated with other neurological symptoms, such as epilepsy or blindness.

Several aspects of FHM3 have long been incompletely understood. Even the question of whether pathological mutations (12 known so far) lead to loss or to gain of function was not resolved. Recently using novel Knock-In (KI) FHM3 mouse models and heterologous expression of FHM3 mutations, several studies have contributed to uncovering the molecular and cellular mechanisms underlying FHM3. Results in vitro are in line with a major gain of function effect as a possible explanation of familial hemiplegic migraine 3; indeed a shift of the steady state inactivation to more positive voltages, an accelerated recovery from inactivation, and an increase of the persistent current were observed in all tested FHM3 mutation (L1649Q, L1670W, F1774S, Q1489H, I1498M, F1499L, M1500V, F1661L) [34,35].

Accordingly, in the KI L1649Q-FHM3 mouse model, hyperactivity of the Nav1.1 channel, which is predominantly expressed in inhibitory GABAergic interneurons was observed resulting in hyperexcitability of interneurons; this element contributes to lower the threshold to elicit the onset of the Cortical Spreading Depression (CSD), a neural pathological mechanism consisting in a wave of neuronal depolarization slowly propagating across cortex and strictly linked to migraine aura [36]. The combination of these investigations shed light on the molecular defects causing FHM3 that can be potentially relevant also for other non-genetic forms of migraine with and without aura with which FHM3 may share molecular pathogenic mechanisms.

3.3. Other Epileptogenic Nav Isoforms

Nav1.2 is mainly expressed in excitatory neurons of the cortex and hippocampus and its mutant variants are mostly related to the benign familial neonatal infantile seizures (BFNIS), infantile West syndromes, Early infantile epileptic encephalopathy (EIEE), epilepsy of infancy with migrating focal seizures (EIMFS) [37]. About 100 Nav1.2 epileptogenic mutations have been identified so far with several hot spots recognized mostly at the level of the selectivity filter, the pore, the voltage sensor and of N- and C-termini [12,38]. Overall there is no consensus on the molecular process causing the forms of epilepsy associated with Nav1.2 since in vitro functional studies have demonstrated that LoF or GoF mutations can be involved in the epileptic phenotype. Several cases have been shown in which Na1.2-related seizures were not controlled by antiepileptic drugs (AED).

The first epileptogenic mutation of Nav1.3, K345Q, was identified in 2010 and associated with a case of partial cryptogenic epilepsy [39,40].

Since then different hereditary or de novo mutations of Nav1.3 have been identified related to epilepsy, generally with a GoF phenotype. Interestingly several features such as epileptic encephalopathy and polymicrogyria, have been associated exclusively with Nav1.3-related epilepsies and have not been reported in other channelopathies [41]. However, so far the clinical data on Nav1.3-associated epilepsies are quite scarce and accordingly efficacious therapeutic treatments still need to be optimized.

It is worth mentioning that in the last decades, a pivotal role of Nav1.3 have emerged in nervous system injury and neuropathic pain and for this reason, many efforts have been spent in this direction [42].

Even later than Nav1.3, the first case of Nav1.6-related epilepsy was diagnosed in 2012 [43]. In the last decades, many mutations of the SCN8 gene have been associated with EIEE, with benign familial infantile seizures-5 (BFIS5) and with several other forms showing a wide spectrum of severity, with symptoms ranging from cognitive and motor regression to cortical blindness [44]. The high number of new variants diagnosed has enabled the identification of several mutational hotspots; interestingly, a mutation localized in the untranslated region of *SCN8A* has been reported that produces a pathogenic variant with a mechanism that could interfere with transcription [45]. Nav1.6 is one of the most common sodium channels in CNS but is also expressed in heart muscle and this can explain the correlation with the increased risk of sudden unexpected death (SUDEP) observed in Na1.6-related epileptic patients [46].

As a general rule, it is established that epileptogenic GoF or LoF variants of the Nav CNS isoforms shift in the opposite direction of neuronal excitability and firing, differently affecting neural network activity, depending on whether the channel is prevalent in excitatory or inhibitory neurons. This implies that different approaches and molecular targets will be reasonably required to optimize the pharmacological treatments. Within this topic, an additional degree of complexity is added by the fact that patients with the same mutation may respond differently to the same therapeutic approach. Moreover, about 30% of patients are resistant to current antiepileptic drugs [47].

When planning to modulate Nav channel malfunctions using drugs, the high degree of conservation among Nav isoforms has to be taken into account: for instance, reducing the kinetics of Nav1.2 may result in a reduction of the cardiac Nav1.5. Importantly, it has to be considered that, because of the prevalence of Nav1.1 in regulating the excitability of GABAergic interneurons, prescription of antiepileptic drugs which non-selectively block Na channels is not recommended in Nav1.1-related epilepsies because this could worsen the crisis provoked by the decrease of inhibitory activity; conversely the best approach to design a drug based therapy could imply an enhancement GABA production or stability. Additionally, the use of antisense oligonucleotide (ASO) could be a promising strategy to treat haploinsufficiency [48].

Many treatments to fight epileptic seizures and epilepsy in general have been developed and applied over time and many efforts are still currently spent to exploit new therapeutic strategies. However, going into more detail on the pharmacology of epilepsies caused by Nav channels is not within the scope of this review; regarding this subject, it is possible to consult more exhaustive reviews [49,50].

4. VGSC in Intellectual Disability

As discussed above, mutations in all CNS-expressed sodium channel genes (*SCN1A*, *SCN2A*, *SCN3A* and *SCN8A*) cause various forms of epileptic phenotypes, and in the case of *SCN1A* also familial hemiplegic migraine, symptoms that by themselves not necessarily imply the presence of neurodevelopmental problems, autism or neurodegenerative phenotypes (Table 1). However, it is becoming increasingly clear that such phenotypes are often associated with variants of these sodium channel genes, with large differences among the different genes affected.

Table 1. List of the main neurological and neurodegenerative diseases correlated with Nav channel dysfunctions.

Nav Isoform	Gene	Neurological Disorder	References	Neurodegenerative Disorder	References
Nav1.1	*SCN1A*				
		Dravet Syndrome	[51,52]	Alzheimer's Disease	[53]
		GEFS+ (genetic epilepsy with febrile seizures plus)	[52,54,55]	Parkinson's Disease	[56]
		Epilepsy of infancy with migrating focal seizures	[52,54]		
		Myoclonic-atonic epilepsy	[52,54]		
		Familial hemiplegic migraine	[32]		
Nav1.2	*SCN2A*				
		Developmental and epileptic encephalopathy (DEE).	[52,57]	Alzheimer's disease	[53]
		Benign Familial Neonatal-Infantile Seizures (BFNIS)	[52,57]		
		West syndrome	[38,58]		
		Epilepsy of infancy with migrating focal seizures (EIMFS)	[52,54]		
		Autism Syndrome Disorder (ASD)	[58]		
		Intellectual Disability	[58]		
		Episodic ataxia	[59]		
Nav1.3	*SCN3A*				
		Developmental and Epileptic Encephalopathy (DEE)	[60]	Parkinson's Disease	[61]
		Polymicrogyria	[60]		
		Intellectual Disability	[60]		
Nav1.4	*SCN4A*				
				Amyotrophic Lateral Sclerosis	[62]
				Huntington's Disease	[63]
Nav1.5	*SCN5A*				
				Multiple Sclerosis	[64,65]

Table 1. *Cont.*

Nav Isoform	Gene	Neurological Disorder	References	Neurodegenerative Disorder	References
NaX	*SCN7A*				
				Amyotrophic Lateral Sclerosis	[66]
Nav1.6	*SCN8A*				
		Developmental and epileptic encephalopathy (DEE).	[52]	Alzheimer Disease	[67,68]
		Autism Syndrome Disorder (ASD)	[52]	Amyotrophic Lateral Sclerosis	[69]
		Intellectual Disability	[70,71]	Multiple Sclerosis	[72]
Nav1.8	*SCN10A*				
				Multiple Sclerosis	[64,73]

Rather severe forms of neurodevelopmental disorders, often associated with brain malformations, are found in patients carrying *SCN3A* variants [74]. Affected individuals frequently have developmental and epileptic encephalopathy (DEE) and all patients show some degree of early childhood developmental delay. As a general rule, variants cause a gain of function of the ion channel similar to GoF *SCN1A* variants that lead to FHM3 [35], i.e., a defective inactivation process with increased persistent currents [60]. The severity of the disease is likely related to the fact that *SCN3A* is prominently expressed during fetal development, such that its over-activity somehow results in a compromised development of the brain [74].

Variants of *SCN2A* are frequently associated with autism spectrum disorder and intellectual disability, in addition to infantile seizures [12,75–77]. Nav1.2 is widely expressed in the CNS, often co-localizing with Nav1.6, predominantly in glutamatergic excitatory neurons [78,79]. Surprisingly Nav1.2 GoF variants are mostly implicated in benign infantile-onset seizures (BIS) or infantile epileptic encephalopathy followed by developmental delay (IEE), while loss of function variants (for example truncations) cause autism spectrum disorder/intellectual disability with or without childhood-onset seizures [12]. Children with IEE can show microcephaly and cerebral and/or cerebellar atrophy. Mutational hotspots include the S4–S5 segment and the pore loop. Some variants, like K1422E alter ion selectivity and fall out of the usual loss- versus gain-of –function classification [78].

Similarly to *SCN2A*, variants of *SCN8A* (encoding Nav1.6) are often associated with developmental impairment and regression [43] in agreement with the overlapping function of these two genes. The first patient discovered with this disease carried the N1768D variant of a highly conserved asparagine [80]. In heterologous expression systems, the variant led to a dramatic increase in persistent currents and incomplete channel inactivation [80]. Similar to *SCN2A*, both GoF as well as LoF are associated with disease but with different outcomes regarding neurodevelopmental phenotypes. GoF variants are associated with seizures and significant developmental impairment and intellectual disability, while LoF variants are not necessarily associated with seizures [43].

SCN1A, the Na^+ channel gene with the largest number of epilepsy-associated variants (mostly Dravet syndrome caused by LoF), is less implicated in neurodevelopmental clinical phenotypes than the other Na^+ channel genes [54]. Beyond epilepsy, patients can present with autism spectrum disorder and SUDEP. Nevertheless, a few rare variants, like the recurrent T226M, can cause severe epilepsy together with profound developmental impairment [81]. In heterologous expression systems, that variant exhibited hyperpolarizing shifts of both activation as well as inactivation and enhanced fast inactivation [82].

In general, the cellular bases underlying neurodevelopmental phenotypes and possible structural brain anomalies caused by Na^+ channel gene variants are largely unknown. Maturation of the brain, necessitating specific electrical activity for proper neuronal differentiation and migration, synapse formation, axonal sprouting etc. [83] critically depends on the correct functioning of voltage-gated sodium channels, in agreement with the rather severe phenotypes associated with *SCN3A* variants. The fact that *SCN1A* variants generally cause less severe developmental phenotypes is possibly related to its later upregulation after birth. In addition to defective brain maturation, which is continuing during infancy and childhood, neural network-independent mechanisms, like direct cytotoxicity due for example to Na^+ overloading could play an additional role [84].

5. VGSC in Neurodegenerative Diseases

5.1. Alzheimer's Disease

Alzheimer's disease (AD) is a heterogeneous neurodegenerative disorder with irreversible progression, characterized by the progressive loss of synapses and neurons and by the formation of amyloid plaques in the brain. Clinically, it is characterized by loss of memory, followed by deterioration of all mental functions, neuronal degeneration of both cerebral and limbic cortices, reactive gliosis and deposition in the brain parenchyma of amyloid aggregates (or plaques) closely associated with dystrophic neurons. Activated phagocytic microglia, and intraneuronal aggregates of disrupted microtubules, known as "neurofibrillary tangles" are also detectable [85]. The molecular mechanisms underlying the development of AD are not well known so far and also the physiological functions of the crucial proteins the amyloid precursor protein (APP) and the presenilins 1 and 2 (PS1 and PS2) are unclear. The toxic extracellular amyloid oligomers detectable in AD plaques are composed of amyloid-Aß peptides (Aß) derived from the sequential proteolytic cleavage of APP by the ß-secretase BACE1 and of the "ß-secretase-complex" in which PS1 has a regulatory role. Mutations of the *APP* gene are responsible for AD as well. In addition, intracellular neurofibrillary tangles, composed of filaments of hyper-phosphorylated Tau protein, are neuropathological hallmarks of AD. Furthermore, presenilins, which are part of the molecular machinery that processes APP, when mutated are responsible for most of the cases of familial AD; more than 70 different mutations in presenilin 1 (PS1) have been associated with inherited early onset Alzheimer's disease [86]. The phenotypical heterogeneity among patients, and even among familial patients with the same genetic mutation, implies that other proteins might have a role in regulating the onset and severity of the neurodegeneration. Recent findings suggest that APP and PSs are the center of a complex network of interactions with many different intracellular adaptors, but the role of these proteins in the physiology or pathology is still unknown [87]. APP contains a YENPTY motif that has been previously described as an internalization motif, which now has been recognized to be involved as a key player in the regulation of multiple interactions with intracellular proteins [88]. The significance of the motif, which is typical of the receptor (TKR) and non-receptor tyrosine kinases (TK), in amyloid formation and in general for AD development is under investigation. Furthermore, the cytoplasmic tail of APP undergoes post-translational modifications; in particular, one of the mechanisms which may regulate APP cleavage and protein–protein interactions is linked to the occurrence of phosphorylation at Ser, Thr and Tyr residues. For example, Thr 668 can be phosphorylated by c-Jun N-terminal kinase-3 [89]. Thus APP is a tightly regulated protein, post-translationally modified by kinases. The pathophysiological significance of APP phosphorylation is unclear and there are even contrasting opinions about the effective influence of such post-translational modifications on APP cleavage, amyloid formation and AD development. Indeed in this context, it has to be underlined that APP and APP-related proteins, APLP1 and APLP2, can interact with several proteins such as X11 and Fe65 [90], c-Abl [91], mDab [92], JIP-1 [93] independently of the phosphorylation of the tyrosine residues of the YENPTY motif.

Interestingly, several studies suggest that the physical interaction of APP and its related proteins, Aß oligomers and BACE1 enzyme with various isoforms of Nav could interfere with neural physiological processes and be involved in the development of the AD.

5.1.1. Alzheimer Disease and Nav Involvement: APP Phosphorylation

Even if a causative dependence of AD from VGSC mutations has not been assessed so far, it has been recently hypothesized that several isoforms may be modulated from APP. In a paper published in 2015, Liu and collaborators demonstrated that in murine cortical neurons, APP co-localizes and interacts with Nav1.6 and that knocking down APP provokes a decrease in Nav1.6 cell surface expression and function [94]. Conversely, APP-induced increases of Nav1.6 cell surface expression have been shown to be dependent on Go protein, the most abundant G protein in the CNS, being enhanced by a constitutively-active mutant Go protein and blocked by a dominant negative mutant Go protein. Interestingly, Nav1.6 sodium channel surface expression was shown to be increased by T668E and decreased by T668A mutations of APP, mimicking and preventing Thr-668 phosphorylation, respectively. In agreement phosphorylation of APP at Thr-668 enhanced its interaction with Nav1.6. Furthermore, APP regulates JNK activity in a Go protein-dependent manner and JNK, in turn, phosphorylates APP. Therefore, APP enhances Nav1.6 sodium channel cell surface expression through a Go-coupled JNK pathway [94]. The interaction between APP and Nav1.6 sodium channel was further studied by Shao li and colleagues. From studies in APP knockout mice they observed that APP molecules aggregated at nodes of Ranvier (NORs) in CNS myelinated axons and interacted with Nav1.6 and described a reduction of sodium current density in hippocampal neurons as well. Coexpression of APP or its intracellular domains (AICD) with Nav1.6 in *Xenopus laevis* oocytes resulted in an increase of peak sodium currents, which also in this case was enhanced by constitutively-active Go mutant and blocked by a Go dominant negative mutant. Similarly to the results of Liu and colleagues, Nav1.6 current was increased by APP mutation T668E and decreased by T668A. Accordingly, the cell surface expression of Nav1.6 sodium channels in the white matter of the spinal cord and the spinal conduction velocity was decreased in APP/JNK3/knockout mice. Thus also in this study, the conclusion was that APP modulates Nav1.6 sodium channels through a Go-coupled JNK pathway and that this modulation is dependent on the phosphorylation of the Thr668 residue of APP [67].

5.1.2. Alzheimer's Disease and Nav Involvement: Aß Oligomers

The correlation among AD pathophysiology, seizures and increased neuronal excitability was established by Busche and colleagues examining an AD mouse model. Indeed in the CA1 hippocampal region of young APP/PS1 mice, they observed an increased number of hyperactive neurons associated with the high level of Aβ oligomers produced, and therefore speculated that soluble Aβ oligomers might directly induce neuronal hyperactivity [95]. Several lines of evidence indicated that amyloid-β1–42 (Aβ1–42) induced neuronal hyperactivity may give rise to cognitive deficits and memory dysfunction in AD. Indeed recent studies on primary hippocampal neurons of another AD mouse model (Tg2576) exposed to Aβ1–42 oligomers, demonstrated that the overexpression of Nav1.6 contributes to membrane depolarization and to the increase of spike frequency, thereby resulting in neuronal hyperexcitability. These findings identify the Nav1.6 channel as a determinant of hippocampal neuronal hyperexcitability induced by Aβ1–42 oligomers [96].

5.1.3. Alzheimer's Disease and Nav Involvement: BACE1

Another relevant molecule for the sequential proteolytic cleavage of APP and plaques formation is the ß-secretase BACE1. Some authors describe a correlation between BACE1 and sodium channel expression. A recent study performed by De-Juan Yuan and colleagues on WT mice and on the APP/PS1 AD mouse model has shown that Nav1.6 is overexpressed in old AD mice. The high expression of Nav1.6 in APP/PS1 mice enhances BACE1 transcription through activation of the NFAT1 factor regulated by the Na(+)/Ca(2+)exchanger (NCX). Interestingly, the authors demonstrated that knocking down Nav1.6 with a bilateral injection of adeno-associated viruses (serotype 8, AAV8) encoding shRNA of Nav1.6 in the hippocampus significantly reduced the density of Aβ plaques through the suppression of the β-secretase-mediated cleavage of APP. As a consequence, the cognitive deficit of the mice and the neural network hyper excitability were both remarkably reduced [68]. The authors went further in defining the molecular mechanisms underlying the role of Nav1.6 in AD pathogenesis. Treating Nav1.6 overexpressing HEK cells with TTX, an unspecific blocker of CNS VGSC, they observed a remarkable reduction of BACE1 expression; the same reduction was less evident when Nav1.6 channel was knocked down by shRNA transfection and TTX was applied, unveiling a molecular mechanism dependent on Na ion flux and not only the presence of the channel itself. Thus for the first time, Nav1.6 has been indicated as a new target to be considered to slow down AD evolution.

Other authors suggest that Nav1.1 and 1.2 might also be involved to some extent in AD; indeed the results of their investigations indicated a reduction of Nav1.1, 1.2 and Nav1.6 α-subunits protein in primary neurons in a culture of wild type BACE1-null mice. They propose an underlying mechanism involving BACE1 activity regulating mRNA levels of the Nav1.1 α-subunit through the cleavage of the Navβ2 subunit expressed on the surface. Interestingly in the hippocampus of the same murine model, Nav1.1 expression appeared significantly reduced, while Nav1.2, perhaps as a compensatory mechanism, was remarkably increased. Thus endogenous BACE1 activity seems to regulate total and surface levels of VGSC in mouse brains [53].

5.2. Parkinson's Disease

Parkinson's disease (PD) is a neurodegenerative disorder characterized by motor disabilities that affects predominantly the dopaminergic neurons of the substantia nigra causing a decrease in dopamine levels in the striatum [97]. The main symptoms are bradykinesia, akinesia, muscle rigidity, postural instability, stiffness and resting tremor which may be due to the high levels of synchronous oscillations in the basal ganglia neurons [98,99]. The causes of PD are unknown, although it is speculated that there may be a contribution from genetic and environmental factors [100]. PD pathogenesis has been associated with a number of factors, including impairments linked to intracellular Ca^{2+} excess, mitochondrial malfunction, oxidative or metabolic stress, and, in particular, a small number of neurotoxins that render neuronal cells more susceptible to death. VGSCs have an important role in the abnormal electrical activity of neurons in the globus pallidus and the subthalamic nucleus in PD [101] and are involved in cognitive impairments. By using the rat PD model infused with 6-OHDA (6-hyroxydopamine), Wang and colleagues showed that the expression of Nav1.1, 1.3 and 1.6 in the hippocampus was dynamically increased at different time points after dopamine depletion [61]. In contrast, treating rats with phenytoin, a sodium channel blocker that slows down the recovery from inactivation [102], remarkably improved cognitive impairments. In MPTP (1-metyl-4-fenyl-1,2,3,6-tetrahydropyridin)-treated PD mice it was found that Nav1.1 expression was increased in the external globus pallidus [103]. Globus pallidus is a central nucleus of the basal ganglia; it receives the majority of the inhibitory GABAergic inputs from the striatum and plays a key role in the propagation of synchronized oscillatory activity of basal ganglia [99]. In particular, the increase of Nav1.1 expression in MPTP-treated mice was evident in parvalbumin (PV) positive GABAergic interneurons that exhibit fast-firing spontaneous activity, and exert their inhibitory control on the activity of innervated neurons in the subthalamic nucleus, substantia nigra and in the striatum [56]. In these cells, Nav1.1 is a determinant for the maintenance of sustained fast spiking more than for its onset [104,105]. Additionally, in this study phenytoin was used to test the effectiveness of blocking VGSCs in reducing PD symptoms. Indeed both motor disability and high synchronous oscillations were reduced in MPTP-treated mice, thus confirming the potential therapeutic role of this compound in PD. Even if the role of GABAergic transmission in PD is still unknown, it is possible to speculate that the observed effect of phenytoin could result from blocking the increased activity of Nav1.1 in the globus pallidus thus restoring the physiological GABAergic activity. Probably, the upregulation of Nav1.1 expression in globus pallidus may be a compensatory molecular mechanism aimed to enhance inhibitory response in the basal ganglia and counteract the abnormal neural activity of PD animals. Nav1.3 seems to be involved in PD as well. This

VGSC isoform generally is robustly expressed during the fetal period and is downregulated after birth. By using the rat PD model infused with 6-OHDA, it has been demonstrated that 49 days after infusion Nav1.3 is re-expressed in dopaminergic neurons of the substantia nigra [61]. Additionally, in this case the authors suggest that the re-expression of Nav1.3 could be a compensatory mechanism for the degeneration of dopaminergic neurons caused by PD progression. Current therapies treat only PD symptoms, but several investigations have also been carried out in order to find putative neuroprotective drugs for dopaminergic neurons. In particular, Sadeghian and colleagues have examined the effects of safinamide on microglial activation and dopaminergic neurons degeneration in a rat model of PD in vivo. In the PD rat model, safinamide reduced the number of activated microglial cells and increased survival of dopaminergic neurons [106]. Safinamide is a sodium and calcium channels modulator and inhibits glutamate release induced by abnormal neuronal activity, promoting its neuroprotective effect [107]. Specifically safinamide interacts with the inactivated state of the VGSCs, keeping most of the channels in the inactive state and preventing their activation. This effect induces a selective depression of the pathological high-frequency firing, leaving physiologic activity unaffected and thus avoiding CNS depressant effects [108].

5.3. *Amyotrophic Lateral Sclerosis (ALS)*

Amyotrophic Lateral Sclerosis (ALS) is an unknown etiology disease, caused by the progressive neurodegeneration of motor neurons [109]. The degenerative process induces a progressive atrophy of the neuromuscular system which causes death from paralysis 3–5 years after the onset of the disease [110]. At the moment there is no effective therapy for ALS on slowing down or arresting the neurodegenerative process [111]. To date, there are two main drugs for the treatment of ALS aimed to prolong the patient's life expectancy: riluzole [109,112] and edaravone [112,113]. Although the mechanisms by which these drugs exert their effects are not well known, various hypotheses have been formulated. Riluzole is believed to modulate the release of glutamate [114,115] and sodium channel activity. In particular, this drug is able to down-regulate neuronal firing and inhibit the persistent current of VGSC [116,117]. Persistent current (see Figure 1) is caused by a particular kinetics of VGSC characterized by a rapid activation followed by a subsequent slow inactivation which maintains the channel in the activated state for hundreds of milliseconds [118]. It is generally more evident in various pathological conditions, when the ionic environment is altered or when a mutation modifies functional properties of VGSCs and contributes to maintaining the neuronal membrane potential near the threshold value triggering spontaneous action potentials [119]. Vucic and Kiernan using the transcranial magnetic stimulation technique on ALS patients have demonstrated that cortical excitability is abnormally increased in an early state of the disease [120]. Similarly experiments performed on ASL animal models have confirmed that a neuronal hyperexcitability of the motor cortex activating the glutamate excitotoxic cascade via trans-synaptic mechanism is at the route of the neurodegenerative process of the motor neuron [121,122].

Although the molecular mechanisms underlying ALS are still not well understood, several pieces of evidence indicate that Nav1.6 channels could be a potential therapeutic target. Using G93A mice, Saba and collaborators showed that the expression levels of the Nav1.6 channels during ASL progression are modified in the primary motor cortex but not in other cortical areas with consequent alteration of the excitability and of persistent current of this neural district [69].

While it is assumed that Nav1.6 dysfunction may be linked to ALS, evidence has recently been provided of the existence of sporadic ALS forms caused by heterozygous point mutations in the *SCN4A* gene that precede the development of the disease. In this context, two mutations, Arg672His and Ser1159Pro, which have opposite effects on neuromuscular excitability have been identified in two different patients. In both cases, the authors hypothesize that the abnormal Nav1.4 channels predisposed to depolarization-induced cellular excitotoxicity, leading to the development of ALS [62]. Whole genome sequencing analysis of ALS patients identified the presence of missense mutations in the *SCN7A* gene, which codes for NaX, a type II sodium channel sensitive to the extracellular [Na^+] [66]. Mutations in this channel result in a loss of function phenotype which provokes a dysregulation of sodium homeostasis and neuronal hyperexcitability. Overall, the findings described above support the hypothesis of the role of VGSC dysfunction in ALS development. Gaining more insights into the molecular mechanisms related to VGSC underlying the disease could lead to the identification of new therapeutic targets and even pave the way for personalized gene therapy.

5.4. Multiple Sclerosis

Multiple sclerosis (MS) is a multifactorial neurodegenerative disease of the central nervous system whose etiology is still mostly unknown. It is a chronic demyelinating disease characterized by an autoimmune response against the tissues of the central nervous system with lymphocytic and macrophage infiltration [123]. The pathological hallmarks of MS are demyelinated plaques in the CNS with inflammation, gliosis, and neurodegeneration [124]. At the beginning of the disease, the lymphocyte infiltration that triggers the axonal and myelin damage can be recovered. Later the inflammatory episodes occur repeatedly and microglia activation causes extensive and chronic neurodegeneration leading to disability.

An experimental autoimmune encephalomyelitis (EAE) mouse model is used to study MS. In this murine model, T cells infiltrate the CNS, initiate demyelination and cause loss of axons [125]. VGSCs have an important role in axonal loss in MS. It has been shown that in demyelinated axons there is a particular distribution of sodium channels, with Nav1.2 and Nav1.6 present in the plaques together with the Na^+/Ca^{2+} exchanger whereas in non-demyelinated control axons Nav1.6 is located only in Ranvier nodes [72]. High sodium flux along Nav1.6 reverses the Na^+/Ca^{2+} exchanger and increases axonal calcium finally leading to axonal damage through the activation Ca^{2+} dependent proteases [72]. Upregulation of Nav1.8 detected in cerebellar Purkinje neurons of MS patients and in the experimental EAE mouse model appears to be a determinant for the cerebellar dysfunction observed in this disease [64,73]. Indeed the administration of a selective Nav1.8 blocker in the cerebrospinal fluid of EAE mice partially improved symptomatology [126]. Overexpression of Nav1.5 has been detected in astrocytes of post-mortem MS brain tissue. It has been suggested that its upregulation is necessary to restore physiological ATPase-dependent Na^+/K^+ homeostasis in damaged neural areas [64]. Experiments performed on an in vitro model of glial injury [127] have assessed that Nav1.5 in non-excitable cells plays the main role of reversing the NCX function. It has been proposed that the application of sodium channel blockers could attenuate the inflammatory effects provoked by microglia injury and activation such as phagocytosis, the release of cytokines interleukin-1 and tumor necrosis factor-α [128]. In vitro studies have shown that Nav1.5 is present in the membrane of maturing endosomes of macrophages suggesting a possible role of the channel in the phagocytic pathway of myelin degradation within MS lesions [65].

6. Conclusions

The purpose of this review was to collect and summarize the main information currently available in the scientific literature on the multifaceted role of Nav channels on pathologies affecting the CNS, whether neurodegenerative or not. We expect that having all the data available in the literature on such a complex subject grouped together can effectively support future scientific activity on this topic.

A picture emerged in which malfunctions of the same brain Nav isoform can lead to very different neural pathologies; for instance, Nav1.1 mutations may cause several forms of epilepsy or genetic migraine, but the channel is suspected to also be involved in autism spectrum disorders; similarly, Nav1.2 is correlated with epilepsy and with several forms of autism; finally Nav1.6 when mutated gives rise to epilepsy, whereas when its expression is impaired is correlated with developmental regression or with the most common neurodegenerative diseases such as Alzheimer's and Parkinson's.

Even if some of the pathologies described are not directly categorized as neurodegenerative, nevertheless some of their manifestations can produce neurodegeneration as well. Epileptic crises with prolonged seizures are clearly capable of injuring the brain, while brief and isolated seizures are likely to cause negative changes in brain function and possibly the loss of specific brain cells.

In order to better dissect and elucidate the molecular basis of Nav-correlated brain pathologies a tight collaboration between clinicians and researchers is essential to obtain new experimental models and techniques.

Many efforts have been spent until now to find new treatments to limit neuronal damage once the pathology has manifested.

Currently, therapies that intervene by blocking the degenerative process, reducing neuronal loss and restoring nerve transmission, are not available. There is numerous clinical evidence which suggests that antioxidant substances, for example, are useful both in preventing and modifying the course of neurodegenerative diseases.

Further studies on neuroprotection as a therapeutic intervention aimed at slowing or even halting the progression of neurodegeneration would be desirable.

Author Contributions: All authors contributed to the original draft preparation, P.G. supervised the final draft. All authors have read and agreed to the published version of the manuscript.

Funding: This research was funded by the Fondazione AIRC per la Ricerca sul Cancro (grant # IG 21558) to M.P.

Institutional Review Board Statement: Not applicable.

Informed Consent Statement: Not applicable.

Data Availability Statement: No new data were created or analyzed in this study. Data sharing is not applicable to this article.

Conflicts of Interest: The authors declare no conflict of interest.

References

1. Catterall, W.A.; Perez-Reyes, E.; Snutch, T.P.; Striessnig, J. International Union of Pharmacology. XLVIII. Nomenclature and structure-function relationships of voltage-gated calcium channels. *Pharmacol. Rev.* **2005**, *57*, 411–425. [CrossRef]
2. Catterall, W.A. Forty Years of Sodium Channels: Structure, Function, Pharmacology, and Epilepsy. *Neurochem. Res.* **2017**, *42*, 2495–2504. [CrossRef] [PubMed]
3. Isom, L.L. The role of sodium channels in cell adhesion. *Front. Biosci. J. Virtual Libr.* **2002**, *7*, 12–23. [CrossRef] [PubMed]
4. Yu, F.H.; Westenbroek, R.E.; Silos-Santiago, I.; McCormick, K.A.; Lawson, D.; Ge, P.; Ferriera, H.; Lilly, J.; DiStefano, P.S.; Catterall, W.A.; et al. Sodium channel beta4, a new disulfide-linked auxiliary subunit with similarity to beta2. *J. Neurosci. Off. J. Soc. Neurosci.* **2003**, *23*, 7577–7585. [CrossRef]
5. O'Malley, H.A.; Isom, L.L. Sodium channel beta subunits: Emerging targets in channelopathies. *Annu. Rev. Physiol.* **2015**, *77*, 481–504. [CrossRef]
6. Goldin, A.L. Diversity of mammalian voltage-gated sodium channels. *Ann. N. Y. Acad. Sci.* **1999**, *868*, 38–50. [CrossRef]
7. Westenbroek, R.E.; Merrick, D.K.; Catterall, W.A. Differential subcellular localization of the RI and RII Na^+ channel subtypes in central neurons. *Neuron* **1989**, *3*, 695–704. [CrossRef]
8. Kearney, J.A.; Buchner, D.A.; De Haan, G.; Adamska, M.; Levin, S.I.; Furay, A.R.; Albin, R.L.; Jones, J.M.; Montal, M.; Stevens, M.J.; et al. Molecular and pathological effects of a modifier gene on deficiency of the sodium channel Scn8a (Na(v)1.6). *Hum. Mol. Genet.* **2002**, *11*, 2765–2775. [CrossRef] [PubMed]
9. Yu, F.H.; Mantegazza, M.; Westenbroek, R.E.; Robbins, C.A.; Kalume, F.; Burton, K.A.; Spain, W.J.; McKnight, G.S.; Scheuer, T.; Catterall, W.A. Reduced sodium current in GABAergic interneurons in a mouse model of severe myoclonic epilepsy in infancy. *Nat. Neurosci.* **2006**, *9*, 1142–1149. [CrossRef]
10. Rubinstein, M.; Han, S.; Tai, C.; Westenbroek, R.E.; Hunker, A.; Scheuer, T.; Catterall, W.A. Dissecting the phenotypes of Dravet syndrome by gene deletion. *Brain J. Neurol.* **2015**, *138*, 2219–2233. [CrossRef]
11. Syrbe, S.; Zhorov, B.S.; Bertsche, A.; Bernhard, M.K.; Hornemann, F.; Mutze, U.; Hoffmann, J.; Hortnagel, K.; Kiess, W.; Hirsch, F.W.; et al. Phenotypic Variability from Benign Infantile Epilepsy to Ohtahara Syndrome Associated with a Novel Mutation in SCN2A. *Mol. Syndromol.* **2016**, *7*, 182–188. [CrossRef]

12. Sanders, S.J.; Campbell, A.J.; Cottrell, J.R.; Moller, R.S.; Wagner, F.F.; Auldridge, A.L.; Bernier, R.A.; Catterall, W.A.; Chung, W.K.; Empfield, J.R.; et al. Progress in Understanding and Treating SCN2A-Mediated Disorders. *Trends Neurosci.* **2018**, *41*, 442–456. [CrossRef] [PubMed]
13. Caldwell, J.H.; Schaller, K.L.; Lasher, R.S.; Peles, E.; Levinson, S.R. Sodium channel Na(v)1.6 is localized at nodes of ranvier, dendrites, and synapses. *Proc. Natl. Acad. Sci. USA* **2000**, *97*, 5616–5620. [CrossRef]
14. Whitaker, W.R.; Faull, R.L.; Waldvogel, H.J.; Plumpton, C.J.; Emson, P.C.; Clare, J.J. Comparative distribution of voltage-gated sodium channel proteins in human brain. *Brain Res. Mol. Brain Res.* **2001**, *88*, 37–53. [CrossRef]
15. Bezanilla, F. Ion channels: From conductance to structure. *Neuron* **2008**, *60*, 456–468. [CrossRef]
16. Stuhmer, W.; Conti, F.; Suzuki, H.; Wang, X.D.; Noda, M.; Yahagi, N.; Kubo, H.; Numa, S. Structural parts involved in activation and inactivation of the sodium channel. *Nature* **1989**, *339*, 597–603. [CrossRef] [PubMed]
17. Hodgkin, A.L.; Huxley, A.F. The components of membrane conductance in the giant axon of Loligo. *J. Physiol.* **1952**, *116*, 473–496. [CrossRef] [PubMed]
18. Cox, B. CHAPTER 1 Ion Channel Drug Discovery: A Historical Perspective. In *Ion Channel Drug Discovery*; The Royal Society of Chemistry: London, UK, 2015; pp. 1–15.
19. Stevens, F.L.; Hurley, R.A.; Taber, K.H. Anterior cingulate cortex: Unique role in cognition and emotion. *J. Neuropsychiatry Clin. Neurosci.* **2011**, *23*, 121–125. [CrossRef]
20. de Lera Ruiz, M.; Kraus, R.L. Voltage-Gated Sodium Channels: Structure, Function, Pharmacology, and Clinical Indications. *J. Med. Chem.* **2015**, *58*, 7093–7118. [CrossRef]
21. Tucker, G.J. Seizure disorders presenting with psychiatric symptomatology. *Psychiatr. Clin. N. Am.* **1998**, *21*, 625–635. [CrossRef]
22. Encinas, A.C.; Watkins, J.C.; Longoria, I.A.; Johnson, J.P., Jr.; Hammer, M.F. Variable patterns of mutation density among NaV1.1, NaV1.2 and NaV1.6 point to channel-specific functional differences associated with childhood epilepsy. *PLoS ONE* **2020**, *15*, e0238121. [CrossRef] [PubMed]
23. Menezes, L.F.S.; Sabia Junior, E.F.; Tibery, D.V.; Carneiro, L.D.A.; Schwartz, E.F. Epilepsy-Related Voltage-Gated Sodium Channelopathies: A Review. *Front. Pharmacol.* **2020**, *11*, 1276. [CrossRef]
24. Dravet, C. Dravet syndrome history. *Dev. Med. Child Neurol.* **2011**, *53* (Suppl. 2), 1–6. [CrossRef]
25. Catterall, W.A. Dravet Syndrome: A Sodium Channel Interneuronopathy. *Curr. Opin. Physiol.* **2018**, *2*, 42–50. [CrossRef]
26. Mantegazza, M.; Broccoli, V. SCN1A/Na(V) 1.1 channelopathies: Mechanisms in expression systems, animal models, and human iPSC models. *Epilepsia* **2019**, *60* (Suppl. 3), S25–S38. [CrossRef] [PubMed]
27. Spampanato, J.; Escayg, A.; Meisler, M.H.; Goldin, A.L. Functional effects of two voltage-gated sodium channel mutations that cause generalized epilepsy with febrile seizures plus type 2. *J. Neurosci. Off. J. Soc. Neurosci.* **2001**, *21*, 7481–7490. [CrossRef]
28. Lossin, C.; Rhodes, T.H.; Desai, R.R.; Vanoye, C.G.; Wang, D.; Carniciu, S.; Devinsky, O.; George, A.L., Jr. Epilepsy-associated dysfunction in the voltage-gated neuronal sodium channel SCN1A. *J. Neurosci. Off. J. Soc. Neurosci.* **2003**, *23*, 11289–11295. [CrossRef]
29. Tang, B.; Dutt, K.; Papale, L.; Rusconi, R.; Shankar, A.; Hunter, J.; Tufik, S.; Yu, F.H.; Catterall, W.A.; Mantegazza, M.; et al. A BAC transgenic mouse model reveals neuron subtype-specific effects of a Generalized Epilepsy with Febrile Seizures Plus (GEFS+) mutation. *Neurobiol. Dis.* **2009**, *35*, 91–102. [CrossRef]
30. Catterall, W.A.; Kalume, F.; Oakley, J.C. NaV1.1 channels and epilepsy. *J. Physiol.* **2010**, *588*, 1849–1859. [CrossRef] [PubMed]
31. Pietrobon, D.; Moskowitz, M.A. Pathophysiology of migraine. *Annu. Rev. Physiol.* **2013**, *75*, 365–391. [CrossRef]
32. Dichgans, M.; Freilinger, T.; Eckstein, G.; Babini, E.; Lorenz-Depiereux, B.; Biskup, S.; Ferrari, M.D.; Herzog, J.; van den Maagdenberg, A.M.; Pusch, M.; et al. Mutation in the neuronal voltage-gated sodium channel SCN1A in familial hemiplegic migraine. *Lancet* **2005**, *366*, 371–377. [CrossRef] [PubMed]
33. Vanmolkot, K.R.; Babini, E.; de Vries, B.; Stam, A.H.; Freilinger, T.; Terwindt, G.M.; Norris, L.; Haan, J.; Frants, R.R.; Ramadan, N.M.; et al. The novel p.L1649Q mutation in the SCN1A epilepsy gene is associated with familial hemiplegic migraine: Genetic and functional studies. Mutation in brief #957. Online. *Hum. Mutat.* **2007**, *28*, 522. [CrossRef] [PubMed]
34. Bertelli, S.; Barbieri, R.; Pusch, M.; Gavazzo, P. Gain of function of sporadic/familial hemiplegic migraine-causing SCN1A mutations: Use of an optimized cDNA. *Cephalalgia Int. J. Headache* **2019**, *39*, 477–488. [CrossRef]
35. Barbieri, R.; Bertelli, S.; Pusch, M.; Gavazzo, P. Late sodium current blocker GS967 inhibits persistent currents induced by familial hemiplegic migraine type 3 mutations of the SCN1A gene. *J. Headache Pain* **2019**, *20*, 107. [CrossRef]
36. Auffenberg, E.; Hedrich, U.B.; Barbieri, R.; Miely, D.; Groschup, B.; Wuttke, T.V.; Vogel, N.; Luhrs, P.; Zanardi, I.; Bertelli, S.; et al. Hyperexcitable interneurons trigger cortical spreading depression in an Scn1a migraine model. *J. Clin. Investig.* **2021**, *131*, 21. [CrossRef]
37. Reynolds, C.; King, M.D.; Gorman, K.M. The phenotypic spectrum of SCN2A-related epilepsy. *Eur. J. Paediatr. Neurol. EJPN Off. J. Eur. Paediatr. Neurol. Soc.* **2020**, *24*, 117–122. [CrossRef] [PubMed]
38. Perucca, P.; Perucca, E. Identifying mutations in epilepsy genes: Impact on treatment selection. *Epilepsy Res.* **2019**, *152*, 18–30. [CrossRef]
39. Holland, K.D.; Kearney, J.A.; Glauser, T.A.; Buck, G.; Keddache, M.; Blankston, J.R.; Glaaser, I.W.; Kass, R.S.; Meisler, M.H. Mutation of sodium channel SCN3A in a patient with cryptogenic pediatric partial epilepsy. *Neurosci. Lett.* **2008**, *433*, 65–70. [CrossRef]

40. Estacion, M.; Gasser, A.; Dib-Hajj, S.D.; Waxman, S.G. A sodium channel mutation linked to epilepsy increases ramp and persistent current of Nav1.3 and induces hyperexcitability in hippocampal neurons. *Exp. Neurol.* **2010**, *224*, 362–368. [CrossRef]
41. Inuzuka, L.M.; Macedo-Souza, L.I.; Della-Ripa, B.; Cabral, K.S.S.; Monteiro, F.; Kitajima, J.P.; de Souza Godoy, L.F.; de Souza Delgado, D.; Kok, F.; Garzon, E. Neurodevelopmental disorder associated with de novo SCN3A pathogenic variants: Two new cases and review of the literature. *Brain Dev.* **2020**, *42*, 211–216, Erratum in *Brain Dev.* **2021**, *43*, 671. [CrossRef]
42. Black, J.A.; Nikolajsen, L.; Kroner, K.; Jensen, T.S.; Waxman, S.G. Multiple sodium channel isoforms and mitogen-activated protein kinases are present in painful human neuromas. *Ann. Neurol.* **2008**, *64*, 644–653. [CrossRef]
43. Veeramah, K.R.; O'Brien, J.E.; Meisler, M.H.; Cheng, X.; Dib-Hajj, S.D.; Waxman, S.G.; Talwar, D.; Girirajan, S.; Eichler, E.E.; Restifo, L.L.; et al. De novo pathogenic SCN8A mutation identified by whole-genome sequencing of a family quartet affected by infantile epileptic encephalopathy and SUDEP. *Am. J. Hum. Genet.* **2012**, *90*, 502–510. [CrossRef] [PubMed]
44. Gardella, E.; Moller, R.S. Phenotypic and genetic spectrum of SCN8A-related disorders, treatment options, and outcomes. *Epilepsia* **2019**, *60* (Suppl. 3), S77–S85. [CrossRef] [PubMed]
45. Johannesen, K.M.; Gardella, E.; Encinas, A.C.; Lehesjoki, A.E.; Linnankivi, T.; Petersen, M.B.; Lund, I.C.B.; Blichfeldt, S.; Miranda, M.J.; Pal, D.K.; et al. The spectrum of intermediate SCN8A-related epilepsy. *Epilepsia* **2019**, *60*, 830–844. [CrossRef]
46. Hammer, M.F.; Xia, M.; Schreiber, J.M. SCN8A-Related Epilepsy and/or Neurodevelopmental Disorders. In *GeneReviews(R)*; Adam, M.P., Mirzaa, G.M., Pagon, R.A., Wallace, S.E., Bean, L.J.H., Gripp, K.W., Amemiya, A., Eds.; University of Washington: Seattle, WA, USA, 1993.
47. Loscher, W.; Potschka, H.; Sisodiya, S.M.; Vezzani, A. Drug Resistance in Epilepsy: Clinical Impact, Potential Mechanisms, and New Innovative Treatment Options. *Pharmacol. Rev.* **2020**, *72*, 606–638. [CrossRef]
48. Hsiao, J.; Yuan, T.Y.; Tsai, M.S.; Lu, C.Y.; Lin, Y.C.; Lee, M.L.; Lin, S.W.; Chang, F.C.; Liu Pimentel, H.; Olive, C.; et al. Upregulation of Haploinsufficient Gene Expression in the Brain by Targeting a Long Non-coding RNA Improves Seizure Phenotype in a Model of Dravet Syndrome. *EBioMedicine* **2016**, *9*, 257–277. [CrossRef]
49. Rodrigues, T.; de Moura, J.P.; Dos Santos, A.M.F.; Monteiro, A.F.M.; Lopes, S.M.; Scotti, M.T.; Scotti, L. Epileptic Targets and Drugs: A Mini-Review. *Curr. Drug Targets* **2023**, *24*, 212–224. [CrossRef] [PubMed]
50. Ghovanloo, M.R.; Ruben, P.C. Cannabidiol and Sodium Channel Pharmacology: General Overview, Mechanism, and Clinical Implications. *Neurosci. A Rev. J. Bringing Neurobiol. Neurol. Psychiatry* **2022**, *28*, 318–334. [CrossRef]
51. Meng, H.; Xu, H.Q.; Yu, L.; Lin, G.W.; He, N.; Su, T.; Shi, Y.W.; Li, B.; Wang, J.; Liu, X.R.; et al. The SCN1A mutation database: Updating information and analysis of the relationships among genotype, functional alteration, and phenotype. *Hum. Mutat.* **2015**, *36*, 573–580. [CrossRef]
52. Meisler, M.H.; Hill, S.F.; Yu, W. Sodium channelopathies in neurodevelopmental disorders. *Nat. Rev. Neurosci.* **2021**, *22*, 152–166. [CrossRef]
53. Kim, D.Y.; Gersbacher, M.T.; Inquimbert, P.; Kovacs, D.M. Reduced Sodium Channel Nav1.1 Levels in BACE1-null Mice. *J Biol Chem.* **2011**, *286*, 8106–8116. [CrossRef] [PubMed]
54. Scheffer, I.E.; Nabbout, R. SCN1A-related phenotypes: Epilepsy and beyond. *Epilepsia* **2019**, *60* (Suppl. 3), S17–S24. [CrossRef]
55. Escayg, A.; Goldin, A.L. Sodium channel SCN1A and epilepsy: Mutations and mechanisms. *Epilepsia* **2010**, *51*, 1650–1658. [CrossRef] [PubMed]
56. Saunders, A.; Huang, K.W.; Sabatini, B.L. Globus Pallidus Externus Neurons Expressing parvalbumin Interconnect the Subthalamic Nucleus and Striatal Interneurons. *PLoS ONE* **2016**, *11*, e0149798. [CrossRef] [PubMed]
57. Begemann, A.; Acuna, M.A.; Zweier, M.; Vincent, M.; Steindl, K.; Bachmann-Gagescu, R.; Hackenberg, A.; Abela, L.; Plecko, B.; Kroell-Seger, J.; et al. Further corroboration of distinct functional features in SCN2A variants causing intellectual disability or epileptic phenotypes. *Mol. Med.* **2019**, *25*, 6. [CrossRef]
58. Wolff, M.; Johannesen, K.M.; Hedrich, U.B.S.; Masnada, S.; Rubboli, G.; Gardella, E.; Lesca, G.; Ville, D.; Milh, M.; Villard, L.; et al. Genetic and phenotypic heterogeneity suggest therapeutic implications in SCN2A-related disorders. *Brain J. Neurol.* **2017**, *140*, 1316–1336. [CrossRef]
59. Schwarz, N.; Bast, T.; Gaily, E.; Golla, G.; Gorman, K.M.; Griffiths, L.R.; Hahn, A.; Hukin, J.; King, M.; Korff, C.; et al. Clinical and genetic spectrum of SCN2A-associated episodic ataxia. *Eur. J. Paediatr. Neurol. EJPN Off. J. Eur. Paediatr. Neurol. Soc.* **2019**, *23*, 438–447. [CrossRef]
60. Zaman, T.; Helbig, K.L.; Clatot, J.; Thompson, C.H.; Kang, S.K.; Stouffs, K.; Jansen, A.E.; Verstraete, L.; Jacquinet, A.; Parrini, E.; et al. SCN3A-Related Neurodevelopmental Disorder: A Spectrum of Epilepsy and Brain Malformation. *Ann. Neurol.* **2020**, *88*, 348–362. [CrossRef]
61. Wang, Z.; Lin, Y.; Liu, W.; Kuang, P.; Lao, W.; Ji, Y.; Zhu, H. Voltage-Gated Sodium Channels Are Involved in Cognitive Impairments in Parkinson's Disease- like Rats. *Neuroscience* **2019**, *418*, 231–243. [CrossRef]
62. Franklin, J.P.; Cooper-Knock, J.; Baheerathan, A.; Moll, T.; Mannikko, R.; Heverin, M.; Hardiman, O.; Shaw, P.J.; Hanna, M.G. Concurrent sodium channelopathies and amyotrophic lateral sclerosis supports shared pathogenesis. *Amyotroph. Lateral Scler. Front. Degener.* **2020**, *21*, 627–630. [CrossRef]
63. Corrochano, S.; Blanco, G.; Acevedo-Arozena, A. Skeletal Muscle Modulates Huntington's Disease Pathogenesis in Mice: Role of Physical Exercise. *J. Exp. Neurosci.* **2018**, *12*, 1179069518809059. [CrossRef] [PubMed]
64. Black, J.A.; Newcombe, J.; Waxman, S.G. Astrocytes within multiple sclerosis lesions upregulate sodium channel Nav1.5. *Brain J. Neurol.* **2010**, *133*, 835–846. [CrossRef] [PubMed]

65. Black, J.A.; Newcombe, J.; Waxman, S.G. Nav1.5 sodium channels in macrophages in multiple sclerosis lesions. *Mult. Scler.* **2013**, *19*, 532–542. [CrossRef]
66. Hiyama, T.Y.; Watanabe, E.; Ono, K.; Inenaga, K.; Tamkun, M.M.; Yoshida, S.; Noda, M. Na(x) channel involved in CNS sodium-level sensing. *Nat. Neurosci.* **2002**, *5*, 511–512. [CrossRef]
67. Li, S.; Wang, X.; Ma, Q.; Yang, W.-L.; Zhang, X.-G.; Dawe, G.S.; Xiao, Z.-C. Amyloid precursor protein modulates Nav1.6 sodium channel currents through a Go-coupled JNK pathway. *Sci. Rep.* **2016**, *6*, 39320. [CrossRef] [PubMed]
68. Yuan, D.; Yang, G.; Wu, W.; Li, Q.; Xu, D.; Ntim, M.; Jiang, C.; Liu, J.; Zhang, Y.; Wang, Y.; et al. Reducing Nav1.6 expression attenuates the pathogenesis of Alzheimer's disease by suppressing BACE1 transcription. *Aging Cell* **2022**, *21*, e13593. [CrossRef]
69. Saba, L.; Viscomi, M.T.; Martini, A.; Caioli, S.; Mercuri, N.B.; Guatteo, E.; Zona, C. Modified age-dependent expression of NaV1.6 in an ALS model correlates with motor cortex excitability alterations. *Neurobiol. Dis.* **2019**, *130*, 104532. [CrossRef]
70. Wagnon, J.L.; Barker, B.S.; Ottolini, M.; Park, Y.; Volkheimer, A.; Valdez, P.; Swinkels, M.E.M.; Patel, M.K.; Meisler, M.H. Loss-of-function variants of SCN8A in intellectual disability without seizures. *Neurology Genet.* **2017**, *3*, e170. [CrossRef]
71. Blanchard, M.G.; Willemsen, M.H.; Walker, J.B.; Dib-Hajj, S.D.; Waxman, S.G.; Jongmans, M.C.; Kleefstra, T.; van de Warrenburg, B.P.; Praamstra, P.; Nicolai, J.; et al. De novo gain-of-function and loss-of-function mutations of SCN8A in patients with intellectual disabilities and epilepsy. *J. Med. Genet.* **2015**, *52*, 330–337. [CrossRef]
72. Craner, M.J.; Newcombe, J.; Black, J.A.; Hartle, C.; Cuzner, M.L.; Waxman, S.G. Molecular changes in neurons in multiple sclerosis: Altered axonal expression of Nav1.2 and Nav1.6 sodium channels and Na^+/Ca^{2+} exchanger. *Proc. Natl. Acad. Sci. USA* **2004**, *101*, 8168–8173. [CrossRef]
73. Damarjian, T.G.; Craner, M.J.; Black, J.A.; Waxman, S.G. Upregulation and colocalization of p75 and Nav1.8 in Purkinje neurons in experimental autoimmune encephalomyelitis. *Neurosci. Lett.* **2004**, *369*, 186–190. [CrossRef]
74. Helbig, K.L.; Goldberg, E.M. SCN3A-Related Neurodevelopmental Disorder. In *GeneReviews(R)*; Adam, M.P., Mirzaa, G.M., Pagon, R.A., Wallace, S.E., Bean, L.J.H., Gripp, K.W., Amemiya, A., Eds.; University of Washington: Seattle, WA, USA, 1993.
75. Human Genomics. The Genotype-Tissue Expression (GTEx) pilot analysis: Multitissue gene regulation in humans. *Science* **2015**, *348*, 648–660. [CrossRef]
76. de Ligt, J.; Willemsen, M.H.; van Bon, B.W.; Kleefstra, T.; Yntema, H.G.; Kroes, T.; Vulto-van Silfhout, A.T.; Koolen, D.A.; de Vries, P.; Gilissen, C.; et al. Diagnostic exome sequencing in persons with severe intellectual disability. *N. Engl. J. Med.* **2012**, *367*, 1921–1929. [CrossRef] [PubMed]
77. Rauch, A.; Wieczorek, D.; Graf, E.; Wieland, T.; Endele, S.; Schwarzmayr, T.; Albrecht, B.; Bartholdi, D.; Beygo, J.; Di Donato, N.; et al. Range of genetic mutations associated with severe non-syndromic sporadic intellectual disability: An exome sequencing study. *Lancet* **2012**, *380*, 1674–1682. [CrossRef]
78. Echevarria-Cooper, D.M.; Hawkins, N.A.; Misra, S.N.; Huffman, A.M.; Thaxton, T.; Thompson, C.H.; Ben-Shalom, R.; Nelson, A.D.; Lipkin, A.M.; George, A.L., Jr.; et al. Cellular and behavioral effects of altered NaV1.2 sodium channel ion permeability in Scn2aK1422E mice. *Hum. Mol. Genet.* **2022**, *31*, 2964–2988. [CrossRef] [PubMed]
79. Shapiro, L.E.; Katz, C.P.; Wasserman, S.H.; DeFesi, C.R.; Surks, M.I. Heat stress and hydrocortisone are independent stimulators of triiodothyronine-induced growth hormone production in cultured rat somatotrophic tumour cells. *Acta Endocrinol.* **1991**, *124*, 417–424. [CrossRef] [PubMed]
80. Talwar, D.; Hammer, M.F. SCN8A Epilepsy, Developmental Encephalopathy, and Related Disorders. *Pediatr. Neurol.* **2021**, *122*, 76–83. [CrossRef] [PubMed]
81. Sadleir, L.G.; Mountier, E.I.; Gill, D.; Davis, S.; Joshi, C.; DeVile, C.; Kurian, M.A.; Mandelstam, S.; Wirrell, E.; Nickels, K.C.; et al. Not all SCN1A epileptic encephalopathies are Dravet syndrome: Early profound Thr226Met phenotype. *Neurology* **2017**, *89*, 1035–1042. [CrossRef]
82. Berecki, G.; Bryson, A.; Terhag, J.; Maljevic, S.; Gazina, E.V.; Hill, S.L.; Petrou, S. SCN1A gain of function in early infantile encephalopathy. *Ann. Neurol.* **2019**, *85*, 514–525. [CrossRef]
83. Zhang, L.I.; Poo, M.M. Electrical activity and development of neural circuits. *Nat. Neurosci.* **2001**, *4*, 1207–1214. [CrossRef]
84. Kawasaki, K.; Suzuki, Y.; Yamamura, H.; Imaizumi, Y. Rapid Na(+) accumulation by a sustained action potential impairs mitochondria function and induces apoptosis in HEK293 cells expressing non-inactivating Na(+) channels. *Biochem. Biophys. Res. Commun.* **2019**, *513*, 269–274. [CrossRef]
85. Selkoe, D.J. Deciphering the genesis and fate of amyloid beta-protein yields novel therapies for Alzheimer disease. *J. Clin. Investig.* **2002**, *110*, 1375–1381. [CrossRef]
86. Gandy, S.; Naslund, J.; Nordstedt, C. Alzheimer's disease. Molecular consequences of presenilin-1 mutation. *Nature* **2001**, *411*, 654–656. [CrossRef]
87. Zhang, Y.-W.; Thompson, R.; Zhang, H.; Xu, H. APP processing in Alzheimer's disease. *Mol. Brain* **2011**, *4*, 3. [CrossRef]
88. Cao, X.; Sudhof, T.C. A transcriptionally [correction of transcriptively] active complex of APP with Fe65 and histone acetyltransferase Tip60. *Science* **2001**, *293*, 115–120. [CrossRef]
89. Inomata, H.; Nakamura, Y.; Hayakawa, A.; Takata, H.; Suzuki, T.; Miyazawa, K.; Kitamura, N. A scaffold protein JIP-1b enhances amyloid precursor protein phosphorylation by JNK and its association with kinesin light chain 1. *J. Biol. Chem.* **2003**, *278*, 22946–22955. [CrossRef]
90. Borg, J.P.; Ooi, J.; Levy, E.; Margolis, B. The phosphotyrosine interaction domains of X11 and FE65 bind to distinct sites on the YENPTY motif of amyloid precursor protein. *Mol. Cell Biol.* **1996**, *16*, 6229–6241. [CrossRef]

91. Zambrano, N.; Bruni, P.; Minopoli, G.; Mosca, R.; Molino, D.; Russo, C.; Schettini, G.; Sudol, M.; Russo, T. The beta-amyloid precursor protein APP is tyrosine-phosphorylated in cells expressing a constitutively active form of the Abl protoncogene. *J. Biol. Chem.* **2001**, *276*, 19787–19792. [CrossRef]
92. Howell, B.W.; Lanier, L.M.; Frank, R.; Gertler, F.B.; Cooper, J.A. The disabled 1 phosphotyrosine-binding domain binds to the internalization signals of transmembrane glycoproteins and to phospholipids. *Mol. Cell Biol.* **1999**, *19*, 5179–5188. [CrossRef]
93. Roncarati, R.; Sestan, N.; Scheinfeld, M.H.; Berechid, B.E.; Lopez, P.A.; Meucci, O.; McGlade, J.C.; Rakic, P.; D'Adamio, L. The gamma-secretase-generated intracellular domain of beta-amyloid precursor protein binds Numb and inhibits Notch signaling. *Proc. Natl. Acad. Sci. USA* **2002**, *99*, 7102–7107. [CrossRef]
94. Liu, C.; Tan, F.C.K.; Xiao, Z.-C.; Dawe, G.S. Amyloid precursor protein enhances Nav1.6 sodium channel cell surface expression. *J. Biol. Chem.* **2015**, *8*, 12048. [CrossRef]
95. Busche, M.A.; Chen, X.; Henning, H.A.; Reichwald, J.; Staufenbiel, M.; Sakmann, B.; Konnerth, A. Henning, +3, and Arthur Konnerth Critical role of soluble amyloid-β for early hippocampal hyperactivity in a mouse model of Alzheimer's disease. *Proc. Natl. Acad. Sci. USA* **2012**, *109*, 8740–8745. [CrossRef]
96. Ciccone, R.; Franco, C.; Piccialli, I.; Boscia, F.; Casamassa, A.; de Rosa, V.; Cepparulo, P.; Cataldi, M.; Annunziato, L.; Pannaccione, A. Amyloid β-Induced Upregulation of Nav1.6 Underlies Neuronal Hyperactivity in Tg2576 Alzheimer's Disease Mouse Model. *Sci. Rep.* **2019**, *9*, 13592. [CrossRef]
97. Hammond, C.; Bergman, H.; Brown, P. Pathological synchronization in Parkinson's disease: Networks, models and treatments. *Trends Neurosci.* **2007**, *30*, 357–364. [CrossRef]
98. Bergman, H.; Deuschl, G. Pathophysiology of Parkinson's disease: From clinical neurology to basic neuroscience and back. *Mov. Disord. Off. J. Mov. Disord. Soc.* **2002**, *17* (Suppl. 3), S28–S40. [CrossRef]
99. Schwab, B.C.; Heida, T.; Zhao, Y.; Marani, E.; van Gils, S.A.; van Wezel, R.J. Synchrony in Parkinson's disease: Importance of intrinsic properties of the external globus pallidus. *Front. Syst. Neurosci.* **2013**, *7*, 60. [CrossRef]
100. Dolgacheva, L.P.; Zinchenko, V.P.; Goncharov, N.V. Molecular and Cellular Interactions in Pathogenesis of Sporadic Parkinson Disease. *Int. J. Mol. Sci.* **2022**, *23*, 13043. [CrossRef]
101. Zhu, H.; Wang, Z.; Jin, J.; Pei, X.; Zhao, Y.; Wu, H.; Lin, W.; Tao, J.; Ji, Y. Parkinson's disease-like forelimb akinesia induced by BmK I, a sodium channel modulator. *Behav. Brain Res.* **2016**, *308*, 166–176. [CrossRef]
102. Rogawski, M.A.; Loscher, W. The neurobiology of antiepileptic drugs. *Nat. Rev. Neurosci.* **2004**, *5*, 553–564. [CrossRef]
103. Liu, W.; Lao, W.; Zhang, R.; Zhu, H. Altered expression of voltage gated sodium channel Nav1.1 is involved in motor ability in MPTP-treated mice. *Brain Res. Bull.* **2021**, *170*, 187–198. [CrossRef]
104. Ogiwara, I.; Miyamoto, H.; Morita, N.; Atapour, N.; Mazaki, E.; Inoue, I.; Takeuchi, T.; Itohara, S.; Yanagawa, Y.; Obata, K.; et al. Nav1.1 localizes to axons of parvalbumin-positive inhibitory interneurons: A circuit basis for epileptic seizures in mice carrying an Scn1a gene mutation. *J. Neurosci. Off. J. Soc. Neurosci.* **2007**, *27*, 5903–5914. [CrossRef]
105. Duflocq, A.; Le Bras, B.; Bullier, E.; Couraud, F.; Davenne, M. Nav1.1 is predominantly expressed in nodes of Ranvier and axon initial segments. *Mol. Cell. Neurosci.* **2008**, *39*, 180–192. [CrossRef] [PubMed]
106. Sadeghian, M.; Mullali, G.; Pocock, J.M.; Piers, T.; Roach, A.; Smith, K.J. Neuroprotection by safinamide in the 6-hydroxydopamine model of Parkinson's disease. *Neuropathol. Appl. Neurobiol.* **2016**, *42*, 423–435. [CrossRef] [PubMed]
107. Chazot, P.L. Safinamide (Newron Pharmaceuticals). *Curr. Opin. Investig. Drugs* **2001**, *2*, 809–813. [PubMed]
108. Caccia, C.; Maj, R.; Calabresi, M.; Maestroni, S.; Faravelli, L.; Curatolo, L.; Salvati, P.; Fariello, R.G. Safinamide: From molecular targets to a new anti-Parkinson drug. *Neurology* **2006**, *67*, S18–S23. [CrossRef]
109. Mathis, S.; Couratier, P.; Julian, A.; Vallat, J.M.; Corcia, P.; Le Masson, G. Management and therapeutic perspectives in amyotrophic lateral sclerosis. *Expert Rev. Neurother.* **2017**, *17*, 263–276. [CrossRef]
110. Hardiman, O.; van den Berg, L.H. Edaravone: A new treatment for ALS on the horizon? *Lancet. Neurol.* **2017**, *16*, 490–491. [CrossRef]
111. Petrov, D.; Mansfield, C.; Moussy, A.; Hermine, O. ALS Clinical Trials Review: 20 Years of Failure. Are We Any Closer to Registering a New Treatment? *Front. Aging Neurosci.* **2017**, *9*, 68. [CrossRef]
112. Hinchcliffe, M.; Smith, A. Riluzole: Real-world evidence supports significant extension of median survival times in patients with amyotrophic lateral sclerosis. *Degener. Neurol. Neuromuscul. Dis.* **2017**, *7*, 61–70. [CrossRef]
113. Silani, V. Therapy in Amyotrophic Lateral Sclerosis (ALS): An unexpected evolving scenario. *Arch. Ital. De Biol.* **2017**, *155*, 118–130. [CrossRef]
114. Blasco, H.; Mavel, S.; Corcia, P.; Gordon, P.H. The glutamate hypothesis in ALS: Pathophysiology and drug development. *Curr. Med. Chem.* **2014**, *21*, 3551–3575. [CrossRef]
115. Lazarevic, V.; Yang, Y.; Ivanova, D.; Fejtova, A.; Svenningsson, P. Riluzole attenuates the efficacy of glutamatergic transmission by interfering with the size of the readily releasable neurotransmitter pool. *Neuropharmacology* **2018**, *143*, 38–48. [CrossRef]
116. Carunchio, I.; Curcio, L.; Pieri, M.; Pica, F.; Caioli, S.; Viscomi, M.T.; Molinari, M.; Canu, N.; Bernardi, G.; Zona, C. Increased levels of p70S6 phosphorylation in the G93A mouse model of Amyotrophic Lateral Sclerosis and in valine-exposed cortical neurons in culture. *Exp. Neurol.* **2010**, *226*, 218–230. [CrossRef]
117. Lamanauskas, N.; Nistri, A. Riluzole blocks persistent Na^+ and Ca^{2+} currents and modulates release of glutamate via presynaptic NMDA receptors on neonatal rat hypoglossal motoneurons in vitro. *Eur. J. Neurosci.* **2008**, *27*, 2501–2514. [CrossRef]

118. Carter, B.C.; Giessel, A.J.; Sabatini, B.L.; Bean, B.P. Transient sodium current at subthreshold voltages: Activation by EPSP waveforms. *Neuron* **2012**, *75*, 1081–1093. [CrossRef] [PubMed]
119. Ceballos, C.C.; Roque, A.C.; Leao, R.M. The role of negative conductances in neuronal subthreshold properties and synaptic integration. *Biophys. Rev.* **2017**, *9*, 827–834. [CrossRef]
120. Vucic, S.; Kiernan, M.C. Transcranial Magnetic Stimulation for the Assessment of Neurodegenerative Disease. *Neurother. J. Am. Soc. Exp. Neuro Ther.* **2017**, *14*, 91–106. [CrossRef] [PubMed]
121. Ozdinler, P.H.; Benn, S.; Yamamoto, T.H.; Guzel, M.; Brown, R.H., Jr.; Macklis, J.D. Corticospinal motor neurons and related subcerebral projection neurons undergo early and specific neurodegeneration in hSOD1G(9)(3)A transgenic ALS mice. *J. Neurosci. Off. J. Soc. Neurosci.* **2011**, *31*, 4166–4177. [CrossRef]
122. Geevasinga, N.; Menon, P.; Ozdinler, P.H.; Kiernan, M.C.; Vucic, S. Pathophysiological and diagnostic implications of cortical dysfunction in ALS. *Nat. Rev. Neurol.* **2016**, *12*, 651–661. [CrossRef]
123. Compston, A.; Coles, A. Multiple sclerosis. *Lancet* **2008**, *372*, 1502–1517. [CrossRef] [PubMed]
124. Popescu, B.F.; Pirko, I.; Lucchinetti, C.F. Pathology of multiple sclerosis: Where do we stand? *Contin* **2013**, *19*, 901–921. [CrossRef] [PubMed]
125. t Hart, B.A.; Luchicchi, A.; Schenk, G.J.; Stys, P.K.; Geurts, J.J.G. Mechanistic underpinning of an inside-out concept for autoimmunity in multiple sclerosis. *Ann. Clin. Transl. Neurol.* **2021**, *8*, 1709–1719. [CrossRef] [PubMed]
126. Shields, S.D.; Ahn, H.S.; Yang, Y.; Han, C.; Seal, R.P.; Wood, J.N.; Waxman, S.G.; Dib-Hajj, S.D. Nav1.8 expression is not restricted to nociceptors in mouse peripheral nervous system. *Pain* **2012**, *153*, 2017–2030. [CrossRef] [PubMed]
127. Pappalardo, L.W.; Samad, O.A.; Black, J.A.; Waxman, S.G. Voltage-gated sodium channel Nav 1.5 contributes to astrogliosis in an in vitro model of glial injury via reverse Na^+/Ca^{2+} exchange. *Glia* **2014**, *62*, 1162–1175. [CrossRef]
128. Black, J.A.; Liu, S.; Waxman, S.G. Sodium channel activity modulates multiple functions in microglia. *Glia* **2009**, *57*, 1072–1081. [CrossRef]

life

MDPI

Review

Voltage-Gated Na^+ Channels in Alzheimer's Disease: Physiological Roles and Therapeutic Potential

Timothy J. Baumgartner, Zahra Haghighijoo, Nana A. Goode, Nolan M. Dvorak, Parsa Arman and Fernanda Laezza *

Department of Pharmacology & Toxicology, The University of Texas Medical Branch, Galveston, TX 77555, USA; tjbaumga@utmb.edu (T.J.B.); zahaghig@utmb.edu (Z.H.); nagoode@utmb.edu (N.A.G.); nmdvorak@utmb.edu (N.M.D.); paarman@utmb.edu (P.A.)

* Correspondence: felaezza@utmb.edu; Tel.: +1-(409)-772-9672; Fax: +1-(409)-772-9642

Abstract: Alzheimer's disease (AD) is the most common cause of dementia and is classically characterized by two major histopathological abnormalities: extracellular plaques composed of amyloid beta (Aβ) and intracellular hyperphosphorylated tau. Due to the progressive nature of the disease, it is of the utmost importance to develop disease-modifying therapeutics that tackle AD pathology in its early stages. Attenuation of hippocampal hyperactivity, one of the earliest neuronal abnormalities observed in AD brains, has emerged as a promising strategy to ameliorate cognitive deficits and abate the spread of neurotoxic species. This aberrant hyperactivity has been attributed in part to the dysfunction of voltage-gated Na^+ (Nav) channels, which are central mediators of neuronal excitability. Therefore, targeting Nav channels is a promising strategy for developing disease-modifying therapeutics that can correct aberrant neuronal phenotypes in early-stage AD. This review will explore the role of Nav channels in neuronal function, their connections to AD pathology, and their potential as therapeutic targets.

Keywords: voltage-gated sodium channels; Alzheimer's disease; excitability; hippocampus; neurodegeneration; plasticity; pharmacology

Citation: Baumgartner, T.J.; Haghighijoo, Z.; Goode, N.A.; Dvorak, N.M.; Arman, P.; Laezza, F. Voltage-Gated Na^+ Channels in Alzheimer's Disease: Physiological Roles and Therapeutic Potential. *Life* **2023**, *13*, 1655. https://doi.org/10.3390/life13081655

Academic Editors: Carlo Musio and Kiminobu Sugaya

Received: 26 May 2023
Revised: 11 July 2023
Accepted: 26 July 2023
Published: 29 July 2023

1. Introduction

Alzheimer's disease (AD) is a progressive neurodegenerative disorder classically characterized by the accumulation of amyloid beta (Aβ) plaques and hyperphosphorylated tau aggregates that disrupt synaptic function, ultimately culminating in synaptic decline and neurodegeneration [1]. Current FDA-approved small-molecule therapeutics for AD include acetylcholinesterase inhibitors [2] and NMDA receptor antagonists [3], which are effective in providing symptomatic relief but lack disease-modifying properties. FDA-approved monoclonal antibodies, such as aducanumab [4] and lecanemab [5], show efficacy in the clearance of Aβ, but there is a lack of evidence that they convincingly slow AD progression among large clinical populations. Thus, there remains an unmet need for the development of disease-modifying therapeutics for AD.

Accumulation of neurotoxic proteins in key brain regions induces neuronal deficits that are widely thought to be the cause of AD symptoms. The precise mechanisms of Aβ- and tau-mediated AD pathology remain to be elucidated, an issue which is further complicated by interpatient variability [6]. Nonetheless, AD is defined by the accumulation of Aβ and tau deposits [7]. Aβ deposition begins in the frontomedial and temporobasal areas, spreading then to the remaining neocortical regions [8]. Tau accumulation is first observed in the entorhinal cortex [9] and spreads successively into the hippocampus [10]. While intricacies of the relationship between Aβ and tau seeding and accumulation remain elusive, several studies suggest that Aβ may facilitate the seeding of tau [11–14].

The Aβ and tau proteins progressively accumulate at synapses, interrupting synaptic communication through the degeneration of dendritic spines [15], leading to axonal degeneration and eventual neuronal loss [16]. These phenomena progressively hinder the function of the hippocampal circuit, inducing deficits in long-term potentiation (LTP) and long-term depression (LTD), two forms of synaptic plasticity widely thought to be the basis of progressive memory loss in AD [17].

2. Hippocampal Hyperactivity in Early-Stage AD

Prior to global neurodegeneration and resultant progressive loss of memory associated with late stages of AD, hippocampal hyperactivity is observed in rodent models [18–20] as well as human patients [21–24]. Functional MRI studies indicate that patients with mild cognitive impairment (MCI) display increased hippocampal activation during memory-related tasks compared to healthy adults [25,26], and this phenomenon has emerged as a potential biomarker of mild cognitive impairment and early-stage AD [27,28].

This hyperactivity occurs prior to amyloid plaque deposition [20,29,30], positioning the phenotype as one of the first neurophysiological alterations in the AD brain. While there remain many questions to be answered regarding the precise mechanisms, origins, and consequences of this phenotype, it has emerged as a common feature in AD that precedes greater cognitive decline [31]. In support of this aberrant elevated neuronal activity as a precursor to AD, it has been has been linked to cognitive dysfunction and decreased memory performance [27,31,32] as well as the production and accumulation of Aβ and tau [27,33–38]. Moreover, amelioration of hippocampal hyperactivity using anti-epileptics, such as levetiracetam, has been shown to improve cognition and memory performance in rodent models and patients with MCI or early-stage AD [32,39,40]. Therefore, given its acute and longitudinal impacts on AD pathophysiology, correcting hippocampal hyperactivity represents a promising and potentially disease-modifying approach for AD treatment.

As described above, the hyperactivity phenotype is linked to various neuronal processes that accelerate the rate of AD progression. Therefore, evaluation of molecular contributors to the phenotype is warranted. On account of their centrality in initiating and propagating the action potential (AP) [41,42], in this review, we discuss the contribution of voltage-gated sodium (Nav) channels to the hyperactivity phenotype observed in early-stage AD, their function as the disease progresses, and their viability as therapeutic targets for the disease.

3. Overview of Nav Channels in AD Pathology

Broadly speaking, proper hippocampal function is mediated by the activity of glutamatergic principal neurons and GABAergic interneurons [43–45]. Thus, in order to ameliorate aberrant hyperactivity, the functional contributions of each neuronal subtype must be considered. Glutamatergic principal cells comprise the vast majority of hippocampal neurons and are densely packed into layers [46], notably the dentate gyrus (DG), CA3, and CA1 regions whose excitatory connections compose the trisynaptic circuit [47,48]. GABAergic inhibitory interneurons, despite only comprising ~10% of hippocampal cells [43,46], dramatically influence the activity of the hippocampal circuit [49]. Unlike principal cells, interneuron subtypes display significant intrinsic and morphological diversity and are disseminated throughout all hippocampal subfields [43]. The diversity of interneuron axonal projections lends them the capacity to synapse onto single cells or neuronal clusters throughout the hippocampal formation, providing essential GABAergic tone [49,50] and preventing excessive excitation of principal neurons through feedforward and feedback inhibition [43,50]. Hippocampal network dynamics rely on the balance between the excitatory and inhibitory signals (E/I balance) [51,52] produced by the aforementioned cell types. Therefore, pathological deviations in hippocampal activity can be ascribed to altered excitability of either principal cells or interneurons.

Nav channels, which are molecular determinants of neuronal excitability, consist of nine distinct isoforms (Nav1.1–Nav1.9) [53] that have varying expression profiles among different cell types [54–56]. Nav1.1 and Nav1.6 are of particular interest as they display distinctive enriched expression in hippocampal interneurons [57,58] and principal neurons [54,59], respectively, enabling the initiation and propagation of action potentials in these cell types. In early-stage AD, Nav1.1 and Nav1.6 display unique expression profiles and functional activity [29,60], contributing centrally to the aberrant excitability of interneurons and principal cells and resulting in hippocampal hyperactivity. Given the dichotomous functional roles of Nav1.1 and Nav1.6 channels in modulating hippocampal activity, the two isoforms will be considered separately. In this section, the physiological roles and disease-associated functions of Nav1.1 and Nav1.6 channels in their respective cell types will be discussed.

3.1. Nav1.1

GABAergic inhibitory interneurons comprise a small percentage of hippocampal neurons [46] but are able to contribute substantially to the regulation of hippocampal excitability due to their distribution and diverse axonal projections [43,44]. Hippocampal interneurons are classified into several major subtypes based on neuronal molecular expression, including somatostatin neurons, parvalbumin neurons, neuropeptide Y neurons, vasoactive intestinal peptide neurons, and cholecystokinin neurons [61]. Of these subtypes, somatostatin (SOM)- and parvalbumin (PV)-positive interneurons comprise a large majority (~70–80%) of the hippocampal interneuron population [62]. The Nav1.1 channel is predominantly expressed in inhibitory interneurons [60,63], where it serves as a key molecular determinant of intrinsic firing. With its subcellular localization in soma, axon initial segments (AIS), and axons, Nav1.1 channels facilitate the initiation and propagation of action potentials in these cells [63,64].

Loss-of-function mutations to Nav1.1 result in reduced Na^+ current, causing reduced action potential firing in interneurons [57]. The imbalance caused by decreased interneuron activity can lead to seizures and is the leading cause of multiple epilepsies and other disorders, most notably Dravet's syndrome [58,65]. It has been shown that epileptiform activity, particularly in the form of nonconvulsive seizures, is common in early-stage AD patients [66]. Multiple lines of evidence indicate that this epileptiform activity accelerates the rate of AD pathology and worsening of cognitive symptoms [66,67]. Along with displaying epileptiform activity, decreased levels of Nav1.1 expression have been observed across multiple transgenic AD mouse models [68–70] as well as AD patients [70], further supporting the pathophysiological role of this channel in AD. Early studies demonstrated that GABAergic interneurons are resistant to Aβ and tau protein deposition [71,72]. Highlighting this finding, in a study of the expression of hyperphosphorylated tau in AD patients, Blazquez-Llorca et al. found that of almost 4000 PV-positive interneurons in the hippocampal formation and entorhinal cortex analyzed, paired-helical filaments of tau were present in only two [73]. Despite GABAergic interneurons being resistant to Aβ and tau deposition [71,72], the activity of these neurons is dysfunctional in AD, and the precise pathophysiological mechanisms are the focus of many recent investigations [74–77].

One proposed mechanism by which GABAergic interneuron activity becomes compromised in AD is through the activity of β-site APP cleaving enzyme 1 (BACE1) [78], a secretase involved in Aβ production. Crucially, BACE1 activity and levels are significantly increased in the brains of AD patients, which is thought to contribute to the progression of AD by increasing Aβ [79–81]. In addition to its role in Aβ production, the Nav channel β2 subunit, which is covalently linked to the Nav1.1 alpha subunit, is a substrate for BACE1 [78]. Moreover, increased BACE1 causes a reduction in Nav1.1 cell surface expression [78], which would be expected to decrease the excitability of GABAergic interneurons. Consistent with such an expectation, PV interneurons in the hippocampus display reduced excitability in rodent models of AD [82]. Therefore, despite their apparent resistance to Aβ and tau pathology [71–73], depletion of Nav1.1 and suppression of its activity in hippocampal interneurons is potentially a major contributor to network dysfunction and cognitive deficits in AD [70].

3.2. Nav1.6

Contrarily to GABAergic interneurons, glutamatergic principal cells comprise a majority of neurons in the hippocampus [46]. Also, in contrast to interneurons, principal cells display a distinct layered organizational pattern with a relatively high level of anatomical uniformity [83,84]. Nav1.6 is extensively expressed in the adult human brain and is the predominant Nav isoform in hippocampal principal cells [85]. Nav1.6 displays a distinctive pattern of subcellular localization, with concentrated expression at the AIS and nodes of Ranvier in hippocampal principal neurons [59]. This pattern allows the channel to centrally regulate action potential initiation, propagation, and saltatory conduction in these cells [86] and govern the spike threshold of glutamatergic principal cells and contribute to synaptic integration [54]. This distinct role of Nav1.6 is due to its accumulation in the distal AIS and hyperpolarized voltage dependence of activation, setting it apart from other Nav channel isoforms at the AIS. Given the central role of Nav1.6 in regulating the action potential in excitatory neurons, both gain-of-function and loss-of-function mutations have been found to have pathogenic consequences [59]. Functional alterations to the Nav1.6 channel have been implicated in a number of neuropsychiatric disorders [59,87–89]. It has been observed that haploinsufficiency of the SCN8A gene results in cognitive impairment and has been linked to intellectual disability [90]. Conversely, gain-of-function mutations to Nav1.6, which promote aberrant high-frequency neuronal firing, are closely linked with epileptiform activity [88,91].

In the context of AD, Nav1.6 has emerged as the focus of many investigations regarding network abnormalities and disease progression [29,34,92,93]. It has been observed that acute exposure to soluble Aβ oligomers results in hyperactivity of CA1 pyramidal neurons [94]. Further, it was observed that soluble Aβ exposure selectively upregulated the expression and function of Nav1.6 [29] and that pyramidal cell hyperactivity was abolished following treatment with Nav channel blockers [94], implicating this Nav channel isoform as a driver of AD-related hyperactivity. Additionally, it is observed that amyloid precursor protein (APP) both directly interacts with Nav1.6 [93] and enhances Nav1.6 surface expression via a G protein-coupled JNK pathway [92]. Functionally, this results in the potentiation of Nav1.6-mediated Na^+ currents [92]. These mechanisms provide convincing evidence for the contribution of Nav1.6 to the hyperexcitation of principal neurons in early AD and reveal novel, potentially synergistic pathways that drive AD pathology.

4. Therapeutic Potential of Nav Channels for AD

Nav1.1 and Nav1.6 channels have distinct roles in hippocampal hyperactivity and are potential therapeutic targets for early-stage AD [39,69]. However, current Nav channel modulators have the potential for off-target effects due to a lack of isoform selectivity. With the development of high-resolution Nav channel structures [72] and a better understanding of the biology of isoform-specific accessory proteins and signaling pathways [95], progress has been made towards developing targeted therapies that are selective for specific isoforms. This section will discuss Nav1.1 and Nav1.6 modulation mechanisms and potential therapeutic strategies for early AD.

4.1. Nav1.1 Activation

Increased BACE1 activity in early-stage AD reduces Nav1.1 cell surface expression, leading to decreased excitability of GABAergic interneurons in the hippocampus and contributing to increased excitatory/inhibitory tone. Restoring proper interneuron activity may help correct this imbalance. One potential therapeutic strategy for overcoming deficits in interneuron activity in the hippocampus is the transplantation of interneuron progenitor cells [96]. This involves the extraction of progenitor cells from the medial ganglionic eminence and transplantation into the brain of the affected individual [96]. Following the procedure, these cells integrate into the existing neuronal circuitry and mature into functional inhibitory interneurons. In AD rodent models, it has been observed that hippocampal transplantation of wild-type progenitor cells provides little to no improvement;

however, transplantation of progenitor cells overexpressing the Nav1.1 channel results in improved behavior and cognitive performance [60]. Conversely, transplantation of Nav1.1-deficient progenitor cells impaired learning and behavioral functions [60]. Despite encouraging results following the transplantation of Nav1.1 interneuron progenitor cells in rodent models, the associated risks may hinder the translatability of this approach. Gene therapy represents an alternative strategy for the restoration of Nav1.1 activity and interneuron activity [97,98]. The feasibility of this approach has been demonstrated in rodent models, where voltage-gated potassium channels were delivered for the treatment of epilepsy or neuropathic pain [99,100]. Recent and promising developments involving engineered prokaryotic Nav channels for cardiac arrhythmias also illustrate the potential of this concept for clinical applications [101] and may inform future investigations regarding the delivery of neuronal Nav channels.

Cumulatively, these studies not only further indicate the importance of interneuron function during early AD but also establish that interneuron-based interventions for hippocampal hyperactivity are reliant on Nav1.1. Thus, pharmacologically potentiating Nav1.1 activity represents a promising therapeutic avenue for early-stage AD. There are several known classes of sodium channel activators, such as toxins from the venom of several organisms [102] or pyrethroid insecticides [103]. While therapeutic development from these agents is challenging due to their overall toxic effects, using these chemicals to selectively activate Nav channels allows for the demonstration of therapeutic benefit. For example, δ-theraphotoxin-Hm1a and δ-theraphotoxin-Hm1b (Hm1a and Hm1b) are toxins derived from the venom of the Heteroscodra maculata tarantula that potentiates the activity of the Nav1.1 channel selectively [104,105]. Hm1a-mediated activation of Nav1.1 is able to restore the function of inhibitory interneurons from Dravet syndrome mice without affecting the activity of excitatory neurons, and intracerebroventricular infusion of Hm1a in the Dravet syndrome rodent model rescued the animals from seizures and premature death [105]. While Hm1a and Hm1b exhibit high structural homology, further investigation revealed that Hm1b is more stable in biological fluids [106]. Electrophysiological studies revealed that the Hm1b peptide causes a hyperpolarizing shift in the voltage dependence of activation and delays fast inactivation of the Nav1.1 channel, increasing both peak and sustained Nav1.1 currents [106] through stabilizing interactions with its domain four voltage sensor [106]. While further optimization is required for clinical applications, these toxins demonstrate the value of Nav1.1 activation for restoring inhibitory interneuron activity in hyperexcitability disorders.

Given the considerable potential of Nav1.1 activators for Dravet syndrome and other excitability disorders, the development of small molecule Nav1.1 agonists has emerged as a rich area of investigation [102,107,108]. A 2019 high-throughput screening campaign conducted by Takeda Pharmaceutical Company led to the identification of a series of 4-phenyl-2-(pyrrolidinyl)-nicotinamide derivatives as potent and selective Nav1.1 activators [108], and structural optimization yielded a compound with favorable druglike properties that achieves brain concentrations comparable to its potency in vitro. While further investigations are required to determine the safety and translational of this compound's effects, it holds potential as a therapeutic for epilepsy disorders and thus may be applied for hyperexcitability in early-stage AD.

4.2. Nav1.6 Inhibition

In contrast to Nav1.1, Nav1.6 is primarily expressed in excitatory neurons, and increased expression and activity of the channel are observed in AD brains as a result of Aβ oligomer exposure [29,92]. Knockdown of the Nav1.6 channel in AD rodents rescues LTP deficits and mitigates hippocampal hyperactivity in AD models, restoring memory-associated beta, gamma, delta, and theta waves to normal levels [34]. Additionally, knocking down Nav1.6 decreased both the number and size of Aβ plaques in these animals, providing evidence for the disease-modifying potential of Nav1.6 modulation [34]. The reduction in Aβ plaque accumulation is caused by reduced transcription of BACE1. When

Aβ oligomers are present, genetic knockdown or pharmacological inhibition of Nav1.6 reduces BACE1 transcription and BACE1-mediated cleavage of APP, resulting in decreased Aβ production and plaque accumulation [34]. Intriguingly, while treatment with levetiracetam, an anti-epileptic with a mechanism independent of Nav channels, improves memory and cognitive performance, it fails to alter Aβ production or plaque accumulation [109]. Therefore, reducing hippocampal hyperactivity, specifically through modulation of Nav1.6, could provide both acute improvements to memory and cognition as well as simultaneously reducing AD pathophysiology.

NBI-921352 is a Nav channel inhibitor being investigated for the treatment of epilepsy caused by Nav1.6 gain-of-function mutations. It displays high selectivity for Nav1.6~130-fold greater potency for Nav1.6 over Nav1.1 and Nav1.2 isoforms [110]. By stabilizing Nav1.6 inactivated state, NBI-921352 inhibits persistent and resurgent currents, which are the source of hyperexcitability in pyramidal neurons [91]. Crucially, while NBI-921352 reduces excitability in principal cells, the activity of fast-spiking interneurons is spared [110]. Nav1.6 inhibition has shown promise in pre-clinical models of AD by correcting synaptic dysfunction, reducing Aβ plaque accumulation, and improving cognitive function [34]. Phase 1 trials indicate that NCI-921352 is well-tolerated in healthy adults and displays evidence of CNS activity [111]. Further studies are needed to assess NBI-921352's efficacy in diseased individuals. However, increased expression and activity of Nav1.6 in early AD provides the opportunity for increased potency in diseased tissues, and functional studies illustrate the potential of Nav1.6 as a therapeutic target for the correction of aberrant hyperexcitability.

5. Alternative Strategies for Modulation of Nav Channels

While the pore-forming α-subunit of Nav channels is primarily responsible for their physiological functions and has been the target for drug development of current Nav channel inhibitors, these channels rely on an array of intracellular channel-associated proteins (ChAPs) and post-translational modifications (PTMs) via various kinases for full function [112,113] (Figure 1, Table 1). These ChAPs have divergent regulatory effects and provide both isoforms as well as tissue specificity due to their unique intermolecular interactions with the channel as well their differential expression profiles in tissues [95,114–116]. In addition to selectivity, modulation of the Nav channel via ChAP complexes allows for bidirectional regulation of channel activity by either facilitating or inhibiting complex formation [115,117,118], therefore offering the potential for precise tuning of Nav channel activity through mechanisms beyond traditional agonism and inhibition. Promising ChAP modulators targeting specific regions, such as the C-terminal domain [113,114,119] of Nav channels, have been identified. These ChAP modulators hold the potential to pharmacologically modulate Nav channels with better precision and specificity. Additionally, many kinases have been intimately linked with intracellular ChAPs and exhibit altered functions or expression patterns in AD. Thus, targeting the functional interactions between the Nav channels, their interactors, and disease-associated kinase signaling pathways represents another novel strategy for AD drug development.

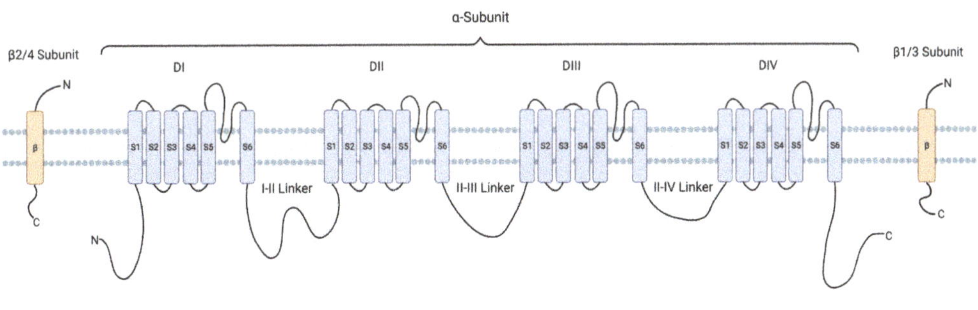

Figure 1. Schematic representation of Nav channel α- and β- subunit structure.

Table 1. Summary of Nav1.1 and Nav1.6 post-translational modifications (PTMs) and regulatory interactions with AD-associated proteins.

Nav1.1 ChAPs and PTMs				
	Type of Interaction	**Key Residues of Nav Channel**	**Functional Outcome**	**Reference**
AKT1	Phosphorylation of I-II Linker	S573, S684, S685, S704	Decreased Nav1.1 Activity	[120]
BACE1	Cleavage of β2 Subunit	C-terminus of β2 Subunit	Decreased Nav1.1 activity and α-subunit surface expression	[78]
Nav1.6 ChAPs and PTMs				
	Type of Interaction	**Key Residues of Nav Channel**	**Functional Outcome**	**Reference**
CaMKII	Phosphorylation of I-II linker	S561, S641, T642	Increased Nav1.6 Activity	[121]
p38 MAPK	Phosphorylation of I-II linker	S553	Decreased Nav1.6 Activity	[122]
FGF14	Binding to C-Terminal Domain	D1833, S1838, H1843, D1846, I1886, T1887, R1892	Increased Nav1.6 Activity	[113,118,123]
GSK3β	Phosphorylation of C-Terminal Domain	T1936	Increased Nav1.6 Activity	[115]

5.1. *Modulation of Nav Channels via Kinase Signaling Pathways*

In the AD brain, several kinase pathways are dysregulated, and the resultant alterations in signaling via these pathways are thought to contribute directly to AD progression [89–91]. Intriguingly, many of the kinases and associated signaling proteins linked with AD pathology are regulators of Nav1.1 or Nav1.6 and exert their functional effects via post-translational modifications (PTMs) and/or direct binding to the channel (Table 1) [92]. Thus, targeting the functional interactions between Nav1.1 or 1.6 and disease-associated proteins represents a novel strategy for AD drug development—and could allow for improved tissue selectivity over agents that modulate Nav channels alone. This section will explore several proteins with pathological links to AD, their regulatory effects on Nav channels, and how their interactions may be modulated to ameliorate hippocampal hyperactivity.

5.1.1. Ca^{2+}/Calmodulin-Dependent Protein Kinase

Calcium dysregulation during AD has long been investigated for its role in facilitating various facets of synaptic dysfunction and AD neuropathology [124–127]. Critically, aberrant alterations in Ca^{2+} signaling are detected in AD rodent models prior to severe memory impairments or histopathological biomarkers [124,128], indicating that correcting aberrant Ca^{2+} signaling and its downstream consequences may have disease-modifying potential. Increased Ca^{2+} levels in AD neurons occur through entry from extracellular space via Ca^{2+} permeable channels such as NMDA receptors [129] and voltage-gated calcium channels [130] or release from intracellular Ca^{2+} stores. These fluctuations in intracellular calcium are processed through the binding of Ca^{2+} ions to proteins such as calmodulin (CaM) [131]. The Ca^{2+}/CaM-dependent protein kinases (CaMKs) are the primary binding partners of Ca^{2+}/CaM complexes and are thus major effectors of alterations to Ca^{2+} ion flux in AD [132]. CaMKII, a member of the CaMK family, is a serine-threonine kinase that has altered expression and activity in the AD brain [133]. Following activation by Ca^{2+}/CaM, CaMKII temporarily retains its kinase activity through autophosphorylation of its Thr286 residue [134]. The coupling of CaMKII with Nav channels has been shown to modulate excitability through the regulation of sodium currents across multiple Nav isoforms and cell types [135,136]. Notably, recent reports indicate that CaMKII phosphorylates Nav1.6, enhancing channel activity and Nav1.6-mediated neuronal excitability [121]. Moreover, inhibition of CaMKII is observed to ameliorate the pathological consequences of Nav1.6 gain-of-function mutations, which result in hyperexcitability disorders [137]. Therefore, inhibiting CaMKII-mediated phosphorylation of the Nav1.6 channel in early AD may represent a novel strategy to ameliorate hippocampal hyperactivity.

5.1.2. Mitogen-Activated Protein Kinase

Mitogen-activated protein kinases (MAPKs) are a family of protein kinases best known for their contribution to various diverse cellular processes, including proliferation, inflammation, and apoptosis [138,139]. The MAPK family is comprised of 3 classes in mammals, extracellular-signal-regulated kinases (ERKS), Jun-amino-terminal kinases (JNKs), and stress-activated protein kinases (p38) [140]. The p38 MAPKs consist of 4 distinct isoforms: p38α, p38β, p38γ, and p38δ. Of particular interest, p38α is highly expressed in the cortex and hippocampus [141] and has been implicated as a regulator of synaptic plasticity [141,142]. In AD, elevated activity of p38 MAPKs is observed in the hippocampus during early phases [143,144], where they contribute to excitotoxicity and tau phosphorylation [145]. However, isoform-selective functional studies have revealed that the p38α isoform is likely the primary driver of AD pathogenesis [146], while other isoforms serve lesser pathogenic roles or potentially exert neuroprotective effects [147]. Nonetheless, given the consensus that p38α is a driver of AD pathology, inhibitors of this isoform have been evaluated and show some promise in ameliorating tau hyperphosphorylation, synaptic decline, AD-associated neuroinflammation, and cognitive impairment [148,149]. p38α has been identified as a direct regulator of Nav1.6 through phosphorylation of a Pro-Gly-Ser^{553}-Pro motif intracellular loop L1, which results in reduced Nav1.6-mediated currents [150]. However, this effect is dependent upon the Pro-Gly-Tyr^{1945} motif of the Nav1.6 C-terminal tail, which is responsible for the binding of Nedd4-2 ubiquitin ligase and subsequent internalization of the channel [122]. Curiously, however, in the absence of a functional Pro-Gly-Try^{1945} motif, it is observed that p38 phosphorylation of Nav 1.6 enhances peak current density [122], indicating that p38 phosphorylation of Nav1.6 may have opposing effects on Nav1.6-mediated neuronal excitability dependent on the function of Nedd4-2. Haploinsufficiency of Nedd4-2 is linked to increased seizure susceptibility [151], which is a hallmark symptom of prodromal AD [30,66,67]. Thus, it is possible that dysfunctional Nedd4-2 reduces Nav1.6 internalization and allows functional upregulation of the channel via p38α, ultimately resulting in aberrant hyperexcitability. The precise functional alterations to p38α and Nedd4-2 and their resultant effects with Nav1.6 during early-stage AD remain to be fully elucidated. Nonetheless, the powerful modulation of Nav1.6 induced by p38α through S^{553} phosphorylation warrants further investigation, and modulation of this system may allow for the regulation of hippocampal activity in early AD.

5.1.3. PI3K/AKT/GSK3β

The phosphatidylinositol 3-kinase (PI3K)/protein kinase B (AKT) pathway contributes to a broad range of physiological processes in the brain, including cell proliferation, differentiation, autophagy, and intraneuronal trafficking [152–154]. PI3K is the most upstream effector of this signaling cascade and is canonically activated by receptor tyrosine kinases in response to extracellular stimuli, followed by recruitment of its catalytic p110 subunit, which enables PI3K-medicated AKT phosphorylation [154,155]. Activated AKT then exerts modulatory effects on a myriad of downstream signaling molecules via phosphorylation and complexation, including mTOR, GABA receptors, and eukaryotic translation initiation factor α (eIF2α) [156,157], which mechanistically link the PI3K/AKT signaling axis to the regulation of synaptic plasticity. In AD brains, dysfunctional PI3K/AKT signaling has been linked to various pathogenic processes [153], including Aβ and tau pathology [158], neuroinflammation, impaired glucose metabolism, and oxidative stress [159], and restoration of its function may have neuroprotective effects [153]. AKT has also been identified as a regulator of Nav channels 1.1 [120] and 1.6 [160]. In the case of Nav1.1, it is observed that AKT directly phosphorylates the channel, resulting in decreased channel activity [120]. However, in CA1 pyramidal cells, Akt inhibition with triciribine results in increased action potential firing, which is likely driven by Nav1.6 [160]. This may be a result of direct interactions between AKT and Nav1.6 or a downstream effect of decreased activation of glycogen synthase kinase 3β (GSK3β), which phosphorylates the Nav1.6 channel at its T1936 residue [115], potentiating its activity. Moreover, GSK3β displays dysregulated function and increased expression in the hippocampus of AD patients and rodent models [161–164]. Therefore, in CA1 pyramidal neurons, activation of AKT may exert neuroprotective effects by reducing GSK3b activity, thereby decreasing Nav1.6 activity. However, considering the dichotomous effects of AKT activation on Nav1.1 [120] and 1.6 [160], inhibition of the downstream interaction between Nav1.6 and GSK3β may serve as a more viable approach than functional modulation of AKT activity.

5.2. Modulation of Nav Channel Macromolecular Complexes

In addition to regulation via phosphorylation, the Nav channel function is also modulated via stable interactions with other ChAPs [113,114,118]. One subclass of these accessory proteins that have been investigated for their ability to regulate Nav channel activity is the intracellular fibroblast growth factors (iFGFs) [95,117,123,165]. These proteins are a subset of the FGF superfamily and include FGF11, FGF12, FGF13, and FGF14 [166]. The iFGFs are expressed in excitable cells, where they are able to modulate the activity of Nav channels through stable interactions with their intracellular C-terminal domains [166]. Intriguingly, interactions of the iFGFs with Nav channels diverge among Nav and iFGF isoforms, resulting in an array of functional outcomes [113]. Despite this complexity, the divergence in cell-type expression among Nav channel isoforms may allow for fine-tuning of channel properties, and thus neuronal excitability, via modulating iFGF/Nav interfaces in target tissues. Of particular interest, FGF14 has been identified as a risk factor for AD [167,168]. In hippocampal neurons, it is observed that FGF14 overexpression results in a significant increase in Nav1.6 current densities, whereas genetic knockdown of FGF14 results in opposing phenotypes and decreased excitability [169]. Moreover, studies from our lab have illustrated that the FGF14/Nav channel complex is a target of GSK3β [170] and that pharmacological inhibition of GSK3β induces dissociation of the FGF14/Nav complex in hippocampal neurons and modifies FGF14-mediated modulation of Nav channel activity [170]. Further investigation revealed that GSK3β phosphorylates FGF14 at S226 and that phosphorylation of this residue is increased in Tg2576 AD rodents [171]. Moreover, alanine mutation of the S226 residue results in decreased complex formation with Nav1.6 [171]. As described in the previous section, GSK3β is a serine-threonine kinase with dysregulated activity and expression in the hippocampus of AD brains [161–164]. Thus, hyperphosphorylation of FGF14 by GSK3β may lead to a significantly increased probability of FGF14/Nav1.6 complex formation in early AD, thereby potentiating Nav1.6 activity and aberrant neuronal hyperexcitability in CA1 neurons. Further investigation is required to determine the precise mechanisms of phosphorylation-driven regulation of the FGF14/Nav1.6 complex, but disruption of FGF14 binding to the channel directly in the hippocampus of AD brains may prevent its potentiation of Nav1.6-mediated neuronal excitability and aid in ameliorating hippocampal network hyperactivity.

6. Conclusions

There remains a major need for AD therapeutics with disease-modifying properties. Therefore, elucidating the neuronal mechanisms that underlie AD progression is a critical step in the development of novel interventions. Hippocampal hyperactivity is one of the first neuronal phenotypes observed in AD patients, and it is linked to the onset of memory deficits as well progression of AD pathophysiology. Correction of this hyperactivity, therefore, is an attractive disease-modifying therapeutic strategy and has emerged as the focus of many recent investigations. While the complete roles of Nav1.1 and Nav1.6 throughout AD progression remain to be elucidated, their contributions to early-stage hyperactivity are under investigation and may prove critical to the development of disease-modifying AD therapeutics. Given their central roles in governing the excitability of these neuronal subtypes, functional modulation of Nav1.1 and Nav1.6 represents a promising therapeutic strategy to regulate hippocampal activity in early-stage AD, and their contributions to early-stage hyperactivity may prove critical to the development of disease-modifying AD therapeutics. In addition to pursuing traditional pharmacological approaches, Nav channel activity may be regulated through modulation of post-translational modifications or stable interactions with auxiliary proteins that alter channel activity. In the AD brain, there are numerous kinase signaling cascades and disease-related proteins that exhibit distinct functions and expression patterns during disease progression, several of which have established functional effects on Nav channels. In many cases, the functional consequences of these interactions are divergent among Nav isoforms. Therefore, functionally modulating Nav channels through altering their regulatory PTMs or protein complexes may provide the opportunity for the development of isoform-specific Nav channel therapeutics with improved specificity for diseased tissues.

Author Contributions: Conceptualization, T.J.B. and F.L.; writing—original draft preparation, T.J.B., N.A.G., Z.H. and P.A.; writing—review and editing, T.J.B., N.M.D. and F.L.; supervision, F.L.; funding acquisition, T.J.B., N.M.D. and F.L. All authors have read and agreed to the published version of the manuscript.

Funding: This research was funded by the National Institutes of Health grants R01MH124351, R01ES031823, R01 MH132226 and R01ES031823-03S1 (F.L.); the training program funded by the National Institute of Aging (NIH Grant # T32AG067952-01; T.J.B.); and the Houston Area Molecular Biophysics Program Grant No. T32 GM008280 (N.M.D).

Data Availability Statement: No new data was created or analyzed in this study. Data sharing is not applicable to this article.

Acknowledgments: Schematics in this review were generated using BioRender.

Conflicts of Interest: F.L. is the founder and president of IonTx Inc., a start-up company focusing on developing regulators of voltage-gated Na^+ channels.

References

1. Nisbet, R.M.; Götz, J. Amyloid-β and Tau in Alzheimer's Disease: Novel Pathomechanisms and Non-Pharmacological Treatment Strategies. *J. Alzheimers Dis.* **2018**, *64*, S517–S527. [CrossRef] [PubMed]
2. U. F. O. Themes. The Practical Pharmacology of Donepezil. *Basicmedical Key*, 21 August 2016. Available online: https://basicmedicalkey.com/the-practical-pharmacology-of-donepezil/ (accessed on 25 July 2023).
3. Guo, J.; Wang, Z.; Liu, R.; Huang, Y.; Zhang, N.; Zhang, R. Memantine, Donepezil, or Combination Therapy—What is the best therapy for Alzheimer's Disease? A Network Meta-Analysis. *Brain Behav.* **2020**, *10*, e01831. [CrossRef] [PubMed]
4. Padda, I.S.; Parmar, M. Aducanumab. In *StatPearls*; StatPearls Publishing: Treasure Island, FL, USA, 2023. Available online: http://www.ncbi.nlm.nih.gov/books/NBK573062/ (accessed on 25 July 2023).
5. van Dyck, C.H.; Swanson, C.J.; Aisen, P.; Bateman, R.J.; Chen, C.; Gee, M.; Kanekiyo, M.; Li, D.; Reyderman, L.; Cohen, S.; et al. Lecanemab in Early Alzheimer's Disease. *N. Engl. J. Med.* **2023**, *388*, 9–21. [CrossRef] [PubMed]
6. Devi, G.; Scheltens, P. Heterogeneity of Alzheimer's disease: Consequence for drug trials? *Alzheimers Res. Ther.* **2018**, *10*, 122. [CrossRef]
7. Jack, C.R.; Bennett, D.A.; Blennow, K.; Carrillo, M.C.; Dunn, B.; Haeberlein, S.B.; Holtzman, D.M.; Jagust, W.; Jessen, F.; Karlawish, J.; et al. NIA-AA Research Framework: Toward a biological definition of Alzheimer's disease. *Alzheimer's Dement.* **2018**, *14*, 535–562. [CrossRef] [PubMed]
8. Grothe, M.J.; Barthel, H.; Sepulcre, J.; Dyrba, M.; Sabri, O.; Teipel, S.J. In vivo staging of regional amyloid deposition. *Neurology* **2017**, *89*, 2031–2038. [CrossRef]
9. Kaufman, S.K.; Del Tredici, K.; Thomas, T.L.; Braak, H.; Diamond, M.I. Tau seeding activity begins in the transentorhinal/entorhinal regions and anticipates phospho-tau pathology in Alzheimer's disease and PART. *Acta Neuropathol.* **2018**, *136*, 57–67. [CrossRef]
10. Iaccarino, L.; Tammewar, G.; Ayakta, N.; Baker, S.L.; Bejanin, A.; Boxer, A.L.; Gorno-Tempini, M.L.; Janabi, M.; Kramer, J.H.; Lazaris, A.; et al. Local and distant relationships between amyloid, tau and neurodegeneration in Alzheimer's Disease. *NeuroImage Clin.* **2017**, *17*, 452–464. [CrossRef]
11. Bolmont, T.; Clavaguera, F.; Meyer-Luehmann, M.; Herzig, M.C.; Radde, R.; Staufenbiel, M.; Lewis, J.; Hutton, M.; Tolnay, M.; Jucker, M. Induction of tau pathology by intracerebral infusion of amyloid-beta -containing brain extract and by amyloid-beta deposition in APP x Tau transgenic mice. *Am. J. Pathol.* **2007**, *171*, 2012–2020. [CrossRef]

12. Vasconcelos, B.; Stancu, I.-C.; Buist, A.; Bird, M.; Wang, P.; Vanoosthuyse, A.; Van Kolen, K.; Verheyen, A.; Kienlen-Campard, P.; Octave, J.-N.; et al. Heterotypic seeding of Tau fibrillization by pre-aggregated Abeta provides potent seeds for prion-like seeding and propagation of Tau-pathology in vivo. *Acta Neuropathol.* **2016**, *131*, 549–569. [CrossRef]
13. Stancu, I.-C.; Vasconcelos, B.; Terwel, D.; Dewachter, I. Models of β-amyloid induced Tau-pathology: The long and "folded" road to understand the mechanism. *Mol. Neurodegener.* **2014**, *9*, 51. [CrossRef] [PubMed]
14. Pooler, A.M.; Polydoro, M.; Maury, E.A.; Nicholls, S.B.; Reddy, S.M.; Wegmann, S.; William, C.; Saqran, L.; Cagsal-Getkin, O.; Pitstick, R.; et al. Amyloid accelerates tau propagation and toxicity in a model of early Alzheimer's disease. *Acta Neuropathol. Commun.* **2015**, *3*, 14. [CrossRef] [PubMed]
15. Mijalkov, M.; Volpe, G.; Fernaud-Espinosa, I.; DeFelipe, J.; Pereira, J.B.; Merino-Serrais, P. Dendritic spines are lost in clusters in Alzheimer's disease. *Sci. Rep.* **2021**, *11*, 12350. [CrossRef]
16. Griffiths, J.; Grant, S.G.N. Synapse pathology in Alzheimer's disease. *Semin. Cell Dev. Biol.* **2023**, *139*, 13–23. [CrossRef]
17. Benarroch, E.E. Glutamatergic synaptic plasticity and dysfunction in Alzheimer disease: Emerging mechanisms. *Neurology* **2018**, *91*, 125–132. [CrossRef] [PubMed]
18. Busche, M.A.; Chen, X.; Henning, H.A.; Reichwald, J.; Staufenbiel, M.; Sakmann, B.; Konnerth, A. Critical role of soluble amyloid-β for early hippocampal hyperactivity in a mouse model of Alzheimer's disease. *Proc. Natl. Acad. Sci. USA* **2012**, *109*, 8740–8745. [CrossRef]
19. Šišková, Z.; Justus, D.; Kaneko, H.; Friedrichs, D.; Henneberg, N.; Beutel, T.; Pitsch, J.; Schoch, S.; Becker, A.; von der Kammer, H.; et al. Dendritic Structural Degeneration Is Functionally Linked to Cellular Hyperexcitability in a Mouse Model of Alzheimer's Disease. *Neuron* **2014**, *84*, 1023–1033. [CrossRef]
20. Kazim, S.F.; Chuang, S.-C.; Zhao, W.; Wong, R.K.S.; Bianchi, R.; Iqbal, K. Early-Onset Network Hyperexcitability in Presymptomatic Alzheimer's Disease Transgenic Mice Is Suppressed by Passive Immunization with Anti-Human APP/Aβ Antibody and by mGluR5 Blockade. *Front. Aging Neurosci.* **2017**, *9*, 71. [CrossRef]
21. Lee-Liu, D.; Gonzalez-Billault, C. Neuron-intrinsic origin of hyperexcitability during early pathogenesis of Alzheimer's disease. *J. Neurochem.* **2021**, *158*, 586–588. [CrossRef]
22. Celone, K.A.; Calhoun, V.D.; Dickerson, B.C.; Atri, A.; Chua, E.F.; Miller, S.L.; DePeau, K.; Rentz, D.M.; Selkoe, D.J.; Blacker, D.; et al. Alterations in Memory Networks in Mild Cognitive Impairment and Alzheimer's Disease: An Independent Component Analysis. *J. Neurosci.* **2006**, *26*, 10222–10231. [CrossRef]
23. Filippini, N.; MacIntosh, B.J.; Hough, M.G.; Goodwin, G.M.; Frisoni, G.B.; Smith, S.M.; Matthews, P.M.; Beckmann, C.F.; Mackay, C.E. Distinct patterns of brain activity in young carriers of the APOE-ε4 allele. *Proc. Natl. Acad. Sci. USA* **2009**, *106*, 7209–7214. [CrossRef] [PubMed]
24. Bassett, S.S.; Yousem, D.M.; Cristinzio, C.; Kusevic, I.; Yassa, M.A.; Caffo, B.S.; Zeger, S.L. Familial risk for Alzheimer's disease alters fMRI activation patterns. *Brain* **2006**, *129*, 1229–1239. [CrossRef] [PubMed]
25. Hämäläinen, A.; Pihlajamäki, M.; Tanila, H.; Hänninen, T.; Niskanen, E.; Tervo, S.; Karjalainen, P.A.; Vanninen, R.L.; Soininen, H. Increased fMRI responses during encoding in mild cognitive impairment. *Neurobiol. Aging* **2007**, *28*, 1889–1903. [CrossRef] [PubMed]
26. Dickerson, B.C.; Salat, D.H.; Greve, D.N.; Chua, E.F.; Rand-Giovannetti, E.; Rentz, D.M.; Bertram, L.; Mullin, K.; Tanzi, R.E.; Blacker, D.; et al. Increased hippocampal activation in mild cognitive impairment compared to normal aging and AD. *Neurology* **2005**, *65*, 404–411. [CrossRef] [PubMed]
27. Toniolo, S.; Sen, A.; Husain, M. Modulation of Brain Hyperexcitability: Potential New Therapeutic Approaches in Alzheimer's Disease. *Int. J. Mol. Sci.* **2020**, *21*, 9318. [CrossRef]
28. Setti, S.E.; Hunsberger, H.C.; Reed, M.N. Alterations in Hippocampal Activity and Alzheimer's Disease. *Transl. Issues Psychol. Sci.* **2017**, *3*, 348–356. [CrossRef]
29. Ciccone, R.; Franco, C.; Piccialli, I.; Boscia, F.; Casamassa, A.; de Rosa, V.; Cepparulo, P.; Cataldi, M.; Annunziato, L.; Pannaccione, A. Amyloid β-Induced Upregulation of Nav1.6 Underlies Neuronal Hyperactivity in Tg2576 Alzheimer's Disease Mouse Model. *Sci. Rep.* **2019**, *9*, 13592. [CrossRef]
30. Vico Varela, E.; Etter, G.; Williams, S. Excitatory-inhibitory imbalance in Alzheimer's disease and therapeutic significance. *Neurobiol. Dis.* **2019**, *127*, 605–615. [CrossRef]
31. Targa Dias Anastacio, H.; Matosin, N.; Ooi, L. Neuronal hyperexcitability in Alzheimer's disease: What are the drivers behind this aberrant phenotype? *Transl. Psychiatry* **2022**, *12*, 257. [CrossRef]
32. Bakker, A.; Krauss, G.L.; Albert, M.S.; Speck, C.L.; Jones, L.R.; Stark, C.E.; Yassa, M.A.; Bassett, S.S.; Shelton, A.L.; Gallagher, M. Reduction of hippocampal hyperactivity improves cognition in amnestic mild cognitive impairment. *Neuron* **2012**, *74*, 467–474. [CrossRef]
33. Leal, S.L.; Landau, S.M.; Bell, R.K.; Jagust, W.J. Hippocampal activation is associated with longitudinal amyloid accumulation and cognitive decline. *Elife* **2017**, *6*, e22978. [CrossRef] [PubMed]
34. Yuan, D.; Yang, G.; Wu, W.; Li, Q.; Xu, D.; Ntim, M.; Jiang, C.; Liu, J.; Zhang, Y.; Wang, Y.; et al. Reducing Nav1.6 expression attenuates the pathogenesis of Alzheimer's disease by suppressing BACE1 transcription. *Aging Cell* **2022**, *21*, e13593. [CrossRef] [PubMed]

35. Gail Canter, R.; Huang, W.-C.; Choi, H.; Wang, J.; Ashley Watson, L.; Yao, C.G.; Abdurrob, F.; Bousleiman, S.M.; Young, J.Z.; Bennett, D.A.; et al. 3D mapping reveals network-specific amyloid progression and subcortical susceptibility in mice. *Commun. Biol.* **2019**, *2*, 360. [CrossRef] [PubMed]
36. Cirrito, J.R.; Yamada, K.A.; Finn, M.B.; Sloviter, R.S.; Bales, K.R.; May, P.C.; Schoepp, D.D.; Paul, S.M.; Mennerick, S.; Holtzman, D.M. Synaptic activity regulates interstitial fluid amyloid-beta levels in vivo. *Neuron* **2005**, *48*, 913–922. [CrossRef]
37. Yamamoto, K.; Tanei, Z.-I.; Hashimoto, T.; Wakabayashi, T.; Okuno, H.; Naka, Y.; Yizhar, O.; Fenno, L.E.; Fukayama, M.; Bito, H.; et al. Chronic optogenetic activation augments aβ pathology in a mouse model of Alzheimer disease. *Cell Rep.* **2015**, *11*, 859–865. [CrossRef]
38. Pooler, A.M.; Phillips, E.C.; Lau, D.H.W.; Noble, W.; Hanger, D.P. Physiological release of endogenous tau is stimulated by neuronal activity. *EMBO Rep.* **2013**, *14*, 389–394. [CrossRef]
39. Tse, M.T. Anti-epileptic drug shows benefit in AD mouse model. *Nat. Rev. Drug Discov.* **2012**, *11*, 748–749. [CrossRef]
40. Sanchez, P.E.; Zhu, L.; Verret, L.; Vossel, K.A.; Orr, A.G.; Cirrito, J.R.; Devidze, N.; Ho, K.; Yu, G.-Q.; Palop, J.J.; et al. Levetiracetam suppresses neuronal network dysfunction and reverses synaptic and cognitive deficits in an Alzheimer's disease model. *Proc. Natl. Acad. Sci. USA* **2012**, *109*, E2895–E2903. [CrossRef]
41. Hu, W.; Tian, C.; Li, T.; Yang, M.; Hou, H.; Shu, Y. Distinct contributions of Nav1.6 and Nav1.2 in action potential initiation and backpropagation. *Nat. Neurosci.* **2009**, *12*, 996–1002. [CrossRef]
42. Catterall, W.A. Forty Years of Sodium Channels: Structure, Function, Pharmacology, and Epilepsy. *Neurochem. Res.* **2017**, *42*, 2495–2504. [CrossRef]
43. Booker, S.A.; Vida, I. Morphological diversity and connectivity of hippocampal interneurons. *Cell Tissue Res.* **2018**, *373*, 619–641. [CrossRef] [PubMed]
44. Gloveli, T.; Kopell, N.; Dugladze, T. Neuronal Activity Patterns During Hippocampal Network Oscillations In Vitro. In *Hippocampal Microcircuits: A Computational Modeler's Resource Book*; Cutsuridis, V., Graham, B., Cobb, S., Vida, I., Eds.; Springer Series in Computational Neuroscience; Springer: New York, NY, USA, 2010; pp. 247–276. ISBN 978-1-4419-0996-1.
45. Kowalski, J.; Gan, J.; Jonas, P.; Pernía-Andrade, A.J. Intrinsic membrane properties determine hippocampal differential firing pattern in vivo in anesthetized rats. *Hippocampus* **2016**, *26*, 668–682. [CrossRef]
46. Chauhan, P.; Jethwa, K.; Rathawa, A.; Chauhan, G.; Mehra, S. The Anatomy of the Hippocampus. In *Cerebral Ischemia*; Pluta, R., Ed.; Exon Publications: Brisbane, Australia, 2021; ISBN 978-0-645-00179-2.
47. Stepan, J.; Dine, J.; Eder, M. Functional optical probing of the hippocampal trisynaptic circuit in vitro: Network dynamics, filter properties, and polysynaptic induction of CA1 LTP. *Front. Neurosci.* **2015**, *9*, 160. [CrossRef] [PubMed]
48. Sosa, M.; Gillespie, A.K.; Frank, L.M. Neural Activity Patterns Underlying Spatial Coding in the Hippocampus. *Curr. Top. Behav. Neurosci.* **2018**, *37*, 43–100. [CrossRef]
49. Pelkey, K.A.; Chittajallu, R.; Craig, M.T.; Tricoire, L.; Wester, J.C.; McBain, C.J. Hippocampal GABAergic Inhibitory Interneurons. *Physiol. Rev.* **2017**, *97*, 1619–1747. [CrossRef] [PubMed]
50. Pouille, F.; Scanziani, M. Enforcement of Temporal Fidelity in Pyramidal Cells by Somatic Feed-Forward Inhibition. *Science* **2001**, *293*, 1159–1163. [CrossRef] [PubMed]
51. Ferguson, B.R.; Gao, W.-J. PV Interneurons: Critical Regulators of E/I Balance for Prefrontal Cortex-Dependent Behavior and Psychiatric Disorders. *Front. Neural Circuits* **2018**, *12*, 37. [CrossRef]
52. Sadeh, S.; Clopath, C. Excitatory-inhibitory balance modulates the formation and dynamics of neuronal assemblies in cortical networks. *Sci. Adv.* **2021**, *7*, eabg8411. [CrossRef]
53. de Lera Ruiz, M.; Kraus, R.L. Voltage-Gated Sodium Channels: Structure, Function, Pharmacology, and Clinical Indications. *J. Med. Chem.* **2015**, *58*, 7093–7118. [CrossRef]
54. Royeck, M.; Horstmann, M.-T.; Remy, S.; Reitze, M.; Yaari, Y.; Beck, H. Role of axonal NaV1.6 sodium channels in action potential initiation of CA1 pyramidal neurons. *J. Neurophysiol.* **2008**, *100*, 2361–2380. [CrossRef]
55. Wang, J.; Ou, S.-W.; Wang, Y.-J. Distribution and function of voltage-gated sodium channels in the nervous system. *Channels* **2017**, *11*, 534–554. [CrossRef] [PubMed]
56. Duflocq, A.; Le Bras, B.; Bullier, E.; Couraud, F.; Davenne, M. Nav1.1 is predominantly expressed in nodes of Ranvier and axon initial segments. *Mol. Cell. Neurosci.* **2008**, *39*, 180–192. [CrossRef] [PubMed]
57. Catterall, W.A.; Kalume, F.; Oakley, J.C. NaV1.1 channels and epilepsy. *J. Physiol.* **2010**, *588*, 1849–1859. [CrossRef] [PubMed]
58. Cheah, C.S.; Yu, F.H.; Westenbroek, R.E.; Kalume, F.K.; Oakley, J.C.; Potter, G.B.; Rubenstein, J.L.; Catterall, W.A. Specific deletion of NaV1.1 sodium channels in inhibitory interneurons causes seizures and premature death in a mouse model of Dravet syndrome. *Proc. Natl. Acad. Sci. USA* **2012**, *109*, 14646–14651. [CrossRef]
59. Zybura, A.; Hudmon, A.; Cummins, T.R. Distinctive Properties and Powerful Neuromodulation of Nav1.6 Sodium Channels Regulates Neuronal Excitability. *Cells* **2021**, *10*, 1595. [CrossRef]
60. Martinez-Losa, M.; Tracy, T.E.; Ma, K.; Verret, L.; Clemente-Perez, A.; Khan, A.S.; Cobos, I.; Ho, K.; Gan, L.; Mucke, L.; et al. Nav1.1-Overexpressing Interneuron Transplants Restore Brain Rhythms and Cognition in a Mouse Model of Alzheimer's Disease. *Neuron* **2018**, *98*, 75–89.e5. [CrossRef]

61. DeFelipe, J.; López-Cruz, P.L.; Benavides-Piccione, R.; Bielza, C.; Larrañaga, P.; Anderson, S.; Burkhalter, A.; Cauli, B.; Fairén, A.; Feldmeyer, D.; et al. New insights into the classification and nomenclature of cortical GABAergic interneurons. *Nat. Rev. Neurosci.* **2013**, *14*, 202–216. [CrossRef]
62. Wonders, C.P.; Anderson, S.A. The origin and specification of cortical interneurons. *Nat. Rev. Neurosci.* **2006**, *7*, 687–696. [CrossRef]
63. Ogiwara, I.; Miyamoto, H.; Morita, N.; Atapour, N.; Mazaki, E.; Inoue, I.; Takeuchi, T.; Itohara, S.; Yanagawa, Y.; Obata, K.; et al. Nav1.1 Localizes to Axons of Parvalbumin-Positive Inhibitory Interneurons: A Circuit Basis for Epileptic Seizures in Mice Carrying an Scn1a Gene Mutation. *J. Neurosci.* **2007**, *27*, 5903–5914. [CrossRef]
64. Wang, W.; Takashima, S.; Segawa, Y.; Itoh, M.; Shi, X.; Hwang, S.-K.; Nabeshima, K.; Takeshita, M.; Hirose, S. The developmental changes of Na(v)1.1 and Na(v)1.2 expression in the human hippocampus and temporal lobe. *Brain Res.* **2011**, *1389*, 61–70. [CrossRef]
65. Ding, J.; Li, X.; Tian, H.; Wang, L.; Guo, B.; Wang, Y.; Li, W.; Wang, F.; Sun, T. SCN1A Mutation—Beyond Dravet Syndrome: A Systematic Review and Narrative Synthesis. *Front. Neurol.* **2021**, *12*, 743726. [CrossRef] [PubMed]
66. Vossel, K.A.; Beagle, A.J.; Rabinovici, G.D.; Shu, H.; Lee, S.E.; Naasan, G.; Hegde, M.; Cornes, S.B.; Henry, M.L.; Nelson, A.B.; et al. Seizures and epileptiform activity in the early stages of Alzheimer disease. *JAMA Neurol.* **2013**, *70*, 1158–1166. [CrossRef] [PubMed]
67. Csernus, E.A.; Werber, T.; Kamondi, A.; Horvath, A.A. The Significance of Subclinical Epileptiform Activity in Alzheimer's Disease: A Review. *Front. Neurol.* **2022**, *13*, 856500. [CrossRef] [PubMed]
68. Hamm, V.; Héraud, C.; Bott, J.-B.; Herbeaux, K.; Strittmatter, C.; Mathis, C.; Goutagny, R. Differential contribution of APP metabolites to early cognitive deficits in a TgCRND8 mouse model of Alzheimer's disease. *Sci. Adv.* **2017**, *3*, e1601068. [CrossRef]
69. Hu, T.; Xiao, Z.; Mao, R.; Chen, B.; Lu, M.-N.; Tong, J.; Mei, R.; Li, S.-S.; Xiao, Z.-C.; Zhang, L.-F.; et al. Navβ2 knockdown improves cognition in APP/PS1 mice by partially inhibiting seizures and APP amyloid processing. *Oncotarget* **2017**, *8*, 99284–99295. [CrossRef] [PubMed]
70. Verret, L.; Mann, E.O.; Hang, G.B.; Barth, A.M.I.; Cobos, I.; Ho, K.; Devidze, N.; Masliah, E.; Kreitzer, A.C.; Mody, I.; et al. Inhibitory Interneuron Deficit Links Altered Network Activity and Cognitive Dysfunction in Alzheimer Model. *Cell* **2012**, *149*, 708–721. [CrossRef]
71. Rossor, M.N.; Garrett, N.J.; Johnson, A.L.; Mountjoy, C.Q.; Roth, M.; Iversen, L.L. A post-mortem study of the cholinergic and GABA systems in senile dementia. *Brain J. Neurol.* **1982**, *105*, 313–330. [CrossRef]
72. Pike, C.J.; Cotman, C.W. Cultured GABA-immunoreactive neurons are resistant to toxicity induced by beta-amyloid. *Neuroscience* **1993**, *56*, 269–274. [CrossRef]
73. Blazquez-Llorca, L. Pericellular innervation of neurons expressing abnormally hyperphosphorylated tau in the hippocampal formation of Alzheimer's disease patients. *Front. Neuroanat.* **2010**, *4*, 20. [CrossRef]
74. Huang, Y.; Mucke, L. Alzheimer Mechanisms and Therapeutic Strategies. *Cell* **2012**, *148*, 1204–1222. [CrossRef] [PubMed]
75. Govindpani, K.; Calvo-Flores Guzmán, B.; Vinnakota, C.; Waldvogel, H.J.; Faull, R.L.; Kwakowsky, A. Towards a Better Understanding of GABAergic Remodeling in Alzheimer's Disease. *Int. J. Mol. Sci.* **2017**, *18*, 1813. [CrossRef] [PubMed]
76. Selkoe, D.J. Early network dysfunction in Alzheimer's disease. *Science* **2019**, *365*, 540–541. [CrossRef] [PubMed]
77. Villette, V.; Dutar, P. GABAergic Microcircuits in Alzheimer's Disease Models. *Curr. Alzheimer Res.* **2017**, *14*, 30–39. [CrossRef]
78. Kim, D.Y.; Carey, B.W.; Wang, H.; Ingano, L.A.M.; Binshtok, A.M.; Wertz, M.H.; Pettingell, W.H.; He, P.; Lee, V.M.-Y.; Woolf, C.J.; et al. BACE1 regulates voltage-gated sodium channels and neuronal activity. *Nat. Cell Biol.* **2007**, *9*, 755–764. [CrossRef] [PubMed]
79. Tyler, S.J.; Dawbarn, D.; Wilcock, G.K.; Allen, S.J. alpha- and beta-secretase: Profound changes in Alzheimer's disease. *Biochem. Biophys. Res. Commun.* **2002**, *299*, 373–376. [CrossRef] [PubMed]
80. Fukumoto, H.; Cheung, B.S.; Hyman, B.T.; Irizarry, M.C. β-Secretase Protein and Activity Are Increased in the Neocortex in Alzheimer Disease. *Arch. Neurol.* **2002**, *59*, 1381–1389. [CrossRef]
81. Yang, L.-B.; Lindholm, K.; Yan, R.; Citron, M.; Xia, W.; Yang, X.-L.; Beach, T.; Sue, L.; Wong, P.; Price, D.; et al. Elevated β-secretase expression and enzymatic activity detected in sporadic Alzheimer disease. *Nat. Med.* **2003**, *9*, 3–4. [CrossRef]
82. Caccavano, A.; Bozzelli, P.L.; Forcelli, P.A.; Pak, D.T.S.; Wu, J.-Y.; Conant, K.; Vicini, S. Inhibitory Parvalbumin Basket Cell Activity is Selectively Reduced during Hippocampal Sharp Wave Ripples in a Mouse Model of Familial Alzheimer's Disease. *J. Neurosci.* **2020**, *40*, 5116–5136. [CrossRef]
83. Briend, F.; Nelson, E.A.; Maximo, O.; Armstrong, W.P.; Kraguljac, N.V.; Lahti, A.C. Hippocampal glutamate and hippocampus subfield volumes in antipsychotic-naive first episode psychosis subjects and relationships to duration of untreated psychosis. *Transl. Psychiatry* **2020**, *10*, 137. [CrossRef]
84. van Dijk, R.M.; Huang, S.-H.; Slomianka, L.; Amrein, I. Taxonomic Separation of Hippocampal Networks: Principal Cell Populations and Adult Neurogenesis. *Front. Neuroanat.* **2016**, *10*, 22. Available online: https://www.frontiersin.org/articles/10.3389/fnana.2016.00022 (accessed on 25 July 2023).
85. Erickson, A.; Deiteren, A.; Harrington, A.M.; Garcia-Caraballo, S.; Castro, J.; Caldwell, A.; Grundy, L.; Brierley, S.M. Voltage-gated sodium channels: (NaV)igating the field to determine their contribution to visceral nociception. *J. Physiol.* **2018**, *596*, 785–807. [CrossRef] [PubMed]
86. Kaplan, M.R.; Cho, M.-H.; Ullian, E.M.; Isom, L.L.; Levinson, S.R.; Barres, B.A. Differential Control of Clustering of the Sodium Channels Nav1.2 and Nav1.6 at Developing CNS Nodes of Ranvier. *Neuron* **2001**, *30*, 105–119. [CrossRef] [PubMed]

87. Tapia, C.M.; Folorunso, O.; Singh, A.K.; McDonough, K.; Laezza, F. Effects of Deltamethrin Acute Exposure on Nav1.6 Channels and Medium Spiny Neurons of the Nucleus Accumbens. *Toxicology* **2020**, *440*, 152488. [CrossRef]
88. Solé, L.; Wagnon, J.L.; Tamkun, M.M. Functional analysis of three Nav1.6 mutations causing early infantile epileptic encephalopathy. *Biochim. Biophys. Acta BBA Mol. Basis Dis.* **2020**, *1866*, 165959. [CrossRef] [PubMed]
89. Alrashdi, B.; Dawod, B.; Schampel, A.; Tacke, S.; Kuerten, S.; Marshall, J.S.; Côté, P.D. Nav1.6 promotes inflammation and neuronal degeneration in a mouse model of multiple sclerosis. *J. Neuroinflamm.* **2019**, *16*, 215. [CrossRef]
90. Wagnon, J.L.; Barker, B.S.; Ottolini, M.; Park, Y.; Volkheimer, A.; Valdez, P.; Swinkels, M.E.M.; Patel, M.K.; Meisler, M.H. Loss-of-function variants of SCN8A in intellectual disability without seizures. *Neurol. Genet.* **2017**, *3*, e170. [CrossRef]
91. Hargus, N.J.; Nigam, A.; Bertram, E.H.; Patel, M.K. Evidence for a role of Nav1.6 in facilitating increases in neuronal hyperexcitability during epileptogenesis. *J. Neurophysiol.* **2013**, *110*, 1144–1157. [CrossRef]
92. Li, S.; Wang, X.; Ma, Q.-H.; Yang, W.; Zhang, X.-G.; Dawe, G.S.; Xiao, Z.-C. Amyloid precursor protein modulates Nav1.6 sodium channel currents through a Go-coupled JNK pathway. *Sci. Rep.* **2016**, *6*, 39320. [CrossRef]
93. Liu, C.; Tan, F.C.K.; Xiao, Z.-C.; Dawe, G.S. Amyloid precursor protein enhances Nav1.6 sodium channel cell surface expression. *J. Biol. Chem.* **2015**, *290*, 12048–12057. [CrossRef]
94. Ren, S.; Chen, P.; Jiang, H.; Mi, Z.; Xu, F.; Hu, B.; Zhang, J.; Zhu, Z. Persistent sodium currents contribute to Aβ1-42-induced hyperexcitation of hippocampal CA1 pyramidal neurons. *Neurosci. Lett.* **2014**, *580*, 62–67. [CrossRef]
95. Dvorak, N.M.; Wadsworth, P.A.; Wang, P.; Zhou, J.; Laezza, F. Development of Allosteric Modulators of Voltage-Gated Na+ Channels: A Novel Approach for an Old Target. *Curr. Top. Med. Chem.* **2021**, *21*, 841–848. [CrossRef] [PubMed]
96. Southwell, D.G.; Nicholas, C.R.; Basbaum, A.I.; Stryker, M.P.; Kriegstein, A.R.; Rubenstein, J.L.; Alvarez-Buylla, A. Interneurons from Embryonic Development to Cell-Based Therapy. *Science* **2014**, *344*, 1240622. [CrossRef] [PubMed]
97. Martier, R.; Konstantinova, P. Gene Therapy for Neurodegenerative Diseases: Slowing Down the Ticking Clock. *Front. Neurosci.* **2020**, *14*, 580179. [CrossRef]
98. Snowball, A.; Schorge, S. Changing channels in pain and epilepsy: Exploiting ion channel gene therapy for disorders of neuronal hyperexcitability. *FEBS Lett.* **2015**, *589*, 1620–1634. [CrossRef] [PubMed]
99. Dey, D.; Eckle, V.-S.; Vitko, I.; Sullivan, K.A.; Lasiecka, Z.M.; Winckler, B.; Stornetta, R.L.; Williamson, J.M.; Kapur, J.; Perez-Reyes, E. A potassium leak channel silences hyperactive neurons and ameliorates status epilepticus. *Epilepsia* **2014**, *55*, 203–213. [CrossRef]
100. Wykes, R.C.; Heeroma, J.H.; Mantoan, L.; Zheng, K.; MacDonald, D.C.; Deisseroth, K.; Hashemi, K.S.; Walker, M.C.; Schorge, S.; Kullmann, D.M. Optogenetic and potassium channel gene therapy in a rodent model of focal neocortical epilepsy. *Sci. Transl. Med.* **2012**, *4*, 161ra152. [CrossRef]
101. Nguyen, H.X.; Wu, T.; Needs, D.; Zhang, H.; Perelli, R.M.; DeLuca, S.; Yang, R.; Tian, M.; Landstrom, A.P.; Henriquez, C.; et al. Engineered bacterial voltage-gated sodium channel platform for cardiac gene therapy. *Nat. Commun.* **2022**, *13*, 620. [CrossRef]
102. Jensen, H.S.; Grunnet, M.; Bastlund, J.F. Therapeutic potential of NaV1.1 activators. *Trends Pharmacol. Sci.* **2014**, *35*, 113–118. [CrossRef]
103. James, T.F.; Nenov, M.N.; Tapia, C.M.; Lecchi, M.; Koshy, S.; Green, T.A.; Laezza, F. Consequences of acute Nav1.1 exposure to deltamethrin. *Neurotoxicology* **2017**, *60*, 150–160. [CrossRef]
104. Osteen, J.D.; Herzig, V.; Gilchrist, J.; Emrick, J.J.; Zhang, C.; Wang, X.; Castro, J.; Garcia-Caraballo, S.; Grundy, L.; Rychkov, G.Y.; et al. Selective spider toxins reveal a role for the Nav1.1 channel in mechanical pain. *Nature* **2016**, *534*, 494–499. [CrossRef]
105. Richards, K.L.; Milligan, C.J.; Richardson, R.J.; Jancovski, N.; Grunnet, M.; Jacobson, L.H.; Undheim, E.A.B.; Mobli, M.; Chow, C.Y.; Herzig, V.; et al. Selective NaV1.1 activation rescues Dravet syndrome mice from seizures and premature death. *Proc. Natl. Acad. Sci. USA* **2018**, *115*, E8077–E8085. [CrossRef]
106. Chow, C.Y.; Chin, Y.K.Y.; Ma, L.; Undheim, E.A.B.; Herzig, V.; King, G.F. A selective NaV1.1 activator with potential for treatment of Dravet syndrome epilepsy. *Biochem. Pharmacol.* **2020**, *181*, 113991. [CrossRef] [PubMed]
107. Crestey, F.; Frederiksen, K.; Jensen, H.S.; Dekermendjian, K.; Larsen, P.H.; Bastlund, J.F.; Lu, D.; Liu, H.; Yang, C.R.; Grunnet, M.; et al. Identification and Electrophysiological Evaluation of 2-Methylbenzamide Derivatives as Nav1.1 Modulators. *ACS Chem. Neurosci.* **2015**, *6*, 1302–1308. [CrossRef] [PubMed]
108. Miyazaki, T.; Kawasaki, M.; Suzuki, A.; Ito, Y.; Imanishi, A.; Maru, T.; Kawamoto, T.; Koike, T. Discovery of novel 4-phenyl-2-(pyrrolidinyl)nicotinamide derivatives as potent Nav1.1 activators. *Bioorg. Med. Chem. Lett.* **2019**, *29*, 815–820. [CrossRef]
109. Klitgaard, H. Levetiracetam: The preclinical profile of a new class of antiepileptic drugs? *Epilepsia* **2001**, *42* (Suppl. S4), 13–18. [CrossRef]
110. Johnson, J.; Focken, T.; Khakh, K.; Tari, P.K.; Dube, C.; Goodchild, S.J.; Andrez, J.-C.; Bankar, G.; Bogucki, D.; Burford, K.; et al. NBI-921352, a first-in-class, NaV1.6 selective, sodium channel inhibitor that prevents seizures in Scn8a gain-of-function mice, and wild-type mice and rats. *Elife* **2022**, *11*, e72468. [CrossRef]
111. Beatch, G.; Namdari, R.; Cadieux, J.; Kato, H.; Aycardi, E. A Phase 1 Study to Assess the Safety, Tolerability and Pharmacokinetics of Two Formulations of a Novel Nav1.6 Sodium Channnel Blocker (XEN901) in Healthy Adult Subjects. *Neurology* **2020**, *94*, 4757.
112. Pitt, G.S.; Lee, S.-Y. Current view on regulation of voltage-gated sodium channels by calcium and auxiliary proteins. *Protein Sci.* **2016**, *25*, 1573–1584. [CrossRef] [PubMed]

113. Laezza, F.; Lampert, A.; Kozel, M.A.; Gerber, B.R.; Rush, A.M.; Nerbonne, J.M.; Waxman, S.G.; Dib-Hajj, S.D.; Ornitz, D.M. FGF14 N-terminal splice variants differentially modulate Nav1.2 and Nav1.6-encoded sodium channels. *Mol. Cell. Neurosci.* **2009**, *42*, 90–101. [CrossRef]
114. Effraim, P.R.; Huang, J.; Lampert, A.; Stamboulian, S.; Zhao, P.; Black, J.A.; Dib-Hajj, S.D.; Waxman, S.G. Fibroblast growth factor homologous factor 2 (FGF-13) associates with Nav1.7 in DRG neurons and alters its current properties in an isoform-dependent manner. *Neurobiol. Pain* **2019**, *6*, 100029. [CrossRef]
115. Scala, F.; Nenov, M.N.; Crofton, E.J.; Singh, A.K.; Folorunso, O.; Zhang, Y.; Chesson, B.C.; Wildburger, N.C.; James, T.F.; Alshammari, M.A.; et al. Environmental Enrichment and Social Isolation Mediate Neuroplasticity of Medium Spiny Neurons through the GSK3 Pathway. *Cell Rep.* **2018**, *23*, 555–567. [CrossRef] [PubMed]
116. Lou, J.-Y.; Laezza, F.; Gerber, B.R.; Xiao, M.; Yamada, K.A.; Hartmann, H.; Craig, A.M.; Nerbonne, J.M.; Ornitz, D.M. Fibroblast growth factor 14 is an intracellular modulator of voltage-gated sodium channels. *J. Physiol.* **2005**, *569*, 179–193. [CrossRef] [PubMed]
117. Dvorak, N.M.; Wadsworth, P.A.; Wang, P.; Chen, H.; Zhou, J.; Laezza, F. Bidirectional Modulation of the Voltage-Gated Sodium (Nav1.6) Channel by Rationally Designed Peptidomimetics. *Molecules* **2020**, *25*, 3365. [CrossRef] [PubMed]
118. Ali, S.R.; Liu, Z.; Nenov, M.N.; Folorunso, O.; Singh, A.; Scala, F.; Chen, H.; James, T.F.; Alshammari, M.; Panova-Elektronova, N.I.; et al. Functional Modulation of Voltage-Gated Sodium Channels by a FGF14-Based Peptidomimetic. *ACS Chem. Neurosci.* **2018**, *9*, 976–987. [CrossRef]
119. Wadsworth, P.A.; Singh, A.K.; Nguyen, N.; Dvorak, N.M.; Tapia, C.M.; Russell, W.K.; Stephan, C.; Laezza, F. JAK2 regulates Nav1.6 channel function via FGF14Y158 phosphorylation. *Biochim. Biophys. Acta Mol. Cell Res.* **2020**, *1867*, 118786. [CrossRef]
120. Arribas-Blázquez, M.; Piniella, D.; Olivos-Oré, L.A.; Bartolomé-Martín, D.; Leite, C.; Giménez, C.; Artalejo, A.R.; Zafra, F. Regulation of the voltage-dependent sodium channel NaV1.1 by AKT1. *Neuropharmacology* **2021**, *197*, 108745. [CrossRef]
121. Zybura, A.S.; Baucum, A.J.; Rush, A.M.; Cummins, T.R.; Hudmon, A. CaMKII enhances voltage-gated sodium channel Nav1.6 activity and neuronal excitability. *J. Biol. Chem.* **2020**, *295*, 11845–11865. [CrossRef]
122. Gasser, A.; Cheng, X.; Gilmore, E.S.; Tyrrell, L.; Waxman, S.G.; Dib-Hajj, S.D. Two Nedd4-binding Motifs Underlie Modulation of Sodium Channel Nav1.6 by p38 MAPK. *J. Biol. Chem.* **2010**, *285*, 26149–26161. [CrossRef]
123. Singh, A.K.; Wadsworth, P.A.; Tapia, C.M.; Aceto, G.; Ali, S.R.; Chen, H.; D'Ascenzo, M.; Zhou, J.; Laezza, F. Mapping of the FGF14:Nav1.6 complex interface reveals FLPK as a functionally active peptide modulating excitability. *Physiol. Rep.* **2020**, *8*, e14505. [CrossRef]
124. Chakroborty, S.; Briggs, C.; Miller, M.B.; Goussakov, I.; Schneider, C.; Kim, J.; Wicks, J.; Richardson, J.C.; Conklin, V.; Cameransi, B.G.; et al. Stabilizing ER Ca^{2+} Channel Function as an Early Preventative Strategy for Alzheimer's Disease. *PLoS ONE* **2012**, *7*, e52056. [CrossRef]
125. Berridge, M.J. Calcium hypothesis of Alzheimer's disease. *Pflügers Arch. Eur. J. Physiol.* **2010**, *459*, 441–449. [CrossRef] [PubMed]
126. Fukushima, H.; Maeda, R.; Suzuki, R.; Suzuki, A.; Nomoto, M.; Toyoda, H.; Wu, L.-J.; Xu, H.; Zhao, M.-G.; Ueda, K.; et al. Upregulation of calcium/calmodulin-dependent protein kinase IV improves memory formation and rescues memory loss with aging. *J. Neurosci.* **2008**, *28*, 9910–9919. [CrossRef] [PubMed]
127. Ghosh, A.; Giese, K.P. Calcium/calmodulin-dependent kinase II and Alzheimer's disease. *Mol. Brain* **2015**, *8*, 78. [CrossRef] [PubMed]
128. Oddo, S.; Caccamo, A.; Shepherd, J.D.; Murphy, M.P.; Golde, T.E.; Kayed, R.; Metherate, R.; Mattson, M.P.; Akbari, Y.; LaFerla, F.M. Triple-Transgenic Model of Alzheimer's Disease with Plaques and Tangles. *Neuron* **2003**, *39*, 409–421. [CrossRef] [PubMed]
129. Foster, T.C.; Kyritsopoulos, C.; Kumar, A. Central role for NMDA receptors in redox mediated impairment of synaptic function during aging and Alzheimer's disease. *Behav. Brain Res.* **2017**, *322*, 223–232. [CrossRef]
130. Anekonda, T.S.; Quinn, J.F.; Harris, C.; Frahler, K.; Wadsworth, T.L.; Woltjer, R.L. L-type voltage-gated calcium channel blockade with isradipine as a therapeutic strategy for Alzheimer's disease. *Neurobiol. Dis.* **2011**, *41*, 62–70. [CrossRef]
131. Junho, C.V.C.; Caio-Silva, W.; Trentin-Sonoda, M.; Carneiro-Ramos, M.S. An Overview of the Role of Calcium/Calmodulin-Dependent Protein Kinase in Cardiorenal Syndrome. *Front. Physiol.* **2020**, *11*, 735. Available online: https://www.frontiersin.org/articles/10.3389/fphys.2020.00735 (accessed on 25 July 2023). [CrossRef]
132. O'Day, D.H.; Eshak, K.; Myre, M.A. Calmodulin Binding Proteins and Alzheimer's Disease. *J. Alzheimers Dis.* **2015**, *46*, 553–569. [CrossRef]
133. Wang, Y.-J.; Chen, G.-H.; Hu, X.-Y.; Lu, Y.-P.; Zhou, J.-N.; Liu, R.-Y. The expression of calcium/calmodulin-dependent protein kinase II-α in the hippocampus of patients with Alzheimer's disease and its links with AD-related pathology. *Brain Res.* **2005**, *1031*, 101–108. [CrossRef]
134. Chang, J.-Y.; Parra-Bueno, P.; Laviv, T.; Szatmari, E.M.; Lee, S.-J.R.; Yasuda, R. CaMKII Autophosphorylation is Necessary for Optimal Integration of Ca^{2+} Signals During LTP Induction but Not Maintenance. *Neuron* **2017**, *94*, 800–808.e4. [CrossRef]
135. Thompson, C.H.; Hawkins, N.A.; Kearney, J.A.; George, A.L. CaMKII modulates sodium current in neurons from epileptic Scn2a mutant mice. *Proc. Natl. Acad. Sci. USA* **2017**, *114*, 1696–1701. [CrossRef] [PubMed]
136. Ashpole, N.M.; Herren, A.W.; Ginsburg, K.S.; Brogan, J.D.; Johnson, D.E.; Cummins, T.R.; Bers, D.M.; Hudmon, A. Ca^{2+}/Calmodulin-dependent Protein Kinase II (CaMKII) Regulates Cardiac Sodium Channel NaV1.5 Gating by Multiple Phosphorylation Sites. *J. Biol. Chem.* **2012**, *287*, 19856–19869. [CrossRef] [PubMed]

137. Zybura, A.S.; Sahoo, F.K.; Hudmon, A.; Cummins, T.R. CaMKII Inhibition Attenuates Distinct Gain-of-Function Effects Produced by Mutant Nav1.6 Channels and Reduces Neuronal Excitability. *Cells* **2022**, *11*, 2108. [CrossRef] [PubMed]
138. Turjanski, A.G.; Vaqué, J.P.; Gutkind, J.S. MAP kinases and the control of nuclear events. *Oncogene* **2007**, *26*, 3240–3253. [CrossRef]
139. Arthur, J.S.C.; Ley, S.C. Mitogen-activated protein kinases in innate immunity. *Nat. Rev. Immunol.* **2013**, *13*, 679–692. [CrossRef] [PubMed]
140. Morrison, D.K. MAP Kinase Pathways. *Cold Spring Harb. Perspect. Biol.* **2012**, *4*, a011254. [CrossRef]
141. Asih, P.R.; Prikas, E.; Stefanoska, K.; Tan, A.R.P.; Ahel, H.I.; Ittner, A. Functions of p38 MAP Kinases in the Central Nervous System. *Front. Mol. Neurosci.* **2020**, *13*, 570586. [CrossRef]
142. Bolshakov, V.Y.; Carboni, L.; Cobb, M.H.; Siegelbaum, S.A.; Belardetti, F. Dual MAP kinase pathways mediate opposing forms of long-term plasticity at CA3-CA1 synapses. *Nat. Neurosci.* **2000**, *3*, 1107–1112. [CrossRef]
143. Bhaskar, K.; Konerth, M.; Kokiko-Cochran, O.N.; Cardona, A.; Ransohoff, R.M.; Lamb, B.T. Regulation of tau pathology by the microglial fractalkine receptor. *Neuron* **2010**, *68*, 19–31. [CrossRef] [PubMed]
144. Zhu, X.; Rottkamp, C.A.; Hartzler, A.; Sun, Z.; Takeda, A.; Boux, H.; Shimohama, S.; Perry, G.; Smith, M.A. Activation of MKK6, an upstream activator of p38, in Alzheimer's disease. *J. Neurochem.* **2001**, *79*, 311–318. [CrossRef]
145. Feijoo, C.; Campbell, D.G.; Jakes, R.; Goedert, M.; Cuenda, A. Evidence that phosphorylation of the microtubule-associated protein Tau by SAPK4/p38delta at Thr50 promotes microtubule assembly. *J. Cell Sci.* **2005**, *118*, 397–408. [CrossRef] [PubMed]
146. Maphis, N.; Jiang, S.; Xu, G.; Kokiko-Cochran, O.N.; Roy, S.M.; Van Eldik, L.J.; Watterson, D.M.; Lamb, B.T.; Bhaskar, K. Selective suppression of the α isoform of p38 MAPK rescues late-stage tau pathology. *Alzheimers Res. Ther.* **2016**, *8*, 54. [CrossRef] [PubMed]
147. Ittner, A.; Chua, S.W.; Bertz, J.; Volkerling, A.; van der Hoven, J.; Gladbach, A.; Przybyla, M.; Bi, M.; van Hummel, A.; Stevens, C.H.; et al. Site-specific phosphorylation of tau inhibits amyloid-β toxicity in Alzheimer's mice. *Science* **2016**, *354*, 904–908. [CrossRef] [PubMed]
148. Rutigliano, G.; Stazi, M.; Arancio, O.; Watterson, D.M.; Origlia, N. An isoform-selective p38α mitogen-activated protein kinase inhibitor rescues early entorhinal cortex dysfunctions in a mouse model of Alzheimer's disease. *Neurobiol. Aging* **2018**, *70*, 86–91. [CrossRef] [PubMed]
149. Roy, S.M.; Grum-Tokars, V.L.; Schavocky, J.P.; Saeed, F.; Staniszewski, A.; Teich, A.F.; Arancio, O.; Bachstetter, A.D.; Webster, S.J.; Van Eldik, L.J.; et al. Targeting human central nervous system protein kinases: An isoform selective p38αMAPK inhibitor that attenuates disease progression in Alzheimer's disease mouse models. *ACS Chem. Neurosci.* **2015**, *6*, 666–680. [CrossRef] [PubMed]
150. Wittmack, E.K.; Rush, A.M.; Hudmon, A.; Waxman, S.G.; Dib-Hajj, S.D. Voltage-Gated Sodium Channel Nav1.6 Is Modulated by p38 Mitogen-Activated Protein Kinase. *J. Neurosci.* **2005**, *25*, 6621–6630. [CrossRef]
151. Liu, X.; Zhang, L.; Zhang, H.; Liang, X.; Zhang, B.; Tu, J.; Zhao, Y. Nedd4-2 Haploinsufficiency in Mice Impairs the Ubiquitination of Rer1 and Increases the Susceptibility to Endoplasmic Reticulum Stress and Seizures. *Front. Mol. Neurosci.* **2022**, *15*, 919718. [CrossRef]
152. Kitagishi, Y.; Kobayashi, M.; Kikuta, K.; Matsuda, S. Roles of PI3K/AKT/GSK3/mTOR Pathway in Cell Signaling of Mental Illnesses. *Depress. Res. Treat.* **2012**, *2012*, 752563. [CrossRef]
153. Kumar, M.; Bansal, N. Implications of Phosphoinositide 3-Kinase-Akt (PI3K-Akt) Pathway in the Pathogenesis of Alzheimer's Disease. *Mol. Neurobiol.* **2022**, *59*, 354–385. [CrossRef]
154. Long, H.-Z.; Cheng, Y.; Zhou, Z.-W.; Luo, H.-Y.; Wen, D.-D.; Gao, L.-C. PI3K/AKT Signal Pathway: A Target of Natural Products in the Prevention and Treatment of Alzheimer's Disease and Parkinson's Disease. *Front. Pharmacol.* **2021**, *12*, 648636. Available online: https://www.frontiersin.org/articles/10.3389/fphar.2021.648636 (accessed on 25 July 2023).
155. Sato, A.; Sunayama, J.; Matsuda, K.; Tachibana, K.; Sakurada, K.; Tomiyama, A.; Kayama, T.; Kitanaka, C. Regulation of neural stem/progenitor cell maintenance by PI3K and mTOR. *Neurosci. Lett.* **2010**, *470*, 115–120. [CrossRef] [PubMed]
156. Horwood, J.M.; Dufour, F.; Laroche, S.; Davis, S. Signalling mechanisms mediated by the phosphoinositide 3-kinase/Akt cascade in synaptic plasticity and memory in the rat. *Eur. J. Neurosci.* **2006**, *23*, 3375–3384. [CrossRef] [PubMed]
157. Wang, Q.; Liu, L.; Pei, L.; Ju, W.; Ahmadian, G.; Lu, J.; Wang, Y.; Liu, F.; Wang, Y.T. Control of synaptic strength, a novel function of Akt. *Neuron* **2003**, *38*, 915–928. [CrossRef]
158. Yin, G.; Li, L.-Y.; Qu, M.; Luo, H.-B.; Wang, J.-Z.; Zhou, X.-W. Upregulation of AKT Attenuates Amyloid-β-Induced Cell Apoptosis. *J. Alzheimers Dis.* **2011**, *25*, 337–345. [CrossRef] [PubMed]
159. Mackenzie, R.W.; Elliott, B.T. Akt/PKB activation and insulin signaling: A novel insulin signaling pathway in the treatment of type 2 diabetes. *Diabetes Metab. Syndr. Obes. Targets Ther.* **2014**, *7*, 55–64. [CrossRef] [PubMed]
160. Marosi, M.; Nenov, M.N.; Di Re, J.; Dvorak, N.M.; Alshammari, M.; Laezza, F. Inhibition of the Akt/PKB Kinase Increases Nav1.6-Mediated Currents and Neuronal Excitability in CA1 Hippocampal Pyramidal Neurons. *Int. J. Mol. Sci.* **2022**, *23*, 1700. [CrossRef]
161. Leroy, K.; Yilmaz, Z.; Brion, J.-P. Increased level of active GSK-3beta in Alzheimer's disease and accumulation in argyrophilic grains and in neurones at different stages of neurofibrillary degeneration. *Neuropathol. Appl. Neurobiol.* **2007**, *33*, 43–55. [CrossRef]
162. Beurel, E.; Grieco, S.F.; Jope, R.S. Glycogen synthase kinase-3 (GSK3): Regulation, actions, and diseases. *Pharmacol. Ther.* **2015**, *148*, 114–131. [CrossRef]
163. D'Mello, S.R. When Good Kinases Go Rogue: GSK3, p38 MAPK and CDKs as Therapeutic Targets for Alzheimer's and Huntington's Disease. *Int. J. Mol. Sci.* **2021**, *22*, 5911. [CrossRef]
164. Hooper, C.; Killick, R.; Lovestone, S. The GSK3 hypothesis of Alzheimer's disease. *J. Neurochem.* **2008**, *104*, 1433–1439. [CrossRef]

165. Hsu, W.-C.; Nenov, M.N.; Shavkunov, A.; Panova, N.; Zhan, M.; Laezza, F. Identifying a Kinase Network Regulating FGF14:Nav1.6 Complex Assembly Using Split-Luciferase Complementation. *PLoS ONE* **2015**, *10*, e0117246. [CrossRef]
166. Pablo, J.L.; Pitt, G.S. Fibroblast growth factor homologous factors (FHFs): New roles in neuronal health and disease. *Neuroscientist* **2016**, *22*, 19–25. [CrossRef]
167. Antonell, A.; Lladó, A.; Altirriba, J.; Botta-Orfila, T.; Balasa, M.; Fernández, M.; Ferrer, I.; Sánchez-Valle, R.; Molinuevo, J.L. A preliminary study of the whole-genome expression profile of sporadic and monogenic early-onset Alzheimer's disease. *Neurobiol. Aging* **2013**, *34*, 1772–1778. [CrossRef] [PubMed]
168. Di Re, J.; Wadsworth, P.A.; Laezza, F. Intracellular Fibroblast Growth Factor 14: Emerging Risk Factor for Brain Disorders. *Front. Cell. Neurosci.* **2017**, *11*, 103. [CrossRef] [PubMed]
169. Goldfarb, M.; Schoorlemmer, J.; Williams, A.; Diwakar, S.; Wang, Q.; Huang, X.; Giza, J.; Tchetchik, D.; Kelley, K.; Vega, A.; et al. Fibroblast growth factor homologous factors control neuronal excitability through modulation of voltage gated sodium channels. *Neuron* **2007**, *55*, 449–463. [CrossRef] [PubMed]
170. Shavkunov, A.S.; Wildburger, N.C.; Nenov, M.N.; James, T.F.; Buzhdygan, T.P.; Panova-Elektronova, N.I.; Green, T.A.; Veselenak, R.L.; Bourne, N.; Laezza, F. The Fibroblast Growth Factor 14·Voltage-gated Sodium Channel Complex is a New Target of Glycogen Synthase Kinase 3 (GSK3). *J. Biol. Chem.* **2013**, *288*, 19370–19385. [CrossRef]
171. Hsu, W.-C.J.; Wildburger, N.C.; Haidacher, S.J.; Nenov, M.N.; Folorunso, O.; Singh, A.K.; Chesson, B.C.; Franklin, W.F.; Cortez, I.; Sadygov, R.G.; et al. PPARgamma agonists rescue increased phosphorylation of FGF14 at S226 in the Tg2576 mouse model of Alzheimer's disease. *Exp. Neurol.* **2017**, *295*, 1–17. [CrossRef]

life

Review

Targeting Ion Channels and Purkinje Neuron Intrinsic Membrane Excitability as a Therapeutic Strategy for Cerebellar Ataxia

Haoran Huang [1] and Vikram G. Shakkottai [2,*]

[1] Medical Scientist Training Program, The Ohio State University College of Medicine, Columbus, OH 43210, USA; haoran.huang@osumc.edu
[2] Department of Neurology, The University of Texas Southwestern Medical Center, Dallas, TX 75390, USA
* Correspondence: vikram.shakkottai@utsouthwestern.edu; Tel.: +1-214-648-6849

Abstract: In degenerative neurological disorders such as Parkinson's disease, a convergence of widely varying insults results in a loss of dopaminergic neurons and, thus, the motor symptoms of the disease. Dopamine replacement therapy with agents such as levodopa is a mainstay of therapy. Cerebellar ataxias, a heterogeneous group of currently untreatable conditions, have not been identified to have a shared physiology that is a target of therapy. In this review, we propose that perturbations in cerebellar Purkinje neuron intrinsic membrane excitability, a result of ion channel dysregulation, is a common pathophysiologic mechanism that drives motor impairment and vulnerability to degeneration in cerebellar ataxias of widely differing genetic etiologies. We further propose that treatments aimed at restoring Purkinje neuron intrinsic membrane excitability have the potential to be a shared therapy in cerebellar ataxia akin to levodopa for Parkinson's disease.

Keywords: cerebellar ataxia; Purkinje neuron; intrinsic membrane excitability; ion channels; therapy

Citation: Huang, H.; Shakkottai, V.G. Targeting Ion Channels and Purkinje Neuron Intrinsic Membrane Excitability as a Therapeutic Strategy for Cerebellar Ataxia. *Life* **2023**, *13*, 1350. https://doi.org/10.3390/life13061350

Academic Editor: Carlo Musio

Received: 30 March 2023
Revised: 3 June 2023
Accepted: 6 June 2023
Published: 8 June 2023

1. Introduction

The cerebellar ataxias are a group of clinically heterogeneous movement disorders that primarily affect the cerebellum. The causes of cerebellar ataxia are diverse, but patients share similar clinical features, including abnormal gait, impaired balance, poor coordination of limbs, and speech impairment. Despite similar clinical manifestations, a shared treatment for cerebellar ataxia akin to levodopa for Parkinson's disease has not been identified.

Among all types of cerebellar ataxias, spinocerebellar ataxias (SCAs) are a major subgroup of dominantly inherited neurological disorders causing cerebellar impairment. Currently, more than 40 genetic mutations leading to SCAs have been identified [1]. Interestingly, though the disease-causing genes vary, Purkinje neuron pathophysiology is a shared feature of the disease.

Purkinje neurons are located in the cerebellar cortex and integrate all the input into the cerebellum [2]. Cerebellar atrophy and Purkinje neuron degeneration are shared features of cerebellar ataxia [3,4]. One unique feature of Purkinje neurons is that they exhibit autonomous spiking independent of synaptic stimulation. This pacemaking ability of Purkinje neurons is thought to play a vital role in motor coordination as those disturbances in Purkinje neuron firing significantly impair motor function in mouse models of SCA [5–12]. More importantly, in these mouse models of ataxia, rescuing Purkinje neuron firing abnormalities improves motor performance (additional models reviewed in [13]).

The average firing frequency of Purkinje neurons is around 40 Hz, with an unvarying inter-spike interval duration under resting conditions. This regularity is primarily maintained with precisely controlled activities of ion channels [14]. In brief, activation of sodium channels, primarily Nav1.6 but also Nav1.1, depolarizes the Purkinje neuron plasma membrane. Activation of voltage-gated sodium channels depolarizes the membrane to a threshold to initiate the action potential. Voltage-gated potassium channels are then activated, mediating the repolarization phase of the action potential. The depolarization also activates voltage-gated calcium channels, primarily Cav2.1 and Cav3.1, allowing calcium into Purkinje neurons. These calcium channels are coupled to calcium-activated potassium channels, which generate an outward potassium current upon calcium entry and hyperpolarize the Purkinje neuron membrane to generate the after-hyperpolarization or AHP. The major calcium-activated potassium channels that are responsible for generating the AHP in Purkinje neurons are large-conductance calcium-activated potassium (BK) channels and small-conductance calcium-activated potassium type 2 (SK2) channels. The AHP is essential for the initiation of the subsequent action potential by allowing sodium channels to recover from inactivation and preventing the cell from undergoing a depolarization block. Importantly, numerous studies have shown that perturbations of expression and function of these ion channels alter the intrinsic membrane excitability of Purkinje neurons, leading to abnormalities in Purkinje neuron firing (reviewed in [13,15]). In mouse models of the disease, the abnormalities in Purkinje neuron spiking underlie motor dysfunction, at least in the early stages of the disease. Given the important role of ion channels in regulating proper Purkinje neuron physiology, targeting ion channel function and thereby modulating Purkinje neuron intrinsic membrane excitability holds promise as a therapeutic strategy for cerebellar ataxia.

It is worth considering another degenerative disorder, namely Parkinson's disease when evaluating whether a shared treatment strategy exists for cerebellar ataxia. Parkinson's disease is a collection of inherited and sporadic disorders defined by motor impairments of rigidity, tremors, bradykinesia, and postural-gait changes. Most importantly, in spite of diverse potential molecular mechanisms for disease, motor impairment is sensitive to improvement with dopamine replacement therapy, with levodopa being a mainstay of therapy. The cerebellar ataxias, too, are clinically defined by the motor impairment that they produce. In this review, we summarize ion channel gene mutations that result in human cerebellar ataxia. We speculate that targeting ion channels pharmacologically to correct aberrant Purkinje neuron intrinsic membrane excitability holds promise in the treatment of cerebellar ataxia more widely.

2. Ion Channel Gene Mutations That Cause Ataxia

2.1. Voltage-Gated Sodium Channels

Sodium channels, Nav1.6 and Nav1.1, are important in generating the action potential upstroke in cerebellar Purkinje neurons [16,17]. Nav1.6 is encoded by the gene *SCN8A*, gain-of-function mutations of which are associated with epileptic encephalopathy with ataxia (Table 1) [18]. In mice, partial or complete loss-of-function mutations of *Scn8a* result in motor impairment, symptoms of which also include ataxia [19]. Cerebellar Purkinje neurons of *Scna8a* null mice display reduced repetitive firing [20]. Ataxia is, therefore, associated with *SCN8A* mutations in both human and mouse models.

Dravet syndrome is caused by mutations in *SCN1A*, the gene that encodes Nav1.1 (Table 1) [21]. The clinical manifestations of Dravet syndrome include seizures, cognitive impairment, motor deficits, and ataxia [22]. Knockout of Nav1.1 in mice also results in an ataxic phenotype [16]. Recently, it has been shown that either activating Nav1.1 or blocking Nav1.6 can reduce seizure occurrence in a zebrafish model of Dravet syndrome [23], demonstrating the therapeutic potential of targeting these sodium channels.

Defects in the sodium channel, Nav1.6, are also associated with SCA27 in humans. SCA27 results from mutations in the *FGF14* gene, which encodes fibroblast growth factor 14 (FGF14) [24]. *Ffg14* null mice exhibit an ataxic phenotype that resembles SCA27 patients [25]. Spontaneous Purkinje neuron firing in the *Fgf14* null mice is attenuated, accompanied by a negative shift of membrane potential [26]. It has been shown that spontaneous firing of Purkinje neurons requires intracellular FGF14 (iFGF14), and expressing iFGF14 specifically in Purkinje neurons of *Fgf14* null mice restores spontaneous firing and improves motor performance [27]. In addition, FGF14 interacts with Nav1.6 [28], and the level of Nav1.6

is reduced in Purkinje neurons of FGF14 null mice [26]. These data suggest that the interaction between FGF14 and Nav1.6 is crucial for Nav1.6 expression and function in the cerebellum [26].

Recently, a new genetic cause of late-onset ataxia has been uncovered. This new type of ataxia, termed late-onset SCA27B, is associated with an intronic GAA repeat expansion in *FGF14* [29]. Patients with SCA27B display both episodic and progressive ataxia, with an average age of onset of 55 years [29]. Examination of brain tissue from patients reveals cerebellar atrophy with significant depletion of Purkinje neurons and reduction in FGF14 expression (Table 1). Given the prior data in *Fgf14* knockout mice, it is likely that reduced FGF14 level induced reduction in Nav1.6 expression contributes to disease symptoms.

Table 1. Summary of ion channels and their associated disease and the cerebellar pathology that results.

Types	Ion Channel	Disease	Cerebellar Pathology
Voltage-gated sodium channels	Nav1.1	Dravet syndrome	Cerebellar atrophy on MRI; Cerebellar atrophy with loss of Purkinje neuron on post-mortem tissue [30].
	Nav1.6	Epileptic encephalopathy	Cerebellar atrophy on MRI in one patient [31].
		SCA27B	Severe cerebellar vermis atrophy on MRI and post-mortem tissue. Loss of Purkinje neurons on post-mortem tissue [29].
Voltage-gated potassium channels	Kv1.1	EA1	No cerebellar pathology was reported.
	Kv1.2	Ataxia associated with epileptic encephalopathy	Cerebellar atrophy on MRI in a subset of patients [32].
	Kv1.6	SCA3	See above.
	Kv3.3	SCA13	Cerebellar atrophy on MRI [33].
		SCA3	Mild cerebellar atrophy with enlarged 4th ventricle [4]; Degeneration of the cerebellar fastigial nucleus [3].
		SCA1,2	See above.
	Kv4.3	SCA19, 22	Severe Purkinje neuron degeneration in cerebellar autopsy. Mild cerebellar atrophy on MRI in some patients [34,35].
		SCA1	See above.
Calcium-activated potassium channels	BK	Liang-Wang syndrome	Cerebral atrophy involving the vermis and hemisphere on MRI [36,37].
		SCA1	Global cerebellar volume loss on MRI [4]. Cerebellar atrophy on biopsy and degeneration of cerebellar Purkinje neurons, and the cerebellar fastigial nucleus [3].
		SCA2	Cerebellar atrophy on MRI [38]; Global cerebellar volume loss involving both the vermis and cerebellar hemispheres on MRI [4]; Degeneration of cerebellar Purkinje neurons and the cerebellar fastigial nucleus [3].
		SCA7	Cerebellar atrophy mainly involves the superior part of the vermis on MRI [4]; Degeneration of cerebellar Purkinje neurons and the cerebellar fastigial nucleus [3].
	SK2	Dominant neurodevelopmental movement disorders; autosomal-dominant tremulous myoclonus-dystonia	Cerebellar atrophy on MRI in one case [39] (Mochel, personal communication)

Table 1. *Cont.*

Types	Ion Channel	Disease	Cerebellar Pathology
Voltage-gated calcium channels	Cav2.1	SCA6	Cerebellar atrophy involving the vermis and the hemisphere [4]; Degeneration of cerebellar Purkinje neurons.
		EA2	Cerebellar vermis atrophy on MRI [40].
	Cav3.1	SCA42	Cerebellar atrophy on MRI [41,42].
		SCA1,2,7	See above.
Other calcium channels and calcium pumps	TRPC3	SCA41	Mild cerebellar vermis atrophy on MRI [43].
	IP_3R1	SCA15	Cerebellar atrophy on MRI and CT [44].
		SCA29	Cerebellar atrophy on MRI [45].
		SCA2, 3	See above.
	PMCA2	Congenital cerebellar ataxia	Cerebellar atrophy on MRI [46].
	PMCA3	X-linked congenital cerebellar ataxia	Volume loss of cerebellar hemisphere and vermis on MRI [47].

2.2. Voltage-Gated Potassium Channels

2.2.1. Kv1 Channel Family

Kv1 channels are a group of voltage-gated potassium channels that play an essential role in regulating membrane excitability in neurons. It has been shown that Kv1.1, Kv1.2, Kv1.3, Kv1.5, and Kv1.6 are highly enriched in the rat cerebellum, with higher densities of Kv1.1 and Kv1.2 in the terminal region of Purkinje neurons and higher densities of Kv1.5 and Kv1.6 in the soma [48]. Among them, Kv1.1, Kv1.2, and Kv1.6 expression changes are reported to be associated with ataxia [49].

Kv1.1

Mutations in Kv1.1 channels result in Episodic ataxia type 1 (EA1), which is characterized by episodes of loss of motor coordination and balance in addition to prominent muscle spasms involving the head and extremities (Table 1) [50]. To date, a number of gene mutations in *KCNA1*, the gene encoding Kv1.1 channels, have been identified in patients with EA1. Functional studies of mutated Kv1.1 reveal loss-of-function effects, including a reduction in current, abnormal gating properties, and/or a positive shift of voltage dependence [51].

Gain-of-function mutations in *KCNA1* have also been reported. A patient with a p. A261T Kv1.1 variant presented with mild, childhood-onset focal epilepsy without ataxia [52]. Another patient with a p.L296F mutation was described to display recurrent onset of seizures [53]. Both mutations cause an alteration in the voltage dependence of mutant Kv1.1 channel activation, resulting in the channel opening at a more hyperpolarized state compared to wild-type channels. In addition, treating the patient with the p.L296F variant with 4-aminopyridine, a non-specific potassium channel blocker, improved motor impairment.

Kv1.2

The Kv1.2 channel is encoded by the gene *KCNA2*. De novo mutations in *KCNA2* are associated with epileptic encephalopathy, with about half of the patients with *KCNA2* mutations showing symptoms of ataxia and atrophy of the cerebellum (Table 1) [32]. Interestingly, both gain- and loss-of-function mutations of *KCNA2* are associated with encephalopathy. Specifically, p.R297Q, p.L298F, and p.E157K variants cause an increase in current amplitude and a more hyperpolarized voltage dependence of activation [32]. In contrast, p.I263T and p.P405L variants result in a loss-of-function, characterized by a significant reduction in current amplitude [54]. Furthermore, p.L290R, p.L293H, p.L328V, and T374A variants demonstrate both gain- and loss-of-function effects, resulting in the permanent opening of the channel with a negative voltage shift of inactivation. Patients harboring either a gain-of-function mutation or a combination of gain- and loss-of-function mutations in *KCNA2* are more likely to develop ataxia and cerebellar atrophy [32,54]. 4-aminopridine has been shown to improve symptoms in patients with gain-of-function *KCNA2* encephalopathy [55].

In mice, a p.I402T mutation in *Kcna2* results in chronic motor incoordination. Patch clamp recordings from cerebellar slices of these mice reveal reduced Purkinje neuron firing frequency. In addition, the protein levels of both Kv1.2 and Kv1.1 channels are reduced. Increasing the expression of the Kv1.2 channel partially rescues impaired motor coordination [56].

Kv1.6

The Kv1.6 channel is encoded by the gene *KCNA6*. A recent study identified de novo gain-of-function mutations in *KCNA6* in patients with epilepsy [57]. Mutant Kv1.6 channels display slowed channel closing time and a negative shift in deactivation voltage. Reduced transcript levels of *Kcna6* are associated with increased Purkinje neuron intrinsic membrane excitability in SCA3 mice (Table 1). Treatment with antisense oligonucleotides targeting mutant *Atxn3* rescues the reduction in *Kcna6* transcript levels in SCA3 mice [49].

2.2.2. Kv3.3

Kv3.3 is a voltage-dependent potassium channel encoded by the gene *KCNC3*. Mutations in *KCNC3* cause SCA13 in humans, which is characterized by cerebellar degeneration, impaired motor function, and abnormal auditory processing (Table 1) [58]. The age of disease onset in SCA13 patients ranges from early after birth to middle age. *Kcnc3* knockout mice exhibit a reduction in Purkinje neuron repolarization, altered generation of complex spikes [59,60], and motor defects [61,62]. Restoring Kv3.3 expression in Purkinje neurons of *Kcnc3* knockout mice rescues spiking and motor function [61]. In addition, depending on the mutations of *KCNC3*, Kv3.3 channels can display either loss-of-function or gain-of-function. The p.G592R variant of Kv3.3 shows a gain-of-function phenotype. It results in the degradation of a cell survival protein (Hax-1) that binds to Kv3.3 directly [63]. Other loss-of-function mutations in Kv3.3 that have been identified in human SCA13 patients result in a reduction in Kv3.3 current amplitude and/or Kv3.3 activation voltage [64].

The transcript levels of *Kcnc3* are reduced in mouse models of SCA1, SCA2, and SCA3 (Table 1) [49,65]. In SCA3 mice, the reduction in Kv3.3 transcripts likely contributes to increased excitability of Purkinje neurons and motor impairment [49,66]. Therapeutically, the administration of antisense oligonucleotides targeting the mutated *Atxn3* gene rescues the transcript levels of *Kcnc3*, restores normal Purkinje neuron excitability, and improves the locomotor activity in the early stages of the disease. However, alterations in Purkinje neuron function and motor impairment persist in later disease stages. Meanwhile, a broader reduction in other Kv channels is observed in aged SCA3 mice [49]. Although these reductions do not seem to worsen the Purkinje neuron pathophysiology seen in younger mice, they may contribute to motor dysfunction by altering the function of other brain structures beyond the cerebellum. The detailed mechanism underlying these changes remains to be understood.

The reduction in *Kcnc3* transcript levels has also been reported in mouse models of SCA1 and 2 [65], although little is known regarding the degree of contribution of Kv3.3 to impaired Purkinje neuron firing in these mice. Nevertheless, Kv3.3 dysfunction and dysregulation are associated with multiple cerebellar ataxias [49,62,65,67].

2.2.3. Kv4.3

Kv4.3, encoded by the gene *KCND3*, is another voltage-gated potassium channel highly expressed in cerebellar Purkinje neurons [68]. In humans, mutations in *KCND3* are identified to be the cause of previously designated SCAs: SCA19 and SCA22 (Table 1) [34,35]. The clinical presentation of SCA19/22 includes mild cerebellar ataxia and cognitive impairment. In SCA19/22, the missense mutations in *KCND3* that have been identified include, but are not limited to, p.C317Y, p.P375S, p.V338E, and p.T377M, and are postulated to cause disease due to the loss of function. The mutant Kv4.3 channels show impaired membrane trafficking and are more prone to degradation.

Gain-of-function mutations in Kv4.3 are also implicated in cerebellar ataxia. Recently, a patient presenting with progressive cerebellar ataxia, parkinsonism, cognitive impairment, and iron accumulation in both basal ganglia and cerebellum have been identified to carry a p.R419H mutation in Kv4.3 [69]. Mechanistically, the mutant Kv4.3 channel displays increased outward potassium current and changes in gating properties. In a SCA1 mouse model, increased Kv4.3 current is observed in pre-symptomatic mice, which corresponds to reduced Purkinje neuron firing frequency [5]. Targeting the increased Kv4.3 current with a potassium channel blocker improves firing frequency and motor performance in SCA1 mice [5]. Taken together, both gain- and loss-of-function mutations in Kv4.3 are linked to cerebellar ataxia.

2.3. *Calcium-Activated Potassium Channels*

2.3.1. Large-Conductance Calcium-Activated Potassium (BK) Channels

The large-conductance calcium-activated potassium (BK) channel is encoded by the gene *KCNMA1*. In humans, loss-of-function mutations in *KCNMA1* cause Liang–Wang syndrome (LIWAS), a polymalformation syndrome characterized by global developmental delay and neurological dysfunction (Table 1) [36]. About half of the patients identified with LIWAS show symptoms of ataxia and cerebellar atrophy [36], indicating that loss-of-function BK channel mutations can cause cerebellar ataxia [70].

Downregulation of *Kcnma1* transcripts or BK channel dysfunction has been reported in multiple models of cerebellar ataxia. In a transgenic mouse model of SCA1 (ATXN1[82Q] mice), both the transcripts of *Kcnma1* and the expression of BK channels are decreased [6,9,65]. The decreased level of BK channels is associated with a reduction in the AHP, impaired Purkinje neuron spiking, and motor dysfunction. Significantly, restoring BK channel expression in ATXN1[82Q] mice is able to improve motor performance, suggesting the therapeutic potential of targeting BK channels in SCA1 [6]. Reduced expression of BK channels is also observed in a genetically more precise mouse model of SCA1 (Atxn1$^{154Q/2Q}$) [9,65]. In Atxn1$^{154Q/2Q}$ mice, reduction in BK channel expression is concurrent with irregular Purkinje neuron spiking and impaired motor performance. Applying a BK channel activator (4-chloro-N-(5-chloro-2-cyanophenyl)-3-(trifluoromethyl) benzene-1-sulfonamide, also termed BK-20) is able to rescue Purkinje neuron spiking regularity in Atxn1$^{154Q/2Q}$ mice in vitro [71].

Decreased BK channel expression is also evident in SCA2. In ATXN2[127Q] mice, reduction in the transcripts of *Kcnma1* is accompanied by impaired Purkinje neuron spiking in the early disease stages [67]. Despite the progressive reduction in *Kcnma1* transcripts, regular Purkinje neuron spiking is preserved but at a much lower rate at a later disease stage. The regular but slower Purkinje neuron firing is achieved via the activation of inwardly rectifying potassium (Kir) channels, which may serve as a compensatory mechanism in response to the loss of BK channels [67]. Although restoring BK channel expression or activating remaining BK channels in ATXN2[127Q] mice has not been attempted as a therapeutic strategy, it would be expected that, similar to SCA1 models, this would improve motor function.

Furthermore, a reduction of *Kcnma1* transcript levels is associated with irregular Purkinje neuron spiking in a mouse model of SCA7. Importantly, increasing BK channel expression in the cerebellum of SCA7 mice restores the regularity of Purkinje neuron firing to wild-type levels [7].

It has been shown that in mice, BK channels in Purkinje neurons play a major role in maintaining normal motor coordination. Mice with Purkinje neuron-specific ablation of

BK (PN-BK$^{-/-}$) channels display ataxia, the ataxic symptoms of which resemble global *Kcnma1* knockout mice [72]. At the cellular level, Purkinje neurons of PN-BK$^{-/-}$ mice display a depolarized membrane potential, reduced firing frequency, and no change in firing regularity. In addition, the number of complex spikes is greatly reduced in the Purkinje neurons of PN-BK$^{-/-}$ mice, indicating an impairment in both Purkinje neurons and the cerebellar circuit [72,73].

In contrast, gain-of-function mutations of *KCNMA1* are associated with paroxysmal dyskinesia [74], epilepsy, and dystonia. Patients with a p.D434G mutation in BK channels have both paroxysmal dyskinesia [75] and epilepsy [76]. Mechanistically, the p.D434G BK channel variant displays increased calcium sensitivity and enhanced potassium current [74]. In comparison, although the p.N995S variant of BK channels also displays increased potassium current, there is no change in calcium sensitivity. The p.N995S BK channel enhances both channel open probability and duration and hence increases potassium current. In humans, the p.N995S (also annotated as p.N999S and p.N1053S) mutation causes epilepsy but not paroxysmal dyskinesia [76,77]. In mice, knockout of the beta-4 subunit of BK channels results in a gain of function. Beta-4 null mice develop spontaneous seizures, potentially resulting from lowered threshold for the generation of action potentials in hippocampal dentate granule cells [78].

2.3.2. Small-Conductance Calcium-Activated Potassium (SK2) Channels

The brain expresses three types of small-conductance calcium-activated potassium channels. The type 2 small-conductance calcium-activated potassium (SK2) channel, encoded by the gene *KCNN2*, is enriched in cerebellar Purkinje neurons. Recently, exome sequencing has revealed that patients with *KCNN2* mutations exhibit delays in motor, intellectual, and language development (Table 1) [39]. Other symptoms include cerebellar ataxia, tremor, seizures, and extrapyramidal symptoms such as dyskinesia and myoclonus-dystonia [39,79]. Patch-clamp recordings from cells transfected with selected human *KCNN2* mutants reveal a reduction in current density [39], indicating that the loss-of-function of SK2 channels may be responsible for the disease symptoms observed.

Mice harboring a deletion of the *Kcnn2* gene exhibit symptoms of motor dysfunction such as tremors, abnormal gait, and impaired rotarod performance [80]. Genetically targeted silencing of SK channels in the brain of mice also results in profound motor impairment [81]. Modulation of the SK2 channel has shown beneficial effects in a mouse model of SCA2. SCA2-58Q mice show a phenotype of age-dependent onset of motor deficits [82] that is associated with burst firing patterns of Purkinje neurons [11]. Significantly, SK channel activators are able to convert the burst firing pattern back to tonic firing patterns [11]. Oral administration of SK activators also improves motor performance and delays Purkinje neuron degeneration in SCA2-58Q mice. Furthermore, a loss of function of the SK2 channel has been described as the underlying abnormality in mouse models of tremor [80,83]. No gain-of-function mutations in *KCNN2* have been reported.

2.4. *Voltage-Gated Calcium Channels*

2.4.1. Cav2.1

CACNA1A encodes the P/Q-type voltage-gated calcium channel Cav2.1, which is highly enriched at synaptic terminals and in Purkinje neurons [84]. Mutations in *CACNA1A* are associated with SCA6 and episodic ataxia type 2 (EA2) (Table 1) [85,86].

In SCA6, a CAG repeat expansion in *CACNA1A* results in a poly-glutamine tract at the C-terminus of Cav2.1 [85]. Patients affected with SCA6 have 20-33 CAG repeats in the *CACNA1A* gene, with slowly progressive and late-onset ataxia symptoms [87,88]. A SCA6 mouse model captures the late-onset disease phenotype as seen in human patients. Heterozygous SCA6$^{84Q/+}$ mice develop motor impairment at 19 months of age, whereas homozygous SCA6$^{84Q/84Q}$ mice have an earlier disease onset at 7 months of age [10,89]. Surprisingly, although the CAG repeat expansion is in the gene encoding Cav2.1, the function of the calcium channel itself in Purkinje neurons does not appear to be impaired [89,90]. The exact disease mechanism accounting for the late onset of symptoms in SCA6 remains unknown.

Mutations in the *CACNA1A* gene are also associated with EA2, characterized by episodes of recurrent ataxic symptoms in humans [91,92]. One widely used mouse model of EA2 is the tottering mouse. Tottering mice, which carry a p.P601L missense mutation in the *Cacna1a* gene [93], display ataxic symptoms such as paroxysmal attacks of motor dysfunction and abnormal eye movements [94,95]. Purkinje neurons of tottering mice exhibit firing abnormalities [94,96]. The irregular Purkinje neuron spiking is associated with a reduction in calcium current density [97].

Gain-of-function mutations in Cav2.1 are also associated with abnormal Purkinje neuron firing. The p.S218L missense mutation in *Cacna1a* results in an increase in calcium influx into Purkinje neurons, which lowers the threshold for action potential generation, causing a disrupted firing pattern [98]. Mice with the p.S218L mutation display symptoms of ataxia [99], indicating that both increased and decreased calcium influx can be associated with ataxia.

2.4.2. Cav3.1

Cav3.1, encoded by the gene *CACNA1G*, is a T-type calcium channel highly expressed in Purkinje neurons. In humans, the p.R1715H missense mutation in the *CACNA1G* gene causes SCA42, symptoms of which include predominantly gait instability (Table 1) [100]. The p.R1715H mutation results in a positive shift in the voltage dependence of Cav3.1 activation [101]. Zonisamide, a T-type calcium channel blocker, reverses the abnormal voltage dependence of current activation of the mutant channel close to wild-type levels in HEK293T cells [102]. In addition, reduced transcript levels of *Cacna1g* have been reported in other SCA mouse models, including SCA1, 2, and 7 [7,9,65,103]. Mice lacking Cav3.1 display impaired motor coordination, Purkinje neuron loss, and cerebellar atrophy [104].

Gain-of-function mutations in Cav3.1 are associated with childhood-onset cerebellar atrophy with additional clinical features, including cognitive impairment and variable facial dysmorphism, microcephaly, and epilepsy [105].

2.5. Other Calcium Channels and Calcium Pumps

2.5.1. TRPC3

TRPC3 is one of the family members of the transient receptor potential (TRP) superfamily of ion channels that are highly enriched in cerebellar Purkinje neurons [106]. Mutations in the *TRPC3* gene cause SCA41 in humans (Table 1). Recently, a p.R762H missense mutation in *TRPC3* has been identified in a patient with a progressive imbalance and ataxic gait [43]. The p.R762H variant induces significant neuronal death, indicating a gain-of-function mechanism. Moonwalker (*Mwk*) mice, harboring a gain-of-function mutation in *Trpc3*, display coordination defects and Purkinje neuron degeneration [107]. Mechanistically, in *Mwk* mice, the TRPC3 channel displays prolonged channel opening upon activation [107]. Patch clamp recordings from *Mwk* mice show that a greater number of Purkinje neurons are inactive compared to wild-type neurons. Additionally, the Purkinje neurons of *Mwk* mice that are active fire at a greater frequency than wild-type neurons, indicating the inactive ones are likely in depolarization block [108].

2.5.2. IP_3R1

IP_3R1, or inositol 1,3,5-trisphophate (IP_3) receptor type 1, is encoded by the gene *ITPR1*. IP_3R1 functions as a ligand-gated ion channel and mediates calcium release from the endoplasmic reticulum. Deletions or mutations in *ITPR1* lead to SCA15 in humans, the clinical presentation of which is characterized by adult-onset and slow progression of cerebellar gait ataxia (Table 1) [97]. Missense mutations in *ITPR1*, including p.V494I and p.P1059L, have also been identified in families with SCA15 [44,109]. Missense mutations in *ITPR1* are also associated with SCA29 in humans (Table 1) [45]. Patients with SCA29 present with infant-onset slowly progressive ataxia, delayed motor development, and mild cognitive impairment [110]. Furthermore, patients with autoantibodies against IP_3R1 also develop cerebellar ataxia [111]. The underlying mechanism of SCA15 and SCA29 pathology is likely due to suppressed cytosolic calcium signaling, which results from the loss-of-function of IP_3R1.

Studies in mice show that increased IP_3R1 activity may also contribute to cerebellar ataxia. It has been shown that in SCA2-58Q mice, mutant ATXN2 interacts with IP_3R1, leading to increased activation sensitivity of IP_3R1 and enhanced calcium release from the endoplasmic reticulum [82]. Aberrant calcium signaling also contributes to the pathogenesis of SCA3. Similar to the mutant ATXN2 protein, mutant ATXN3 (Q77 and Q127) can also bind to IP_3R1 and increase the sensitivity of IP_3R1 to IP_3 [112].

2.5.3. PMCAs

The calmodulin-activated plasma membrane calcium-ATPases (PMCAs), especially the brain-enriched isoforms, type 2 (PMCA2) and type 3 (PMCA3), function to maintain cellular calcium homeostasis by pumping calcium out of the cell [113]. In humans, mutations in the gene encoding PMCA2 are linked to hearing loss. Recently, a missense mutation in PCMA2, V1143F, has been identified in a patient with congenital cerebellar ataxia but no sign of deafness (Table 1) [46]. The mutation causes a loss-of-function mutation of PMCA2, leading to impaired calcium ejection from the cytoplasm. In mice, PMCA2 is highly expressed in various tissue types, including cerebellar Purkinje neurons. *Atp2b2* (encodes PMCA2) knockout mice exhibit overt cerebellar ataxia [114]. Recordings from Purkinje neurons of *Atp2b2* knockout mice reveal slower and irregular firing and a more hyperpolarized membrane potential [115]. Missense mutations in PMCA3 are associated with ataxia. The missense mutation G1107D in *ATP2B3* has been identified in patients with X-linked congenital cerebellar ataxia [47,116] (Table 1). Functional analysis of the mutant PMCA3 demonstrates impaired calcium pumping ability, indicating a loss-of-function mutation. Taken together, both PMCA2 and PMCA3 play a role in ataxia pathogenesis.

3. Emerging Therapies for Cerebellar Ataxia Impinge on Ion Channels

3.1. Non-Pharmacological Approaches in Cerebellar Ataxia

3.1.1. Rehabilitation

Currently, there is no definitive treatment for cerebellar ataxia. Most of the current clinically used treatments rely on strategies for rehabilitation that include physical therapy, occupational therapy, and speech therapy. Rehabilitation in cerebellar ataxia aims to improve balance and coordination in patients through intensive exercise [117]. It has been shown that rehabilitation training can improve the impairment measured on an ataxia rating scale [118–120], with evidence that the improvement can persist for up to 1 year. The exact mechanism of how exercise improves motor function in cerebellar ataxia patients with progressive cerebellar atrophy is still unknown. In $SCA6^{84Q/84Q}$ mice, exercise has been shown to improve Purkinje neuron firing defects and thereby improve ataxia, potentially through increasing cerebellar expression of brain-derived neurotrophic factor [12]. So far, this is the only evidence showing that exercise can modify Purkinje neuron firing in cerebellar ataxia. It is likely that exercise modifies potassium channel expression and function to mediate this improvement in spiking.

3.1.2. Gene Suppression Strategies

Oligonucleotide-based therapy is an emerging field aimed at treating genetic disorders. Oligonucleotide-based therapy that has been implicated in treating cerebellar ataxias includes antisense oligonucleotide and RNA-based therapy [121]. Antisense oligonucleotides (ASOs) are small, single-stranded DNA molecules that have the capability to reduce the expression of specific proteins by binding to complementary mRNA transcripts [122]. The therapeutic effects of ASOs have been shown in several mouse models of cerebellar ataxia, including SCA1, SCA2, SCA3, and SCA13 [63,123–125]. In the Knockin mouse model of SCA1 ($Atxn1^{154Q/2Q}$) that displays ataxia and premature lethality, ASO delivery to the right lateral ventricle improves motor performance and prolongs survival [123]. Injection of ASOs into the cerebellum of SCA2 mice can successfully reduce Atxn2mRNA and protein levels in Purkinje neurons, resulting in delayed motor symptom onset [124]. In addition, ASO treatment improves Purkinje cell firing in parallel with behavioral improvement. Longitudinal ASO delivery targeting mutant *Atxn3* in SCA3 mice restores transcript levels of Kv3.3 and Kv1.6 associated with improved Purkinje neuron firing and motor performance [49]. In the Kv3.3-G592R mouse model of SCA13, intracerebroventricular injection of ASOs targeting mutant *Kcnc3* mRNA restores motor performance to a level comparable with wild-type controls [63].

Gene silencing through viral-mediated RNAi is an alternative method to ASOs to reduce the level of targeted proteins. Intracerebellar injection of an adeno-associated virus that expresses short hairpin RNAs against ATXN1 has been shown to significantly improve motor coordination in SCA1 mice [126,127].

3.2. *Pharmacological Approaches in Cerebellar Ataxia*

Treatment for cerebellar ataxia is needed. Few medications have proven effective in managing symptoms of cerebellar ataxia. The majority of agents that have been proposed for treatment have been unsuccessful in clinical trials. Here, we discuss the clinical data and the mechanism of action of the agents that are currently prescribed for treating cerebellar ataxia. These include omaveloxolone, 4-aminopyridine, and chlorzoxazone-baclofen. Importantly, many of these agents target ion channels, further illustrating the therapeutic potential of modulating ion channels for cerebellar ataxia.

3.2.1. Omaveloxolone

Recently, the US FDA approved omaveloxolone as the first and only therapy for Friedreich ataxia, an autosomal recessive cerebellar ataxia. Clinical symptoms in patients with Friedreich ataxia include gait and limb ataxia, discoordination, loss of lower limb reflexes, dysarthria, diabetes, cardiomyopathy, scoliosis, and vision loss [128]. Friedreich ataxia results from a GAA repeat expansion in the *FXN* gene, which encodes the protein frataxin that is localized to mitochondria [129]. The function of frataxin is not entirely understood, but it has been shown to be involved in iron–sulfur cluster assembly [130]. A GAA repeat expansion in *FXN* results in a reduction of functional frataxin protein. Consequences of frataxin deficiency include mitochondrial dysfunction, increased sensitivity to oxidative stress, and dysregulation of iron-sulfur cluster assembly [131–133]. Friedreich ataxia is, therefore, considered to be a mitochondrial disease.

The proposed mechanism of action of omaveloxolone is the activation of the nuclear factor erythroid 2-related factor 2 (NRF2) [134,135]. Under normal conditions, oxidative stress induces NRF2 translocation to the nucleus and increases the expression of antioxidant genes. In Friedreich ataxia, however, NRF2 fails to respond to oxidative stress [134]. The impaired antioxidant defense system may contribute to neurodegeneration in Friedreich ataxia since neurons are sensitive to cellular redox status [136]. Omaveloxolone, as a potent activator of NRF2, has been shown to restore mitochondrial function in multiple mouse models of Friedreich ataxia. In clinical trials, patients receiving omaveloxolone show significant improvement in motor performance than patients receiving a placebo [137]. Given omaveloxolone's ability to rescue antioxidant defense systems, it may also be therapeutically beneficial in other cerebellar ataxias that are associated with mitochondrial dysfunction.

SCA28 is the only known dominantly inherited cerebellar ataxia caused by defects in a mitochondrial protein [138]. The mutated gene responsible for SCA28, *AFG3L2*, encodes a subunit of a human mitochondrial ATPase associated with various cellular protease activities (m-AAA) [139]. Members of the m-AAA protease family participate in protein quality control within mitochondria [140,141]. Clinically, patients affected by SCA28 demonstrate slowly progressive gait and limb ataxia. In a SCA28 mouse model that is haploinsufficient for *Afg3l2*, progressive loss of cerebellar Purkinje neurons, thinning of the cerebellar molecular layer, and impaired motor performance are noted [138]. Purkinje neuron neurodegeneration is likely to be a result of enhanced cytoplasmic calcium concentrations due to defects in mitochondrial calcium buffering capacity [140]. Reducing cellular calcium influx is able to improve ataxia in the *Afg3l2* haploinsufficient mice [142]. Additionally, Purkinje neurons in mice that are heterozygous for the p.M665R-*Afg3l2* allele appear to be more excitable than wild-type mice [143].

Mitochondrial defects have also been shown to contribute to disease in autosomal recessive spastic ataxia of Charlevoix–Saguenay (ARSACS), a childhood-onset disorder with pyramidal spasticity and cerebellar ataxia. ARSACS results from mutations in the *SACS* gene encoding the sacsin protein, which has been shown to regulate the connectivity of the mitochondrial network [144]. Purkinje neurons of *Sacs* knockout mice demonstrate disordered dendritic morphology and slow but regular firing [145] before degeneration.

Taken together, it is likely that impairments in the mitochondrial network can lead to alterations in Purkinje neuron intrinsic membrane excitability and promote Purkinje neuron degeneration. Omaveloxolone, therefore, may have a broader therapeutic potential beyond Friedreich ataxia.

3.2.2. 4-Aminopyridine

4-aminopyridine (4-AP) functions as a non-selective voltage-gated potassium (Kv) channel blocker. Currently, it is prescribed for patients with EA2, multiple sclerosis, and for the treatment of symptomatic downbeat nystagmus. Several studies have shown that 4-AP reduces ataxia attack frequency and decreases attack duration in EA2 patients [146,147]. In a mouse model of EA2, abnormal Purkinje neuron spiking is associated with reduced calcium current through the mutant Cav2.1 channel. 4-AP has been shown to restore the precision of Purkinje neuron pacemaking via the prolongation of action potentials and increasing the amplitude of the AHP [148]. 4-AP appears to restore Purkinje neuron pacemaking with a mechanism similar to Chlorzoxazone, a calcium-activated potassium channel (Kca) activator. By inhibiting Kv channels, 4-AP appears to activate Kca channels. Besides EA2, 4-AP has been shown to reduce the frequency and/or severity of ataxia symptoms in patients with SCA27B [29]. Chronic administration of 4-AP to $SCA6^{84Q/84Q}$ mice restores the precision of Purkinje neuron spiking and improves motor performance [10]. Acute treatment of early symptomatic SCA1 mice with 3,4-diaminopyridine, an analog of 4-AP, restores normal Purkinje neuron firing and improves motor performance. Chronic treatment of SCA1 mice with 3,4-diaminopyridine not only normalizes firing frequency and improves motor function but also partially protects against neuronal degeneration [5].

Blockade of Kv channels by 4-AP as a means to increase neuronal conduction has been used as a therapeutic approach in multiple sclerosis (MS). Clinically, MS patients benefit from 4-AP with improved motor function and vision [149,150]. Studies from experimental mouse models of MS show that demyelination causes an axonal redistribution of potassium channels (especially Kv1.1 and Kv1.2) [151] that are normally clustered near the nodes of Ranvier [152]. Demyelination also leads to enhanced exposure of potassium channels. The misdistribution and increased exposure of potassium channels cause an increase in the threshold for successful action potential generation, which impairs action potential progression and results in motor dysfunction in MS [153]. It has been shown that by blocking potassium channels with 4-AP, action potential conduction is restored.

3.2.3. Chlorzoxazone and Baclofen

Chlorzoxazone (CHZ) is an FDA-approved skeletal muscle relaxant used to treat muscle spasms and pain. Although the exact mechanism of action is unknown, it activates SK and BK channels [154,155]. Clinically, CHZ has been suggested to improve eye movements and visual acuity in patients with cerebellar downbeat nystagmus [156]. In mouse studies, CHZ improves Purkinje neuron firing and motor performance in multiple mouse models of cerebellar ataxia. Intraperitoneal injection of CHZ in SCA2-58Q mice converts irregular bursting Purkinje neuron firing patterns into regular tonic firing in vivo [157]. In the same study, SCA2-58Q mice CHZ also improved performance on a beam-walk test. In EA2 mice, CHZ restores irregular Purkinje neuron firing in vitro and significantly reduces the frequency and duration of ataxia attacks when given orally [148].

Baclofen is another FDA-approved skeletal muscle relaxant that has been used to treat muscle spasms, stiffness, and pain in patients with multiple sclerosis and/or spinal cord injury/disease. Intrathecal baclofen (ITB) infusion is accepted as a treatment for spasticity (increased rigidity of muscle secondary to injury), where it has been shown to improve positioning and decrease muscle tone in targeted patients [158]. One study has shown ITB treatment has beneficial effects in patients with SCA3, SCA7, and Friedreich ataxia who have spasticity and muscle spasms [159]. From a molecular standpoint, baclofen functions as a $GABA_B$ receptor agonist in the central nervous system [160]. In the cerebellum, $GABA_B$ receptors are highly enriched at pre- and post-synaptic terminals [161,162]. It has been shown that activation of $GABA_B$ receptors modulates neuronal excitability via activation of G protein-gated inwardly rectifying K^+ (GIRK or Kir3) channels in cerebellar Purkinje neurons [163]. Under normal conditions, inwardly rectifying K^+ (K_{ir}) channels function in maintaining resting membrane potential in neurons [164]. However, their function may be altered in the disease. In a transgenic mouse model of SCA2, a significant number of non-firing Purkinje neurons is associated with loss of BK and Kv3.3 channels expression at an early disease stage. However, Purkinje neuron firing in the mutant mice is restored at a later disease stage, although at a significantly slower frequency. It has been found that rescued slow Purkinje neuron firing frequency is a result of a novel AHP produced by K_{ir} channels [67], indicating that K_{ir} channels may have a compensatory role in restoring Purkinje neuron physiology when the AHP mediated by BK channels is impaired. A similar loss of the AHP is observed in SCA1 mouse models [6,9]. Given Baclofen's ability to create a novel AHP, a combination of CHZ and baclofen was identified to restore the Purkinje neuron AHP in these models of SCA1 [8,9]. Administration of a combination of baclofen and CHZ in vivo improves motor impairment in several models of SCA1 [8,9].

Clinically, CHZ and baclofen are often prescribed together to patients with cerebellar ataxias. In patients with SCA1, SCA2, SCA6, SCA8, and SCA13, co-administered CHZ and baclofen are suggested to improve motor impairment [8].

3.3. Other Approaches

Cerebellar Stimulation: Deep Brain Stimulation and Non-Invasive Cerebellar Stimulation

Deep brain stimulation (DBS) is currently used to treat patients with Parkinson's disease, dystonia, and tremors. Although the use of DBS in patients with ataxia is limited, ataxia patients may potentially benefit from this therapy. Several studies have shown that DBS can relieve tremors and/or dystonia associated with ataxia. One study showed that high-frequency thalamic ventralis intermedius nucleus (VIM) stimulation and low-frequency stimulation of subthalamic projections attenuated tremors in two patients with SCA27 with no effect on ataxia [165]. In another study, VIM stimulation improved tremor in patients with SCA2 and idiopathic tremor-ataxia syndrome, whereas Globus pallidus interna (Gpi) DBS improved dystonia in one SCA17 patient and one patient with a senataxin mutation, with no evidence of improvement in ataxia-related symptoms [166]. Long-term VIM stimulation regularized the gait cycle and improved tremor in three patients with Fragile X-associated tremor/ataxia syndrome (FXTAS) [167]. However, the outcome of VIM stimulation was reported as poor in another FXTAS patient [166]. Dentate nucleus DBS (DN DBS) was also shown to effectively alleviate cerebellar tremors in patients with SCA3 or post-lesion ataxia [168]. In the same patients, DN DBS moderately improved ataxia symptoms, although the effect was not significant. Despite the lack of evidence for the improvement of ataxia via DBS in human studies, DBS targeting the cerebellum has shown promising therapeutic effects in mouse models of ataxia. DBS stimulation of the interposed cerebellar nucleus induces long-lasting motor benefits and significantly improves motor performance in mouse models of ataxia [169,170].

Another promising therapeutic modality for cerebellar disorders is non-invasive cerebellar stimulation, which includes transcranial magnetic stimulation (TMS) and transcranial direct current stimulation (tDCS). The therapeutic effects of non-invasive cerebellar stimulation on 151 cerebellar ataxia patients from eight studies were summarized in the systemic review by Di Nuzzo et al. [171].

A detailed mechanism of how cerebellar stimulation improves motor dysfunction remains to be understood. A recent modeling study suggests that cerebellar DBS can restore the reduced inhibitory input resulting from Purkinje neuron degeneration in ataxic mice [172], indicating cerebellar stimulation has the potential to modulate synaptic activity and downstream neuronal pathways for motor control. Proper synaptic activity at both pre- and postsynaptic levels relies heavily on ion channels [173]. Additionally, it has been proposed that low-frequency stimulation likely excites neurons at both soma and axon levels, whereas high-frequency stimulation exhibits inhibitory effects [174], potentially via the induction of depolarization block via inactivation of voltage-gated ion channels. Therefore, cerebellar stimulation may elicit its therapeutic effects through the modulation of synaptic ion homeostasis.

4. Discussion: Is There a "Levodopa" for Cerebellar Ataxia?

Cerebellar ataxias are a group of devastating diseases that severely affect patients' quality of life. Unfortunately, effective treatments for cerebellar ataxia are limited. The recently FDA-approved medication for Friedreich ataxia, omaveloxolone, is an exciting development and holds promise for the development of other therapeutic modalities in the future. A therapy that could benefit a broader range of cerebellar ataxia patients and with a larger effect size is still needed.

Based on current studies in ataxia mouse models, it is clear that abnormal Purkinje neuron spiking is associated with impaired motor function. Treatments aimed at restoring normal Purkinje neuron spiking have been shown to improve motor performance. Normal Purkinje neuron spiking is tightly dependent on membrane excitability regulated by ion channels. Changes in ion channel expression and/or function in disease conditions are likely to impair Purkinje neuron intrinsic membrane excitability, leading to irregular and/or slow Purkinje neuron firing patterns. The proper ion channel function is, therefore, vital for regulating Purkinje neuron spiking and, hence, precise motor coordination.

Mutations in a variety of genes have been linked to cerebellar ataxias. Recent studies in mouse models of ataxia of differing etiology suggest that a shared feature among cerebellar ataxias is defects in ion channels (altered expression and/or function). Clinically, agents that modulate ion channels may improve symptoms in ataxia patients. For example, the non-selective potassium channel blocker 4-AP appears to improve motor performance in patients with EA2 and SCA27B. Co-administration of chlorzoxazone and baclofen may improve symptoms in patients with SCA1, SCA2, SCA6, SCA8, and SCA13. In ataxia mouse models, these ion channel modulators have also been shown to improve motor function. Mechanistically, improvements in motor behavior in mice by ion channel modulators are associated with improvements in abnormal Purkinje neuron spiking. From a therapeutic perspective, ion channel modulators that can rescue irregular Purkinje neuron spiking or increase Purkinje neuron firing frequency when it is reduced would theoretically have the ability to restore normal Purkinje neuron intrinsic membrane excitability and normalize spiking in disease states and thus improve motor function. Therapies for cerebellar ataxia should, therefore, aim to restore Purkinje neuron intrinsic membrane excitability by normalizing Purkinje neuron spiking regularity when it is irregular and improving Purkinje neuron firing rate when the firing frequency is reduced. Since ion channels are crucial regulators of Purkinje neuron intrinsic membrane excitability, we propose that agents targeting ion channels have the potential to be a shared therapy for cerebellar ataxia independent of the disease-causing gene (Figure 1).

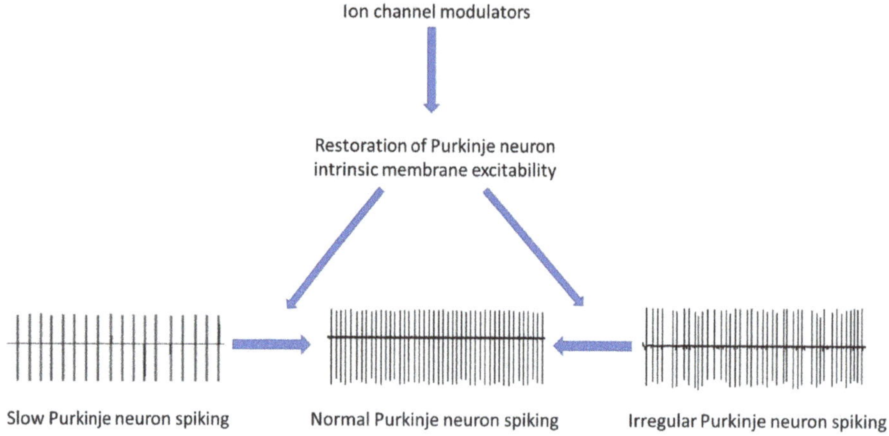

Figure 1. Scheme showing Purkinje neuron pathophysiology in cerebellar ataxia and how modulation of Purkinje neuron intrinsic membrane excitability is a potential therapy for cerebellar ataxia.

4.1. Voltage-Gated Sodium Channels

Both Nav1.6 and Nav1.1 play important roles in the generation of the Purkinje neuron action potential upstroke. Mutations in the genes encoding Nav1.6 and Nav1.1 produce ataxia symptoms in both humans and mice. Specifically, both gain- and loss-of-function mutations in the gene encoding Nav1.6 and loss-of-function mutations of the gene encoding Nav1.1 can result in ataxia. Therefore, activating or inhibiting these sodium channels and tailored to the underlying mechanism may have therapeutic potential in ataxia. It is important to note that although the reduction in sodium current is associated with ataxia, the degree of activation of sodium channels should be tightly controlled. Increased sodium channel activity has been shown to play a role in contributing to neurodegeneration in multiple sclerosis [175]. It has been shown that sustained sodium influx can overwhelm the function of the Na/K ATPase, leading to the accumulation of sodium in axons and thereby activating the Na/Ca exchanger [176]. Activation of the Na/Ca exchanger moves sodium outside of axons at the expense of elevating intra-axonal calcium levels, which contributes to axonal degeneration [177]. Increased calcium concentrations can have detrimental effects on neurons. Separately, an increased inward calcium current can also cause ataxia [98]. Similarly, the dose of sodium channel inhibitors in treating gain-of-function-mutation-associated ataxia should also be tightly controlled as the sodium current is crucial for the generation of action potentials.

4.2. Voltage-Gated Potassium Channels

Defects in voltage-gated potassium channels, including Kv1 family members, Kv3.3, and Kv4.3, are associated with cerebellar ataxia. 4-AP, a non-selective voltage-gated potassium channel inhibitor, has shown promising therapeutic effects in certain ataxia patient populations, including patients with EA2 and SCA27B. 4-AP can restore Purkinje neuron spiking regularity without affecting firing frequency in mouse models of SCA6 and EA2. 3,4-diaminopyridine, an analog of 4-AP, has been shown to improve the reduced Purkinje neuron firing frequency in a mouse model of SCA1 [5]. However, one potential downside of 4-AP being non-selective is that it blocks different types of Kv channels at similar concentrations [178]. Importantly, loss-of-function mutations in certain Kv channels cause ataxia. Loss-of-function mutations of Kv3.3 and Kv4.3 result in SCA13 and SCA19/22, respectively. Reduced transcript levels of the gene encoding Kv1.6 is evident in a mouse model of SCA3. Loss-of-function mutations in Kv1.1 and Kv1.2 are also associated with ataxia phenotypes. Therefore, when considering 4-AP as a shared therapy for cerebellar ataxias, one must take into account that 4-AP has the potential to cause ataxia. Furthermore, a clinical trial examining the short-term effect of 4-AP on patients with SCA1, SCA3, and SCA6 showed no change in the score of an ataxia rating scale compared to a placebo [179].

4.3. Calcium-Activated Potassium Channels

Activation of BK channels can improve Purkinje neuron spiking irregularity and increase Purkinje neuron firing frequency. In Atxn1$^{154Q/2Q}$ mice, irregular Purkinje neuron spiking is associated with motor impairment. BK-20, a BK channel activator, significantly improves the regularity of Purkinje neuron spiking in vitro. Additionally, BK-20 increases firing frequency in Atxn1$^{154Q/2Q}$ Purkinje neurons [71]. We, therefore, speculate that in ataxia, where phenotype results from either irregular or slow Purkinje neuron firing, activating BK channels would help normalize firing. However, the dose of BK channel activators should be tightly controlled as BK channel overactivation is associated with paroxysmal dyskinesia, epilepsy, and dystonia [74].

SK2 channels may also be targeted for treating cerebellar ataxia. Although mutations in SK2 channels have only been identified in a few patients with ataxia, agents activating SK2 channels have been shown to improve Purkinje neuron firing regularity and motor performance in mouse models of cerebellar ataxia. The effect of SK2 activation on Purkinje neuron firing frequency varies. For example, SK2 activators increase Purkinje neuron firing frequency in vivo but decrease firing frequency in vitro in Cav2.1-S218L mice [98]. Additionally, higher concentrations of chlorzoxazone, an SK2 channel activator, significantly decrease Purkinje neuron firing frequency in Atxn1$^{154Q/2Q}$ mice [9].

4.4. Calcium Channels and Calcium Pumps

Calcium channels, including Cav2.1, Cav3.1, TRPC3, and IP$_3$R1, and calcium pumps, including PMCA2 and PMCA3, have been implicated in cerebellar ataxia in both humans and mice. Both gain- and loss-of-function mutations in these calcium channels can cause abnormal Purkinje neuron spiking and motor dysfunction in mice, indicating a vital role of calcium currents and intracellular calcium concentrations in maintaining Purkinje neuron intrinsic membrane excitability. Similarly, mutations in other calcium channels (for example, calcium-release activated calcium channels) that lead to increased or decreased calcium influx may, therefore, also be associated with Purkinje neuron pathophysiology and ataxia. Furthermore, the proper function of calcium-activated potassium channels depends on appropriate calcium flux and intracellular calcium concentrations. Taken together, targeting calcium channels may represent another therapeutic approach to treating cerebellar ataxia.

5. Conclusions

Ion channel dysfunction is a shared disease mechanism in cerebellar ataxias despite the various disease-causing genes. Dysfunction in ion channels disrupts Purkinje neuron intrinsic membrane excitability, which impairs Purkinje neuron spiking and motor function. Therefore, we propose that targeting Purkinje neuron intrinsic membrane excitability holds promise to restore spiking irregularity and increase slowed Purkinje neuron firing. Therapies aiming at restoring normal Purkinje neuron intrinsic membrane excitability, therefore, have the greatest potential to be the "levodopa" for cerebellar ataxia.

The long-term benefits and potential complications of the use of ion channel modulators in cerebellar ataxia are currently unknown. In addition, in Parkinson's disease, levodopa becomes less effective as new symptoms develop and the disease progresses [180]. Similarly, in cerebellar ataxia, it is likely that combining ion channel modulator therapy with other treatments may be needed for the long-term management of progressive symptoms.

Author Contributions: Conceptualization, H.H. and V.G.S.; writing—original draft preparation, H.H. and V.G.S.; writing—review and editing, H.H. and V.G.S.; funding acquisition, V.G.S. All authors have read and agreed to the published version of the manuscript.

Funding: Research was funded by the National Institutes of Health R01 NS085054 (VS).

Institutional Review Board Statement: Not applicable.

Informed Consent Statement: Not applicable.

Data Availability Statement: Data sharing not applicable.

Conflicts of Interest: The authors declare no conflict of interest.

References

1. Durr, A. Autosomal dominant cerebellar ataxias: Polyglutamine expansions and beyond. *Lancet Neurol.* **2010**, *9*, 885–894. [CrossRef] [PubMed]
2. Hirano, T. Purkinje Neurons: Development, Morphology, and Function. *Cerebellum* **2018**, *17*, 699–700. [CrossRef] [PubMed]
3. Seidel, K.; Siswanto, S.; Brunt, E.R.; den Dunnen, W.; Korf, H.W.; Rub, U. Brain pathology of spinocerebellar ataxias. *Acta Neuropathol.* **2012**, *124*, 1–21. [CrossRef] [PubMed]
4. Cocozza, S.; Pontillo, G.; De Michele, G.; Di Stasi, M.; Guerriero, E.; Perillo, T.; Pane, C.; De Rosa, A.; Ugga, L.; Brunetti, A. Conventional MRI findings in hereditary degenerative ataxias: A pictorial review. *Neuroradiology* **2021**, *63*, 983–999. [CrossRef]
5. Hourez, R.; Servais, L.; Orduz, D.; Gall, D.; Millard, I.; de Kerchove d'Exaerde, A.; Cheron, G.; Orr, H.T.; Pandolfo, M.; Schiffmann, S.N. Aminopyridines correct early dysfunction and delay neurodegeneration in a mouse model of spinocerebellar ataxia type 1. *J. Neurosci. Off. J. Soc. Neurosci.* **2011**, *31*, 11795–11807. [CrossRef]
6. Dell'Orco, J.M.; Wasserman, A.H.; Chopra, R.; Ingram, M.A.; Hu, Y.S.; Singh, V.; Wulff, H.; Opal, P.; Orr, H.T.; Shakkottai, V.G. Neuronal Atrophy Early in Degenerative Ataxia Is a Compensatory Mechanism to Regulate Membrane Excitability. *J. Neurosci. Off. J. Soc. Neurosci.* **2015**, *35*, 11292–11307. [CrossRef]
7. Stoyas, C.A.; Bushart, D.D.; Switonski, P.M.; Ward, J.M.; Alaghatta, A.; Tang, M.B.; Niu, C.; Wadhwa, M.; Huang, H.; Savchenko, A.; et al. Nicotinamide Pathway-Dependent Sirt1 Activation Restores Calcium Homeostasis to Achieve Neuroprotection in Spinocerebellar Ataxia Type 7. *Neuron* **2020**, *105*, 630–644.e9. [CrossRef]
8. Bushart, D.D.; Chopra, R.; Singh, V.; Murphy, G.G.; Wulff, H.; Shakkottai, V.G. Targeting potassium channels to treat cerebellar ataxia. *Ann. Clin. Transl. Neurol.* **2018**, *5*, 297–314. [CrossRef]
9. Bushart, D.D.; Huang, H.; Man, L.J.; Morrison, L.M.; Shakkottai, V.G. A Chlorzoxazone-Baclofen Combination Improves Cerebellar Impairment in Spinocerebellar Ataxia Type 1. *Mov. Disord. Off. J. Mov. Disord. Soc.* **2021**, *36*, 622–631. [CrossRef]
10. Jayabal, S.; Chang, H.H.; Cullen, K.E.; Watt, A.J. 4-aminopyridine reverses ataxia and cerebellar firing deficiency in a mouse model of spinocerebellar ataxia type 6. *Sci. Rep.* **2016**, *6*, 29489. [CrossRef]
11. Kasumu, A.W.; Hougaard, C.; Rode, F.; Jacobsen, T.A.; Sabatier, J.M.; Eriksen, B.L.; Strobaek, D.; Liang, X.; Egorova, P.; Vorontsova, D.; et al. Selective positive modulator of calcium-activated potassium channels exerts beneficial effects in a mouse model of spinocerebellar ataxia type 2. *Chem. Biol.* **2012**, *19*, 1340–1353. [CrossRef]
12. Cook, A.A.; Jayabal, S.; Sheng, J.; Fields, E.; Leung, T.C.S.; Quilez, S.; McNicholas, E.; Lau, L.; Huang, S.; Watt, A.J. Activation of TrkB-Akt signaling rescues deficits in a mouse model of SCA6. *Sci. Adv.* **2022**, *8*, eabh3260. [CrossRef]
13. Cook, A.A.; Fields, E.; Watt, A.J. Losing the Beat: Contribution of Purkinje Cell Firing Dysfunction to Disease, and Its Reversal. *Neuroscience* **2021**, *462*, 247–261. [CrossRef]
14. Raman, I.M.; Bean, B.P. Ionic currents underlying spontaneous action potentials in isolated cerebellar Purkinje neurons. *J. Neurosci. Off. J. Soc. Neurosci.* **1999**, *19*, 1663–1674. [CrossRef]
15. Bushart, D.D.; Shakkottai, V.G. Ion channel dysfunction in cerebellar ataxia. *Neurosci. Lett.* **2019**, *688*, 41–48. [CrossRef]
16. Kalume, F.; Yu, F.H.; Westenbroek, R.E.; Scheuer, T.; Catterall, W.A. Reduced sodium current in Purkinje neurons from Nav1.1 mutant mice: Implications for ataxia in severe myoclonic epilepsy in infancy. *J. Neurosci. Off. J. Soc. Neurosci.* **2007**, *27*, 11065–11074. [CrossRef]
17. Khaliq, Z.M.; Gouwens, N.W.; Raman, I.M. The contribution of resurgent sodium current to high-frequency firing in Purkinje neurons: An experimental and modeling study. *J. Neurosci. Off. J. Soc. Neurosci.* **2003**, *23*, 4899–4912. [CrossRef]
18. Veeramah, K.R.; O'Brien, J.E.; Meisler, M.H.; Cheng, X.; Dib-Hajj, S.D.; Waxman, S.G.; Talwar, D.; Girirajan, S.; Eichler, E.E.; Restifo, L.L.; et al. De novo pathogenic SCN8A mutation identified by whole-genome sequencing of a family quartet affected by infantile epileptic encephalopathy and SUDEP. *Am. J. Hum. Genet.* **2012**, *90*, 502–510. [CrossRef]
19. Kohrman, D.C.; Harris, J.B.; Meisler, M.H. Mutation detection in the med and medJ alleles of the sodium channel Scn8a. Unusual splicing due to a minor class AT-AC intron. *J. Biol. Chem.* **1996**, *271*, 17576–17581. [CrossRef]
20. Raman, I.M.; Sprunger, L.K.; Meisler, M.H.; Bean, B.P. Altered subthreshold sodium currents and disrupted firing patterns in Purkinje neurons of Scn8a mutant mice. *Neuron* **1997**, *19*, 881–891. [CrossRef]
21. Dravet, C.; Bureau, M.; Oguni, H.; Fukuyama, Y.; Cokar, O. Severe myoclonic epilepsy in infancy: Dravet syndrome. *Adv. Neurol.* **2005**, *95*, 71–102. [PubMed]
22. Selmer, K.K.; Eriksson, A.S.; Brandal, K.; Egeland, T.; Tallaksen, C.; Undlien, D.E. Parental SCN1A mutation mosaicism in familial Dravet syndrome. *Clin. Genet.* **2009**, *76*, 398–403. [CrossRef] [PubMed]
23. Weuring, W.J.; Singh, S.; Volkers, L.; Rook, M.B.; van 't Slot, R.H.; Bosma, M.; Inserra, M.; Vetter, I.; Verhoeven-Duif, N.M.; Braun, K.P.J.; et al. NaV1.1 and NaV1.6 selective compounds reduce the behavior phenotype and epileptiform activity in a novel zebrafish model for Dravet Syndrome. *PLoS ONE* **2020**, *15*, e0219106. [CrossRef] [PubMed]

24. Brusse, E.; de Koning, I.; Maat-Kievit, A.; Oostra, B.A.; Heutink, P.; van Swieten, J.C. Spinocerebellar ataxia associated with a mutation in the fibroblast growth factor 14 gene (SCA27): A new phenotype. *Mov. Disord. Off. J. Mov. Disord. Soc.* **2006**, *21*, 396–401. [CrossRef] [PubMed]
25. Wang, Q.; Bardgett, M.E.; Wong, M.; Wozniak, D.F.; Lou, J.; McNeil, B.D.; Chen, C.; Nardi, A.; Reid, D.C.; Yamada, K.; et al. Ataxia and paroxysmal dyskinesia in mice lacking axonally transported FGF14. *Neuron* **2002**, *35*, 25–38. [CrossRef]
26. Shakkottai, V.G.; Xiao, M.; Xu, L.; Wong, M.; Nerbonne, J.M.; Ornitz, D.M.; Yamada, K.A. FGF14 regulates the intrinsic excitability of cerebellar Purkinje neurons. *Neurobiol. Dis.* **2009**, *33*, 81–88. [CrossRef]
27. Bosch, M.K.; Carrasquillo, Y.; Ransdell, J.L.; Kanakamedala, A.; Ornitz, D.M.; Nerbonne, J.M. Intracellular FGF14 (iFGF14) Is Required for Spontaneous and Evoked Firing in Cerebellar Purkinje Neurons and for Motor Coordination and Balance. *J. Neurosci. Off. J. Soc. Neurosci.* **2015**, *35*, 6752–6769. [CrossRef]
28. Laezza, F.; Gerber, B.R.; Lou, J.Y.; Kozel, M.A.; Hartman, H.; Craig, A.M.; Ornitz, D.M.; Nerbonne, J.M. The FGF14(F145S) mutation disrupts the interaction of FGF14 with voltage-gated Na+ channels and impairs neuronal excitability. *J. Neurosci. Off. J. Soc. Neurosci.* **2007**, *27*, 12033–12044. [CrossRef]
29. Pellerin, D.; Danzi, M.C.; Wilke, C.; Renaud, M.; Fazal, S.; Dicaire, M.J.; Scriba, C.K.; Ashton, C.; Yanick, C.; Beijer, D.; et al. Deep Intronic FGF14 GAA Repeat Expansion in Late-Onset Cerebellar Ataxia. *N. Engl. J. Med.* **2023**, *388*, 128–141. [CrossRef]
30. Catarino, C.B.; Liu, J.Y.; Liagkouras, I.; Gibbons, V.S.; Labrum, R.W.; Ellis, R.; Woodward, C.; Davis, M.B.; Smith, S.J.; Cross, J.H.; et al. Dravet syndrome as epileptic encephalopathy: Evidence from long-term course and neuropathology. *Brain* **2011**, *134*, 2982–3010. [CrossRef]
31. Ohba, C.; Kato, M.; Takahashi, S.; Lerman-Sagie, T.; Lev, D.; Terashima, H.; Kubota, M.; Kawawaki, H.; Matsufuji, M.; Kojima, Y.; et al. Early onset epileptic encephalopathy caused by de novo SCN8A mutations. *Epilepsia* **2014**, *55*, 994–1000. [CrossRef]
32. Masnada, S.; Hedrich, U.B.S.; Gardella, E.; Schubert, J.; Kaiwar, C.; Klee, E.W.; Lanpher, B.C.; Gavrilova, R.H.; Synofzik, M.; Bast, T.; et al. Clinical spectrum and genotype-phenotype associations of KCNA2-related encephalopathies. *Brain* **2017**, *140*, 2337–2354. [CrossRef]
33. Figueroa, K.P.; Minassian, N.A.; Stevanin, G.; Waters, M.; Garibyan, V.; Forlani, S.; Strzelczyk, A.; Burk, K.; Brice, A.; Durr, A.; et al. KCNC3: Phenotype, mutations, channel biophysics-a study of 260 familial ataxia patients. *Hum. Mutat.* **2010**, *31*, 191–196. [CrossRef]
34. Lee, Y.C.; Durr, A.; Majczenko, K.; Huang, Y.H.; Liu, Y.C.; Lien, C.C.; Tsai, P.C.; Ichikawa, Y.; Goto, J.; Monin, M.L.; et al. Mutations in KCND3 cause spinocerebellar ataxia type 22. *Ann. Neurol.* **2012**, *72*, 859–869. [CrossRef]
35. Duarri, A.; Jezierska, J.; Fokkens, M.; Meijer, M.; Schelhaas, H.J.; den Dunnen, W.F.; van Dijk, F.; Verschuuren-Bemelmans, C.; Hageman, G.; van de Vlies, P.; et al. Mutations in potassium channel kcnd3 cause spinocerebellar ataxia type 19. *Ann. Neurol.* **2012**, *72*, 870–880. [CrossRef]
36. Liang, L.; Li, X.; Moutton, S.; Schrier Vergano, S.A.; Cogne, B.; de Saint-Martin, A.; Hurst, A.C.E.; Hu, Y.; Bodamer, O.; Thevenon, J.; et al. De novo loss-of-function KCNMA1 variants are associated with a new multiple malformation syndrome and a broad spectrum of developmental and neurological phenotypes. *Hum. Mol. Genet.* **2019**, *28*, 2937–2951. [CrossRef]
37. Liang, L.; Liu, H.; Bartholdi, D.; van Haeringen, A.; Fernandez-Jaen, A.; Peeters, E.E.A.; Xiong, H.; Bai, X.; Xu, C.; Ke, T.; et al. Identification and functional analysis of two new de novo KCNMA1 variants associated with Liang-Wang syndrome. *Acta Physiol.* **2022**, *235*, e13800. [CrossRef]
38. Nigri, A.; Sarro, L.; Mongelli, A.; Pinardi, C.; Porcu, L.; Castaldo, A.; Ferraro, S.; Grisoli, M.; Bruzzone, M.G.; Gellera, C.; et al. Progression of Cerebellar Atrophy in Spinocerebellar Ataxia Type 2 Gene Carriers: A Longitudinal MRI Study in Preclinical and Early Disease Stages. *Front. Neurol.* **2020**, *11*, 616419. [CrossRef]
39. Mochel, F.; Rastetter, A.; Ceulemans, B.; Platzer, K.; Yang, S.; Shinde, D.N.; Helbig, K.L.; Lopergolo, D.; Mari, F.; Renieri, A.; et al. Variants in the SK2 channel gene (KCNN2) lead to dominant neurodevelopmental movement disorders. *Brain* **2020**, *143*, 3564–3573. [CrossRef]
40. Yuan, X.; Zheng, Y.; Gao, F.; Sun, W.; Wang, Z.; Zhao, G. Case Report: A Novel CACNA1A Mutation Caused Flunarizine-Responsive Type 2 Episodic Ataxia and Hemiplegic Migraine with Abnormal MRI of Cerebral White Matter. *Front. Neurol.* **2022**, *13*, 899813. [CrossRef]
41. Li, X.; Zhou, C.; Cui, L.; Zhu, L.; Du, H.; Liu, J.; Wang, C.; Fang, S. A case of a novel CACNA1G mutation from a Chinese family with SCA42: A case report and literature review. *Medicine* **2018**, *97*, e12148. [CrossRef] [PubMed]
42. Son, H.; Yoon, J.G.; Kim, M.J.; Moon, J.; Kim, H.J. First Cases of Spinocerebellar Ataxia 42 in Two Korean Families. *J. Mov. Disord.* **2023**, *16*, 110–113. [CrossRef] [PubMed]
43. Fogel, B.L.; Hanson, S.M.; Becker, E.B. Do mutations in the murine ataxia gene TRPC3 cause cerebellar ataxia in humans? *Mov. Disord. Off. J. Mov. Disord. Soc.* **2015**, *30*, 284–286. [CrossRef] [PubMed]
44. Marelli, C.; van de Leemput, J.; Johnson, J.O.; Tison, F.; Thauvin-Robinet, C.; Picard, F.; Tranchant, C.; Hernandez, D.G.; Huttin, B.; Boulliat, J.; et al. SCA15 due to large ITPR1 deletions in a cohort of 333 white families with dominant ataxia. *Arch. Neurol.* **2011**, *68*, 637–643. [CrossRef]
45. Sasaki, M.; Ohba, C.; Iai, M.; Hirabayashi, S.; Osaka, H.; Hiraide, T.; Saitsu, H.; Matsumoto, N. Sporadic infantile-onset spinocerebellar ataxia caused by missense mutations of the inositol 1,4,5-triphosphate receptor type 1 gene. *J. Neurol.* **2015**, *262*, 1278–1284. [CrossRef]

46. Vicario, M.; Zanni, G.; Vallese, F.; Santorelli, F.; Grinzato, A.; Cieri, D.; Berto, P.; Frizzarin, M.; Lopreiato, R.; Zonta, F.; et al. A V1143F mutation in the neuronal-enriched isoform 2 of the PMCA pump is linked with ataxia. *Neurobiol. Dis.* **2018**, *115*, 157–166. [CrossRef]
47. Feyma, T.; Ramsey, K.; Group, C.R.R.; Huentelman, M.J.; Craig, D.W.; Padilla-Lopez, S.; Narayanan, V.; Kruer, M.C. Dystonia in ATP2B3-associated X-linked spinocerebellar ataxia. *Mov. Disord. Off. J. Mov. Disord. Soc.* **2016**, *31*, 1752–1753. [CrossRef]
48. Chung, Y.H.; Shin, C.M.; Kim, M.J.; Cha, C.I. Immunohistochemical study on the distribution of six members of the Kv1 channel subunits in the rat basal ganglia. *Brain Res.* **2000**, *875*, 164–170. [CrossRef]
49. Bushart, D.D.; Zalon, A.J.; Zhang, H.; Morrison, L.M.; Guan, Y.; Paulson, H.L.; Shakkottai, V.G.; McLoughlin, H.S. Antisense Oligonucleotide Therapy Targeted Against ATXN3 Improves Potassium Channel-Mediated Purkinje Neuron Dysfunction in Spinocerebellar Ataxia Type 3. *Cerebellum* **2021**, *20*, 41–53. [CrossRef]
50. Graves, T.D.; Cha, Y.H.; Hahn, A.F.; Barohn, R.; Salajegheh, M.K.; Griggs, R.C.; Bundy, B.N.; Jen, J.C.; Baloh, R.W.; Hanna, M.G.; et al. Episodic ataxia type 1: Clinical characterization, quality of life and genotype-phenotype correlation. *Brain* **2014**, *137*, 1009–1018. [CrossRef]
51. D'Adamo, M.C.; Hasan, S.; Guglielmi, L.; Servettini, I.; Cenciarini, M.; Catacuzzeno, L.; Franciolini, F. New insights into the pathogenesis and therapeutics of episodic ataxia type 1. *Front. Cell. Neurosci.* **2015**, *9*, 317. [CrossRef]
52. Miceli, F.; Guerrini, R.; Nappi, M.; Soldovieri, M.V.; Cellini, E.; Gurnett, C.A.; Parmeggiani, L.; Mei, D.; Taglialatela, M. Distinct epilepsy phenotypes and response to drugs in KCNA1 gain- and loss-of function variants. *Epilepsia* **2022**, *63*, e7–e14. [CrossRef]
53. Muller, P.; Takacs, D.S.; Hedrich, U.B.S.; Coorg, R.; Masters, L.; Glinton, K.E.; Dai, H.; Cokley, J.A.; Riviello, J.J.; Lerche, H.; et al. KCNA1 gain-of-function epileptic encephalopathy treated with 4-aminopyridine. *Ann. Clin. Transl. Neurol.* **2023**, *10*, 656–663. [CrossRef]
54. Syrbe, S.; Hedrich, U.B.S.; Riesch, E.; Djemie, T.; Muller, S.; Moller, R.S.; Maher, B.; Hernandez-Hernandez, L.; Synofzik, M.; Caglayan, H.S.; et al. De novo loss- or gain-of-function mutations in KCNA2 cause epileptic encephalopathy. *Nat. Genet.* **2015**, *47*, 393–399. [CrossRef]
55. Hedrich, U.B.S.; Lauxmann, S.; Wolff, M.; Synofzik, M.; Bast, T.; Binelli, A.; Serratosa, J.M.; Martinez-Ulloa, P.; Allen, N.M.; King, M.D.; et al. 4-Aminopyridine is a promising treatment option for patients with gain-of-function KCNA2-encephalopathy. *Sci. Transl. Med.* **2021**, *13*, eaaz4957. [CrossRef]
56. Xie, G.; Harrison, J.; Clapcote, S.J.; Huang, Y.; Zhang, J.Y.; Wang, L.Y.; Roder, J.C. A new Kv1.2 channelopathy underlying cerebellar ataxia. *J. Biol. Chem.* **2010**, *285*, 32160–32173. [CrossRef]
57. Salpietro, V.; Galassi Deforie, V.; Efthymiou, S.; O'Connor, E.; Marce-Grau, A.; Maroofian, R.; Striano, P.; Zara, F.; Morrow, M.M.; Group, S.S.; et al. De novo KCNA6 variants with attenuated K(V) 1.6 channel deactivation in patients with epilepsy. *Epilepsia* **2023**, *64*, 443–455. [CrossRef]
58. Waters, M.F.; Minassian, N.A.; Stevanin, G.; Figueroa, K.P.; Bannister, J.P.; Nolte, D.; Mock, A.F.; Evidente, V.G.; Fee, D.B.; Muller, U.; et al. Mutations in voltage-gated potassium channel KCNC3 cause degenerative and developmental central nervous system phenotypes. *Nat. Genet.* **2006**, *38*, 447–451. [CrossRef]
59. Zagha, E.; Manita, S.; Ross, W.N.; Rudy, B. Dendritic Kv3.3 potassium channels in cerebellar purkinje cells regulate generation and spatial dynamics of dendritic Ca^{2+} spikes. *J. Neurophysiol.* **2010**, *103*, 3516–3525. [CrossRef]
60. Hurlock, E.C.; McMahon, A.; Joho, R.H. Purkinje-cell-restricted restoration of Kv3.3 function restores complex spikes and rescues motor coordination in Kcnc3 mutants. *J. Neurosci. Off. J. Soc. Neurosci.* **2008**, *28*, 4640–4648. [CrossRef]
61. Hurlock, E.C.; Bose, M.; Pierce, G.; Joho, R.H. Rescue of motor coordination by Purkinje cell-targeted restoration of Kv3.3 channels in Kcnc3-null mice requires Kcnc1. *J. Neurosci. Off. J. Soc. Neurosci.* **2009**, *29*, 15735–15744. [CrossRef]
62. Joho, R.H.; Street, C.; Matsushita, S.; Knopfel, T. Behavioral motor dysfunction in Kv3-type potassium channel-deficient mice. *Genes Brain Behav.* **2006**, *5*, 472–482. [CrossRef] [PubMed]
63. Zhang, Y.; Quraishi, I.H.; McClure, H.; Williams, L.A.; Cheng, Y.; Kale, S.; Dempsey, G.T.; Agrawal, S.; Gerber, D.J.; McManus, O.B.; et al. Suppression of Kv3.3 channels by antisense oligonucleotides reverses biochemical effects and motor impairment in spinocerebellar ataxia type 13 mice. *FASEB J.* **2021**, *35*, e22053. [CrossRef]
64. Zhang, Y.; Kaczmarek, L.K. Kv3.3 potassium channels and spinocerebellar ataxia. *J. Physiol.* **2016**, *594*, 4677–4684. [CrossRef] [PubMed]
65. Chopra, R.; Bushart, D.D.; Cooper, J.P.; Yellajoshyula, D.; Morrison, L.M.; Huang, H.; Handler, H.P.; Man, L.J.; Dansithong, W.; Scoles, D.R.; et al. Altered Capicua expression drives regional Purkinje neuron vulnerability through ion channel gene dysregulation in spinocerebellar ataxia type 1. *Hum. Mol. Genet.* **2020**, *29*, 3249–3265. [CrossRef] [PubMed]
66. Shakkottai, V.G.; do Carmo Costa, M.; Dell'Orco, J.M.; Sankaranarayanan, A.; Wulff, H.; Paulson, H.L. Early changes in cerebellar physiology accompany motor dysfunction in the polyglutamine disease spinocerebellar ataxia type 3. *J. Neurosci. Off. J. Soc. Neurosci.* **2011**, *31*, 13002–13014. [CrossRef]
67. Dell'Orco, J.M.; Pulst, S.M.; Shakkottai, V.G. Potassium channel dysfunction underlies Purkinje neuron spiking abnormalities in spinocerebellar ataxia type 2. *Hum. Mol. Genet.* **2017**, *26*, 3935–3945. [CrossRef]
68. Alfaro-Ruiz, R.; Aguado, C.; Martin-Belmonte, A.; Moreno-Martinez, A.E.; Lujan, R. Cellular and Subcellular Localisation of Kv4-Associated KChIP Proteins in the Rat Cerebellum. *Int. J. Mol. Sci.* **2020**, *21*, 6403. [CrossRef]

69. Hsiao, C.T.; Tropea, T.F.; Fu, S.J.; Bardakjian, T.M.; Gonzalez-Alegre, P.; Soong, B.W.; Tang, C.Y.; Jeng, C.J. Rare Gain-of-Function KCND3 Variant Associated with Cerebellar Ataxia, Parkinsonism, Cognitive Dysfunction, and Brain Iron Accumulation. *Int. J. Mol. Sci.* **2021**, *22*, 8247. [CrossRef]
70. Du, X.; Carvalho-de-Souza, J.L.; Wei, C.; Carrasquel-Ursulaez, W.; Lorenzo, Y.; Gonzalez, N.; Kubota, T.; Staisch, J.; Hain, T.; Petrossian, N.; et al. Loss-of-function BK channel mutation causes impaired mitochondria and progressive cerebellar ataxia. *Proc. Natl. Acad. Sci. USA* **2020**, *117*, 6023–6034. [CrossRef]
71. Srinivasan, S.R.; Huang, H.; Chang, W.C.; Nasburg, J.A.; Nguyen, H.M.; Strassmaier, T.; Wulff, H.; Shakkottai, V.G. Discovery of Novel Activators of Large-Conductance Calcium-Activated Potassium Channels for the Treatment of Cerebellar Ataxia. *Mol. Pharm.* **2022**, *102*, 438–449. [CrossRef]
72. Chen, X.; Kovalchuk, Y.; Adelsberger, H.; Henning, H.A.; Sausbier, M.; Wietzorrek, G.; Ruth, P.; Yarom, Y.; Konnerth, A. Disruption of the olivo-cerebellar circuit by Purkinje neuron-specific ablation of BK channels. *Proc. Natl. Acad. Sci. USA* **2010**, *107*, 12323–12328. [CrossRef]
73. Cheron, G.; Marquez-Ruiz, J.; Cheron, J.; Prigogine, C.; Ammann, C.; Lukowski, R.; Ruth, P.; Dan, B. Purkinje cell BKchannel ablation induces abnormal rhythm in deep cerebellar nuclei and prevents LTD. *Sci. Rep.* **2018**, *8*, 4220. [CrossRef]
74. Du, W.; Bautista, J.F.; Yang, H.; Diez-Sampedro, A.; You, S.A.; Wang, L.; Kotagal, P.; Luders, H.O.; Shi, J.; Cui, J.; et al. Calcium-sensitive potassium channelopathy in human epilepsy and paroxysmal movement disorder. *Nat. Genet.* **2005**, *37*, 733–738. [CrossRef]
75. Zhang, G.; Gibson, R.A.; McDonald, M.; Liang, P.; Kang, P.W.; Shi, J.; Yang, H.; Cui, J.; Mikati, M.A. A Gain-of-Function Mutation in KCNMA1 Causes Dystonia Spells Controlled with Stimulant Therapy. *Mov. Disord. Off. J. Mov. Disord. Soc.* **2020**, *35*, 1868–1873. [CrossRef]
76. Li, X.; Poschmann, S.; Chen, Q.; Fazeli, W.; Oundjian, N.J.; Snoeijen-Schouwenaars, F.M.; Fricke, O.; Kamsteeg, E.J.; Willemsen, M.; Wang, Q.K. De novo BK channel variant causes epilepsy by affecting voltage gating but not Ca(2+) sensitivity. *Eur. J. Hum. Genet.* **2018**, *26*, 220–229. [CrossRef]
77. Moldenhauer, H.J.; Matychak, K.K.; Meredith, A.L. Comparative gain-of-function effects of the KCNMA1-N999S mutation on human BK channel properties. *J. Neurophysiol.* **2020**, *123*, 560–570. [CrossRef]
78. Brenner, R.; Chen, Q.H.; Vilaythong, A.; Toney, G.M.; Noebels, J.L.; Aldrich, R.W. BK channel beta4 subunit reduces dentate gyrus excitability and protects against temporal lobe seizures. *Nat. Neurosci.* **2005**, *8*, 1752–1759. [CrossRef]
79. Balint, B.; Guerreiro, R.; Carmona, S.; Dehghani, N.; Latorre, A.; Cordivari, C.; Bhatia, K.P.; Bras, J. KCNN2 mutation in autosomal-dominant tremulous myoclonus-dystonia. *Eur. J. Neurol.* **2020**, *27*, 1471–1477. [CrossRef]
80. Szatanik, M.; Vibert, N.; Vassias, I.; Guenet, J.L.; Eugene, D.; de Waele, C.; Jaubert, J. Behavioral effects of a deletion in Kcnn2, the gene encoding the SK2 subunit of small-conductance Ca^{2+}-activated K^+ channels. *Neurogenetics* **2008**, *9*, 237–248. [CrossRef]
81. Shakkottai, V.G.; Chou, C.H.; Oddo, S.; Sailer, C.A.; Knaus, H.G.; Gutman, G.A.; Barish, M.E.; LaFerla, F.M.; Chandy, K.G. Enhanced neuronal excitability in the absence of neurodegeneration induces cerebellar ataxia. *J. Clin. Investig.* **2004**, *113*, 582–590. [CrossRef] [PubMed]
82. Liu, J.; Tang, T.S.; Tu, H.; Nelson, O.; Herndon, E.; Huynh, D.P.; Pulst, S.M.; Bezprozvanny, I. Deranged calcium signaling and neurodegeneration in spinocerebellar ataxia type 2. *J. Neurosci. Off. J. Soc. Neurosci.* **2009**, *29*, 9148–9162. [CrossRef] [PubMed]
83. Kuramoto, T.; Yokoe, M.; Kunisawa, N.; Ohashi, K.; Miyake, T.; Higuchi, Y.; Yoshimi, K.; Mashimo, T.; Tanaka, M.; Kuwamura, M.; et al. Tremor dominant Kyoto (Trdk) rats carry a missense mutation in the gene encoding the SK2 subunit of small-conductance Ca(2+)-activated K(+) channel. *Brain Res.* **2017**, *1676*, 38–45. [CrossRef] [PubMed]
84. Jun, K.; Piedras-Renteria, E.S.; Smith, S.M.; Wheeler, D.B.; Lee, S.B.; Lee, T.G.; Chin, H.; Adams, M.E.; Scheller, R.H.; Tsien, R.W.; et al. Ablation of P/Q-type Ca(2+) channel currents, altered synaptic transmission, and progressive ataxia in mice lacking the alpha(1A)-subunit. *Proc. Natl. Acad. Sci. USA* **1999**, *96*, 15245–15250. [CrossRef]
85. Zhuchenko, O.; Bailey, J.; Bonnen, P.; Ashizawa, T.; Stockton, D.W.; Amos, C.; Dobyns, W.B.; Subramony, S.H.; Zoghbi, H.Y.; Lee, C.C. Autosomal dominant cerebellar ataxia (SCA6) associated with small polyglutamine expansions in the alpha 1A-voltage-dependent calcium channel. *Nat. Genet.* **1997**, *15*, 62–69. [CrossRef]
86. Jen, J.C.; Baloh, R.W. Genetics of episodic ataxia. *Adv. Neurol.* **2002**, *89*, 459–461.
87. Ashizawa, T.; Figueroa, K.P.; Perlman, S.L.; Gomez, C.M.; Wilmot, G.R.; Schmahmann, J.D.; Ying, S.H.; Zesiewicz, T.A.; Paulson, H.L.; Shakkottai, V.G.; et al. Clinical characteristics of patients with spinocerebellar ataxias 1, 2, 3 and 6 in the US; a prospective observational study. *Orphanet J. Rare Dis.* **2013**, *8*, 177. [CrossRef]
88. Stevanin, G.; Durr, A.; David, G.; Didierjean, O.; Cancel, G.; Rivaud, S.; Tourbah, A.; Warter, J.M.; Agid, Y.; Brice, A. Clinical and molecular features of spinocerebellar ataxia type 6. *Neurology* **1997**, *49*, 1243–1246. [CrossRef]
89. Watase, K.; Barrett, C.F.; Miyazaki, T.; Ishiguro, T.; Ishikawa, K.; Hu, Y.; Unno, T.; Sun, Y.; Kasai, S.; Watanabe, M.; et al. Spinocerebellar ataxia type 6 knockin mice develop a progressive neuronal dysfunction with age-dependent accumulation of mutant CaV2.1 channels. *Proc. Natl. Acad. Sci. USA* **2008**, *105*, 11987–11992. [CrossRef]
90. Saegusa, H.; Wakamori, M.; Matsuda, Y.; Wang, J.; Mori, Y.; Zong, S.; Tanabe, T. Properties of human Cav2.1 channel with a spinocerebellar ataxia type 6 mutation expressed in Purkinje cells. *Mol. Cell. Neurosci.* **2007**, *34*, 261–270. [CrossRef]
91. Jen, J.; Kim, G.W.; Baloh, R.W. Clinical spectrum of episodic ataxia type 2. *Neurology* **2004**, *62*, 17–22. [CrossRef]
92. Jen, J.C.; Wan, J. Episodic ataxias. *Handb. Clin. Neurol.* **2018**, *155*, 205–215. [CrossRef]

93. Fletcher, C.F.; Tottene, A.; Lennon, V.A.; Wilson, S.M.; Dubel, S.J.; Paylor, R.; Hosford, D.A.; Tessarollo, L.; McEnery, M.W.; Pietrobon, D.; et al. Dystonia and cerebellar atrophy in Cacna1a null mice lacking P/Q calcium channel activity. *FASEB J.* **2001**, *15*, 1288–1290. [CrossRef]
94. Hoebeek, F.E.; Stahl, J.S.; van Alphen, A.M.; Schonewille, M.; Luo, C.; Rutteman, M.; van den Maagdenberg, A.M.; Molenaar, P.C.; Goossens, H.H.; Frens, M.A.; et al. Increased noise level of purkinje cell activities minimizes impact of their modulation during sensorimotor control. *Neuron* **2005**, *45*, 953–965. [CrossRef]
95. Stahl, J.S.; James, R.A.; Oommen, B.S.; Hoebeek, F.E.; De Zeeuw, C.I. Eye movements of the murine P/Q calcium channel mutant tottering, and the impact of aging. *J. Neurophysiol.* **2006**, *95*, 1588–1607. [CrossRef]
96. Walter, J.T.; Alvina, K.; Womack, M.D.; Chevez, C.; Khodakhah, K. Decreases in the precision of Purkinje cell pacemaking cause cerebellar dysfunction and ataxia. *Nat. Neurosci.* **2006**, *9*, 389–397. [CrossRef]
97. Spacey, S.D.; Hildebrand, M.E.; Materek, L.A.; Bird, T.D.; Snutch, T.P. Functional implications of a novel EA2 mutation in the P/Q-type calcium channel. *Ann. Neurol.* **2004**, *56*, 213–220. [CrossRef]
98. Gao, Z.; Todorov, B.; Barrett, C.F.; van Dorp, S.; Ferrari, M.D.; van den Maagdenberg, A.M.; De Zeeuw, C.I.; Hoebeek, F.E. Cerebellar ataxia by enhanced Ca(V)2.1 currents is alleviated by Ca^{2+}-dependent K^+-channel activators in Cacna1a(S218L) mutant mice. *J. Neurosci. Off. J. Soc. Neurosci.* **2012**, *32*, 15533–15546. [CrossRef]
99. van den Maagdenberg, A.M.; Pizzorusso, T.; Kaja, S.; Terpolilli, N.; Shapovalova, M.; Hoebeek, F.E.; Barrett, C.F.; Gherardini, L.; van de Ven, R.C.; Todorov, B.; et al. High cortical spreading depression susceptibility and migraine-associated symptoms in Ca(v)2.1 S218L mice. *Ann. Neurol.* **2010**, *67*, 85–98. [CrossRef]
100. Coutelier, M.; Blesneac, I.; Monteil, A.; Monin, M.L.; Ando, K.; Mundwiller, E.; Brusco, A.; Le Ber, I.; Anheim, M.; Castrioto, A.; et al. A Recurrent Mutation in CACNA1G Alters Cav3.1 T-Type Calcium-Channel Conduction and Causes Autosomal-Dominant Cerebellar Ataxia. *Am. J. Hum. Genet.* **2015**, *97*, 726–737. [CrossRef]
101. Morino, H.; Matsuda, Y.; Muguruma, K.; Miyamoto, R.; Ohsawa, R.; Ohtake, T.; Otobe, R.; Watanabe, M.; Maruyama, H.; Hashimoto, K.; et al. A mutation in the low voltage-gated calcium channel CACNA1G alters the physiological properties of the channel, causing spinocerebellar ataxia. *Mol. Brain* **2015**, *8*, 89. [CrossRef] [PubMed]
102. Hara, N.; Morino, H.; Matsuda, Y.; Satoh, K.; Hashimoto, K.; Maruyama, H.; Kawakami, H. Zonisamide can ameliorate the voltage-dependence alteration of the T-type calcium channel Ca(V)3.1 caused by a mutation responsible for spinocerebellar ataxia. *Mol. Brain* **2020**, *13*, 163. [CrossRef] [PubMed]
103. Dansithong, W.; Paul, S.; Figueroa, K.P.; Rinehart, M.D.; Wiest, S.; Pflieger, L.T.; Scoles, D.R.; Pulst, S.M. Ataxin-2 regulates RGS8 translation in a new BAC-SCA2 transgenic mouse model. *PLoS Genet.* **2015**, *11*, e1005182. [CrossRef] [PubMed]
104. Chang, K.Y.; Park, Y.G.; Park, H.Y.; Homanics, G.E.; Kim, J.; Kim, D. Lack of CaV3.1 channels causes severe motor coordination defects and an age-dependent cerebellar atrophy in a genetic model of essential tremor. *Biochem. Biophys. Res. Commun.* **2011**, *410*, 19–23. [CrossRef] [PubMed]
105. Chemin, J.; Siquier-Pernet, K.; Nicouleau, M.; Barcia, G.; Ahmad, A.; Medina-Cano, D.; Hanein, S.; Altin, N.; Hubert, L.; Bole-Feysot, C.; et al. De novo mutation screening in childhood-onset cerebellar atrophy identifies gain-of-function mutations in the CACNA1G calcium channel gene. *Brain* **2018**, *141*, 1998–2013. [CrossRef]
106. Huang, W.C.; Young, J.S.; Glitsch, M.D. Changes in TRPC channel expression during postnatal development of cerebellar neurons. *Cell Calcium* **2007**, *42*, 1–10. [CrossRef]
107. Becker, E.B.; Oliver, P.L.; Glitsch, M.D.; Banks, G.T.; Achilli, F.; Hardy, A.; Nolan, P.M.; Fisher, E.M.; Davies, K.E. A point mutation in TRPC3 causes abnormal Purkinje cell development and cerebellar ataxia in moonwalker mice. *Proc. Natl. Acad. Sci. USA* **2009**, *106*, 6706–6711. [CrossRef]
108. Sekerkova, G.; Kim, J.A.; Nigro, M.J.; Becker, E.B.; Hartmann, J.; Birnbaumer, L.; Mugnaini, E.; Martina, M. Early onset of ataxia in moonwalker mice is accompanied by complete ablation of type II unipolar brush cells and Purkinje cell dysfunction. *J. Neurosci. Off. J. Soc. Neurosci.* **2013**, *33*, 19689–19694. [CrossRef]
109. Hara, K.; Shiga, A.; Nozaki, H.; Mitsui, J.; Takahashi, Y.; Ishiguro, H.; Yomono, H.; Kurisaki, H.; Goto, J.; Ikeuchi, T.; et al. Total deletion and a missense mutation of ITPR1 in Japanese SCA15 families. *Neurology* **2008**, *71*, 547–551. [CrossRef]
110. Huang, L.; Chardon, J.W.; Carter, M.T.; Friend, K.L.; Dudding, T.E.; Schwartzentruber, J.; Zou, R.; Schofield, P.W.; Douglas, S.; Bulman, D.E.; et al. Missense mutations in ITPR1 cause autosomal dominant congenital nonprogressive spinocerebellar ataxia. *Orphanet J. Rare Dis.* **2012**, *7*, 67. [CrossRef]
111. Jarius, S.; Scharf, M.; Begemann, N.; Stocker, W.; Probst, C.; Serysheva, I.I.; Nagel, S.; Graus, F.; Psimaras, D.; Wildemann, B.; et al. Antibodies to the inositol 1,4,5-trisphosphate receptor type 1 (ITPR1) in cerebellar ataxia. *J. Neuroinflamm.* **2014**, *11*, 206. [CrossRef]
112. Chen, X.; Tang, T.S.; Tu, H.; Nelson, O.; Pook, M.; Hammer, R.; Nukina, N.; Bezprozvanny, I. Deranged calcium signaling and neurodegeneration in spinocerebellar ataxia type 3. *J. Neurosci. Off. J. Soc. Neurosci.* **2008**, *28*, 12713–12724. [CrossRef]
113. Di Leva, F.; Domi, T.; Fedrizzi, L.; Lim, D.; Carafoli, E. The plasma membrane Ca^{2+} ATPase of animal cells: Structure, function and regulation. *Arch. Biochem. Biophys.* **2008**, *476*, 65–74. [CrossRef]
114. Empson, R.M.; Huang, H.; Nagaraja, R.Y.; Roome, C.J.; Knopfel, T. Enhanced synaptic inhibition in the cerebellar cortex of the ataxic PMCA2(−/−) knockout mouse. *Cerebellum* **2013**, *12*, 667–675. [CrossRef]
115. Empson, R.M.; Akemann, W.; Knopfel, T. The role of the calcium transporter protein plasma membrane calcium ATPase PMCA2 in cerebellar Purkinje neuron function. *Funct. Neurol.* **2010**, *25*, 153–158.

116. Zanni, G.; Cali, T.; Kalscheuer, V.M.; Ottolini, D.; Barresi, S.; Lebrun, N.; Montecchi-Palazzi, L.; Hu, H.; Chelly, J.; Bertini, E.; et al. Mutation of plasma membrane Ca^{2+} ATPase isoform 3 in a family with X-linked congenital cerebellar ataxia impairs Ca^{2+} homeostasis. *Proc. Natl. Acad. Sci. USA* **2012**, *109*, 14514–14519. [CrossRef]
117. Marquer, A.; Barbieri, G.; Perennou, D. The assessment and treatment of postural disorders in cerebellar ataxia: A systematic review. *Ann. Phys. Rehabil. Med.* **2014**, *57*, 67–78. [CrossRef]
118. Ilg, W.; Synofzik, M.; Brotz, D.; Burkard, S.; Giese, M.A.; Schols, L. Intensive coordinative training improves motor performance in degenerative cerebellar disease. *Neurology* **2009**, *73*, 1823–1830. [CrossRef]
119. Ilg, W.; Brotz, D.; Burkard, S.; Giese, M.A.; Schols, L.; Synofzik, M. Long-term effects of coordinative training in degenerative cerebellar disease. *Mov. Disord. Off. J. Mov. Disord. Soc.* **2010**, *25*, 2239–2246. [CrossRef]
120. Miyai, I.; Ito, M.; Hattori, N.; Mihara, M.; Hatakenaka, M.; Yagura, H.; Sobue, G.; Nishizawa, M.; Cerebellar Ataxia Rehabilitation Trialists, C. Cerebellar ataxia rehabilitation trial in degenerative cerebellar diseases. *Neurorehabil. Neural Repair* **2012**, *26*, 515–522. [CrossRef]
121. Sullivan, R.; Yau, W.Y.; O'Connor, E.; Houlden, H. Spinocerebellar ataxia: An update. *J. Neurol.* **2019**, *266*, 533–544. [CrossRef]
122. Evers, M.M.; Toonen, L.J.; van Roon-Mom, W.M. Antisense oligonucleotides in therapy for neurodegenerative disorders. *Adv. Drug Deliv. Rev.* **2015**, *87*, 90–103. [CrossRef] [PubMed]
123. Friedrich, J.; Kordasiewicz, H.B.; O'Callaghan, B.; Handler, H.P.; Wagener, C.; Duvick, L.; Swayze, E.E.; Rainwater, O.; Hofstra, B.; Benneyworth, M.; et al. Antisense oligonucleotide-mediated ataxin-1 reduction prolongs survival in SCA1 mice and reveals disease-associated transcriptome profiles. *JCI Insight* **2018**, *3*, e123193. [CrossRef] [PubMed]
124. Scoles, D.R.; Meera, P.; Schneider, M.D.; Paul, S.; Dansithong, W.; Figueroa, K.P.; Hung, G.; Rigo, F.; Bennett, C.F.; Otis, T.S.; et al. Antisense oligonucleotide therapy for spinocerebellar ataxia type 2. *Nature* **2017**, *544*, 362–366. [CrossRef] [PubMed]
125. McLoughlin, H.S.; Moore, L.R.; Chopra, R.; Komlo, R.; McKenzie, M.; Blumenstein, K.G.; Zhao, H.; Kordasiewicz, H.B.; Shakkottai, V.G.; Paulson, H.L. Oligonucleotide therapy mitigates disease in spinocerebellar ataxia type 3 mice. *Ann. Neurol.* **2018**, *84*, 64–77. [CrossRef]
126. Xia, H.; Mao, Q.; Eliason, S.L.; Harper, S.Q.; Martins, I.H.; Orr, H.T.; Paulson, H.L.; Yang, L.; Kotin, R.M.; Davidson, B.L. RNAi suppresses polyglutamine-induced neurodegeneration in a model of spinocerebellar ataxia. *Nat. Med.* **2004**, *10*, 816–820. [CrossRef]
127. Keiser, M.S.; Boudreau, R.L.; Davidson, B.L. Broad therapeutic benefit after RNAi expression vector delivery to deep cerebellar nuclei: Implications for spinocerebellar ataxia type 1 therapy. *Mol. Ther. J. Am. Soc. Gene Ther.* **2014**, *22*, 588–595. [CrossRef]
128. Johnson, W.G. Friedreich ataxia. *Clin. Neurosci.* **1995**, *3*, 33–38.
129. Campuzano, V.; Montermini, L.; Molto, M.D.; Pianese, L.; Cossee, M.; Cavalcanti, F.; Monros, E.; Rodius, F.; Duclos, F.; Monticelli, A.; et al. Friedreich's ataxia: Autosomal.l recessive disease caused by an intronic GAA triplet repeat expansion. *Science* **1996**, *271*, 1423–1427. [CrossRef]
130. Gonzalez-Cabo, P.; Vazquez-Manrique, R.P.; Garcia-Gimeno, M.A.; Sanz, P.; Palau, F. Frataxin interacts functionally with mitochondrial electron transport chain proteins. *Hum. Mol. Genet.* **2005**, *14*, 2091–2098. [CrossRef]
131. Clark, E.; Johnson, J.; Dong, Y.N.; Mercado-Ayon, E.; Warren, N.; Zhai, M.; McMillan, E.; Salovin, A.; Lin, H.; Lynch, D.R. Role of frataxin protein deficiency and metabolic dysfunction in Friedreich ataxia, an autosomal recessive mitochondrial disease. *Neuronal Signal.* **2018**, *2*, NS20180060. [CrossRef]
132. Bolinches-Amoros, A.; Molla, B.; Pla-Martin, D.; Palau, F.; Gonzalez-Cabo, P. Mitochondrial dysfunction induced by frataxin deficiency is associated with cellular senescence and abnormal calcium metabolism. *Front. Cell. Neurosci.* **2014**, *8*, 124. [CrossRef]
133. Pandolfo, M. Frataxin deficiency and mitochondrial dysfunction. *Mitochondrion* **2002**, *2*, 87–93. [CrossRef]
134. Abeti, R.; Baccaro, A.; Esteras, N.; Giunti, P. Novel Nrf2-Inducer Prevents Mitochondrial Defects and Oxidative Stress in Friedreich's Ataxia Models. *Front. Cell. Neurosci.* **2018**, *12*, 188. [CrossRef]
135. Lynch, D.R.; Farmer, J.; Hauser, L.; Blair, I.A.; Wang, Q.Q.; Mesaros, C.; Snyder, N.; Boesch, S.; Chin, M.; Delatycki, M.B.; et al. Safety, pharmacodynamics, and potential benefit of omaveloxolone in Friedreich ataxia. *Ann. Clin. Transl. Neurol.* **2019**, *6*, 15–26. [CrossRef]
136. Kennedy, K.A.; Sandiford, S.D.; Skerjanc, I.S.; Li, S.S. Reactive oxygen species and the neuronal fate. *Cell. Mol. Life Sci. CMLS* **2012**, *69*, 215–221. [CrossRef]
137. Lynch, D.R.; Chin, M.P.; Delatycki, M.B.; Subramony, S.H.; Corti, M.; Hoyle, J.C.; Boesch, S.; Nachbauer, W.; Mariotti, C.; Mathews, K.D.; et al. Safety and Efficacy of Omaveloxolone in Friedreich Ataxia (MOXIe Study). *Ann. Neurol.* **2021**, *89*, 212–225. [CrossRef]
138. Maltecca, F.; Magnoni, R.; Cerri, F.; Cox, G.A.; Quattrini, A.; Casari, G. Haploinsufficiency of AFG3L2, the gene responsible for spinocerebellar ataxia type 28, causes mitochondria-mediated Purkinje cell dark degeneration. *J. Neurosci. Off. J. Soc. Neurosci.* **2009**, *29*, 9244–9254. [CrossRef]
139. Pierson, T.M.; Adams, D.; Bonn, F.; Martinelli, P.; Cherukuri, P.F.; Teer, J.K.; Hansen, N.F.; Cruz, P.; James C. Mullikin For The Nisc Comparative Sequencing Program; Blakesley, R.W.; et al. Whole-exome sequencing identifies homozygous AFG3L2 mutations in a spastic ataxia-neuropathy syndrome linked to mitochondrial m-AAA proteases. *PLoS Genet.* **2011**, *7*, e1002325. [CrossRef]
140. Nolden, M.; Ehses, S.; Koppen, M.; Bernacchia, A.; Rugarli, E.I.; Langer, T. The m-AAA protease defective in hereditary spastic paraplegia controls ribosome assembly in mitochondria. *Cell* **2005**, *123*, 277–289. [CrossRef]
141. Steglich, G.; Neupert, W.; Langer, T. Prohibitins regulate membrane protein degradation by the m-AAA protease in mitochondria. *Mol. Cell. Biol.* **1999**, *19*, 3435–3442. [CrossRef] [PubMed]

142. Maltecca, F.; Baseggio, E.; Consolato, F.; Mazza, D.; Podini, P.; Young, S.M., Jr.; Drago, I.; Bahr, B.A.; Puliti, A.; Codazzi, F.; et al. Purkinje neuron Ca^{2+} influx reduction rescues ataxia in SCA28 model. *J. Clin. Investig.* **2015**, *125*, 263–274. [CrossRef] [PubMed]
143. Mancini, C.; Hoxha, E.; Iommarini, L.; Brussino, A.; Richter, U.; Montarolo, F.; Cagnoli, C.; Parolisi, R.; Gondor Morosini, D.I.; Nicolo, V.; et al. Mice harbouring a SCA28 patient mutation in AFG3L2 develop late-onset ataxia associated with enhanced mitochondrial proteotoxicity. *Neurobiol. Dis.* **2019**, *124*, 14–28. [CrossRef] [PubMed]
144. Girard, M.; Lariviere, R.; Parfitt, D.A.; Deane, E.C.; Gaudet, R.; Nossova, N.; Blondeau, F.; Prenosil, G.; Vermeulen, E.G.; Duchen, M.R.; et al. Mitochondrial dysfunction and Purkinje cell loss in autosomal recessive spastic ataxia of Charlevoix-Saguenay (ARSACS). *Proc. Natl. Acad. Sci. USA* **2012**, *109*, 1661–1666. [CrossRef]
145. Ady, V.; Toscano-Marquez, B.; Nath, M.; Chang, P.K.; Hui, J.; Cook, A.; Charron, F.; Lariviere, R.; Brais, B.; McKinney, R.A.; et al. Altered synaptic and firing properties of cerebellar Purkinje cells in a mouse model of ARSACS. *J. Physiol.* **2018**, *596*, 4253–4267. [CrossRef]
146. Strupp, M.; Kalla, R.; Claassen, J.; Adrion, C.; Mansmann, U.; Klopstock, T.; Freilinger, T.; Neugebauer, H.; Spiegel, R.; Dichgans, M.; et al. A randomized trial of 4-aminopyridine in EA2 and related familial episodic ataxias. *Neurology* **2011**, *77*, 269–275. [CrossRef]
147. Feil, K.; Bremova, T.; Muth, C.; Schniepp, R.; Teufel, J.; Strupp, M. Update on the Pharmacotherapy of Cerebellar Ataxia and Nystagmus. *Cerebellum* **2016**, *15*, 38–42. [CrossRef]
148. Alvina, K.; Khodakhah, K. The therapeutic mode of action of 4-aminopyridine in cerebellar ataxia. *J. Neurosci. Off. J. Soc. Neurosci.* **2010**, *30*, 7258–7268. [CrossRef]
149. Davis, F.A.; Stefoski, D.; Rush, J. Orally administered 4-aminopyridine improves clinical signs in multiple sclerosis. *Ann. Neurol.* **1990**, *27*, 186–192. [CrossRef]
150. Stefoski, D.; Davis, F.A.; Fitzsimmons, W.E.; Luskin, S.S.; Rush, J.; Parkhurst, G.W. 4-Aminopyridine in multiple sclerosis: Prolonged administration. *Neurology* **1991**, *41*, 1344–1348. [CrossRef]
151. Dietrich, M.; Hartung, H.P.; Albrecht, P. Neuroprotective Properties of 4-Aminopyridine. *Neurol.-Neuroimmunol. Neuroinflamm.* **2021**, *8*, e976. [CrossRef]
152. Trimmer, J.S.; Rhodes, K.J. Localization of voltage-gated ion channels in mammalian brain. *Annu. Rev. Physiol.* **2004**, *66*, 477–519. [CrossRef]
153. Bostock, H.; Sears, T.A.; Sherratt, R.M. The effects of 4-aminopyridine and tetraethylammonium ions on normal and demyelinated mammalian nerve fibres. *J. Physiol.* **1981**, *313*, 301–315. [CrossRef]
154. Cao, Y.; Dreixler, J.C.; Roizen, J.D.; Roberts, M.T.; Houamed, K.M. Modulation of recombinant small-conductance Ca(2+)-activated K(+) channels by the muscle relaxant chlorzoxazone and structurally related compounds. *J. Pharmacol. Exp. Ther.* **2001**, *296*, 683–689.
155. Benton, M.D.; Lewis, A.H.; Bant, J.S.; Raman, I.M. Iberiotoxin-sensitive and -insensitive BK currents in Purkinje neuron somata. *J. Neurophysiol.* **2013**, *109*, 2528–2541. [CrossRef]
156. Feil, K.; Claassen, J.; Bardins, S.; Teufel, J.; Krafczyk, S.; Schneider, E.; Schniepp, R.; Jahn, K.; Kalla, R.; Strupp, M. Effect of chlorzoxazone in patients with downbeat nystagmus: A pilot trial. *Neurology* **2013**, *81*, 1152–1158. [CrossRef]
157. Egorova, P.A.; Zakharova, O.A.; Vlasova, O.L.; Bezprozvanny, I.B. In vivo analysis of cerebellar Purkinje cell activity in SCA2 transgenic mouse model. *J. Neurophysiol.* **2016**, *115*, 2840–2851. [CrossRef]
158. Lake, W.; Shah, H. Intrathecal Baclofen Infusion for the Treatment of Movement Disorders. *Neurosurg. Clin. N. Am.* **2019**, *30*, 203–209. [CrossRef]
159. Berntsson, S.G.; Gauffin, H.; Melberg, A.; Holtz, A.; Landtblom, A.M. Inherited Ataxia and Intrathecal Baclofen for the Treatment of Spasticity and Painful Spasms. *Stereotact. Funct. Neurosurg.* **2019**, *97*, 18–23. [CrossRef]
160. Misgeld, U.; Bijak, M.; Jarolimek, W. A physiological role for GABAB receptors and the effects of baclofen in the mammalian central nervous system. *Prog. Neurobiol.* **1995**, *46*, 423–462. [CrossRef]
161. Lujan, R.; Shigemoto, R. Localization of metabotropic GABA receptor subunits GABAB1 and GABAB2 relative to synaptic sites in the rat developing cerebellum. *Eur. J. Neurosci.* **2006**, *23*, 1479–1490. [CrossRef] [PubMed]
162. Kulik, A.; Vida, I.; Lujan, R.; Haas, C.A.; Lopez-Bendito, G.; Shigemoto, R.; Frotscher, M. Subcellular localization of metabotropic GABA(B) receptor subunits GABA(B1a/b) and GABA(B2) in the rat hippocampus. *J. Neurosci. Off. J. Soc. Neurosci.* **2003**, *23*, 11026–11035. [CrossRef] [PubMed]
163. Fernandez-Alacid, L.; Aguado, C.; Ciruela, F.; Martin, R.; Colon, J.; Cabanero, M.J.; Gassmann, M.; Watanabe, M.; Shigemoto, R.; Wickman, K.; et al. Subcellular compartment-specific molecular diversity of pre- and post-synaptic GABA-activated GIRK channels in Purkinje cells. *J. Neurochem.* **2009**, *110*, 1363–1376. [CrossRef] [PubMed]
164. Hibino, H.; Inanobe, A.; Furutani, K.; Murakami, S.; Findlay, I.; Kurachi, Y. Inwardly rectifying potassium channels: Their structure, function, and physiological roles. *Physiol. Rev.* **2010**, *90*, 291–366. [CrossRef]
165. Loeffler, M.A.; Synofzik, M.; Cebi, I.; Klocke, P.; Hormozi, M.; Gasser, T.; Gharabaghi, A.; Weiss, D. Case Report: Deep brain stimulation improves tremor in FGF-14 associated spinocerebellar ataxia. *Front. Neurol.* **2022**, *13*, 1048530. [CrossRef]
166. Oyama, G.; Thompson, A.; Foote, K.D.; Limotai, N.; Abd-El-Barr, M.; Maling, N.; Malaty, I.A.; Rodriguez, R.L.; Subramony, S.H.; Ashizawa, T.; et al. Deep brain stimulation for tremor associated with underlying ataxia syndromes: A case series and discussion of issues. *Tremor Other Hyperkinetic Mov.* **2014**, *4*, 228. [CrossRef]

167. Weiss, D.; Mielke, C.; Wachter, T.; Bender, B.; Liscic, R.M.; Scholten, M.; Naros, G.; Plewnia, C.; Gharabaghi, A.; Kruger, R. Long-term outcome of deep brain stimulation in fragile X-associated tremor/ataxia syndrome. *Park. Relat. Disord.* **2015**, *21*, 310–313. [CrossRef]
168. Cury, R.G.; Franca, C.; Duarte, K.P.; Paraguay, I.; Diniz, J.M.; Cunha, P.; Galhardoni, R.; Silva, V.; Iglesio, R.; Bissoli, A.B.; et al. Safety and Outcomes of Dentate Nucleus Deep Brain Stimulation for Cerebellar Ataxia. *Cerebellum* **2022**, *21*, 861–865. [CrossRef]
169. Kumar, G.; Asthana, P.; Yung, W.H.; Kwan, K.M.; Tin, C.; Ma, C.H.E. Deep Brain Stimulation of the Interposed Nucleus Reverses Motor Deficits and Stimulates Production of Anti-inflammatory Cytokines in Ataxia Mice. *Mol. Neurobiol.* **2022**, *59*, 4578–4592. [CrossRef]
170. Miterko, L.N.; Lin, T.; Zhou, J.; van der Heijden, M.E.; Beckinghausen, J.; White, J.J.; Sillitoe, R.V. Neuromodulation of the cerebellum rescues movement in a mouse model of ataxia. *Nat. Commun.* **2021**, *12*, 1295. [CrossRef]
171. Di Nuzzo, C.; Ruggiero, F.; Cortese, F.; Cova, I.; Priori, A.; Ferrucci, R. Non-invasive Cerebellar Stimulation in Cerebellar Disorders. *CNS Neurol. Disord. Drug Targets* **2018**, *17*, 193–198. [CrossRef]
172. Kumar, G.; Ma, C.H.E. Toward a cerebello-thalamo-cortical computational model of spinocerebellar ataxia. *Neural Netw.* **2023**, *in press*. [CrossRef]
173. Voglis, G.; Tavernarakis, N. The role of synaptic ion channels in synaptic plasticity. *EMBO Rep.* **2006**, *7*, 1104–1110. [CrossRef]
174. Breit, S.; Schulz, J.B.; Benabid, A.L. Deep brain stimulation. *Cell Tissue Res.* **2004**, *318*, 275–288. [CrossRef]
175. Alrashdi, B.; Dawod, B.; Schampel, A.; Tacke, S.; Kuerten, S.; Marshall, J.S.; Cote, P.D. Nav1.6 promotes inflammation and neuronal degeneration in a mouse model of multiple sclerosis. *J. Neuroinflamm.* **2019**, *16*, 215. [CrossRef]
176. Waxman, S.G. Mechanisms of disease: Sodium channels and neuroprotection in multiple sclerosis-current status. *Nat. Clin. Pract. Neurol.* **2008**, *4*, 159–169. [CrossRef]
177. Guyton, M.K.; Wingrave, J.M.; Yallapragada, A.V.; Wilford, G.G.; Sribnick, E.A.; Matzelle, D.D.; Tyor, W.R.; Ray, S.K.; Banik, N.L. Upregulation of calpain correlates with increased neurodegeneration in acute experimental auto-immune encephalomyelitis. *J. Neurosci. Res.* **2005**, *81*, 53–61. [CrossRef]
178. Grissmer, S.; Nguyen, A.N.; Aiyar, J.; Hanson, D.C.; Mather, R.J.; Gutman, G.A.; Karmilowicz, M.J.; Auperin, D.D.; Chandy, K.G. Pharmacological characterization of five cloned voltage-gated K^+ channels, types Kv1.1, 1.2, 1.3, 1.5, and 3.1, stably expressed in mammalian cell lines. *Mol. Pharm.* **1994**, *45*, 1227–1234.
179. Giordano, I.; Bogdanow, M.; Jacobi, H.; Jahn, K.; Minnerop, M.; Schoels, L.; Synofzik, M.; Teufel, J.; Klockgether, T. Experience in a short-term trial with 4-aminopyridine in cerebellar ataxia. *J. Neurol.* **2013**, *260*, 2175–2176. [CrossRef]
180. Marsden, C.D. Problems with long-term levodopa therapy for Parkinson's disease. *Clin. Neuropharmacol.* **1994**, *17* (Suppl. S2), S32–S44. [CrossRef]

life

MDPI

Article

Familial Alzheimer's Disease Neurons Bearing Mutations in *PSEN1* Display Increased Calcium Responses to AMPA as an Early Calcium Dysregulation Phenotype

Helena Targa Dias Anastacio [1], Natalie Matosin [2] and Lezanne Ooi [1,*]

[1] Molecular Horizons and School of Chemistry and Molecular Bioscience, University of Wollongong, Northfields Avenue, Wollongong, NSW 2522, Australia; htda204@uowmail.edu.au
[2] School of Medical Sciences, Faculty of Medicine and Health, The University of Sydney, Sydney, NSW 2050, Australia; natalie.matosin@sydney.edu.au
* Correspondence: lezanne@uow.edu.au; Tel.: +61-2-42215865

Abstract: Familial Alzheimer's disease (FAD) can be caused by mutations in *PSEN1* that encode presenilin-1, a component of the gamma-secretase complex that cleaves amyloid precursor protein. Alterations in calcium (Ca^{2+}) homeostasis and glutamate signaling are implicated in the pathogenesis of FAD; however, it has been difficult to assess in humans whether or not these phenotypes are the result of amyloid or tau pathology. This study aimed to assess the early calcium and glutamate phenotypes of FAD by measuring the Ca^{2+} response of induced pluripotent stem cell (iPSC)-derived neurons bearing PSEN1 mutations to glutamate and the ionotropic glutamate receptor agonists NMDA, AMPA, and kainate compared to isogenic control and healthy lines. The data show that in early neurons, even in the absence of amyloid and tau phenotypes, FAD neurons exhibit increased Ca^{2+} responses to glutamate and AMPA, but not NMDA or kainate. Together, this suggests that *PSEN1* mutations alter Ca^{2+} and glutamate signaling as an early phenotype of FAD.

Keywords: Alzheimer's disease; induced pluripotent stem cells; neurodegeneration; AMPA; glutamate

Citation: Targa Dias Anastacio, H.; Matosin, N.; Ooi, L. Familial Alzheimer's Disease Neurons Bearing Mutations in *PSEN1* Display Increased Calcium Responses to AMPA as an Early Calcium Dysregulation Phenotype. *Life* **2024**, *14*, 625. https://doi.org/10.3390/life14050625

Academic Editors: Andres Daniel Maturana and Carlo Musio

Received: 28 February 2024
Revised: 18 April 2024
Accepted: 8 May 2024
Published: 12 May 2024

1. Introduction

Alzheimer's disease (AD) is the most common neurodegenerative disorder in the world, with memory loss and cognitive dysfunction being common symptoms. The majority of cases are termed sporadic, caused by a combination of genetic and environmental factors, while less than 2% of cases are termed familial Alzheimer's disease (FAD) [1], caused by mutations in one of three genes: *PSEN1*, *PSEN2*, or *APP*, which each play a role in the generation of amyloid-β peptides [2]. Amyloid-β plaques in the extracellular space [3] and hyperphosphorylated tau in neurons [4] are two of the hallmarks of the disease. Current treatments available alleviate behavioral symptoms but do not stop neuronal degeneration. Hence, there is a great need to identify early pathways altered by the disease.

Calcium (Ca^{2+}) dyshomeostasis is implicated in the pathogenesis of AD [5,6]. Several studies have reported mutations in the presenilin genes that alter intracellular Ca^{2+} levels by modifying endoplasmic reticulum (ER) and lysosomal Ca^{2+} release [7–11]. Apart from Ca^{2+} release from intracellular stores, the elevation of cytosolic Ca^{2+} concentration can occur through the influx of extracellular Ca^{2+}. The activation of glutamate receptors is a major source of Ca^{2+} influx [5]. Glutamate is the most abundant excitatory neurotransmitter in the mammalian central nervous system, and it mediates excitatory synaptic transmission via activation of ionotropic (iGluRs) and metabotropic (mGluRs) glutamate receptors in the post-synaptic membrane. Ionotropic receptors are ligand-gated ion channels that mediate excitatory synaptic transmission via the influx of Na^{+} and Ca^{2+} into the cytoplasm. They are divided into 3 groups: α-amino-3-hydroxy-5-methyl-4-isoaxazolepropionic acid (AMPA), *N*-methyl-d-aspartate (NMDA), and kainate (KA) receptors [12]. AMPARs are composed of a combination of four subunits (GluA1-GluA4). In the rodent cortex, GluA1 comprises almost 50% of all AMPAR subunits, followed by similar levels of GluA2 and GluA3 [13]. Soluble oligomeric Aβ binds to AMPARs and promotes their internalization, leading to depressed synaptic transmission [14]. However, contrasting results have also found that Aβ increases expression of synaptic AMPAR, increasing synaptic transmission [15]. Hyperphosphorylated tau mislocalised to dendritic spines also disrupts synaptic function by dysregulating AMPAR trafficking [16]. NMDARs have seven different subunits: one GluN1, four GluN2 (GluN2A, GluN2B, GluN2C, and GluN2D), and two GluN3 subunits (GluN3A and GluN3B). Functional NMDARs are heterotetramers composed of two GluN1 and two GluN2 or GluN3 subunits [17]. Activation of extrasynaptic NMDARs increases production of tau [18] and Aβ [19,20]. In turn, amyloid-β increases extrasynaptic NMDAR activation by decreasing neuronal glutamate uptake [21–23], but it can also promote NMDAR endocytosis, causing synaptic depression [24]. The NMDAR antagonist memantine is used to treat moderate to severe cases of AD [25], thus targeting NMDA signalling can address some of the symptoms of AD. Kainate receptors (KAR) are composed of a tetrameric assembly of subunits GluK1-5. These receptors are strongly implicated in epilepsy and seizures [26], but little is known about their role in AD. Recent work, however, has found decreased protein expression of the GluK2 subunit and impaired KAR-mediated synaptic transmission in a mouse model of AD [27]. The function and expression of iGluRs play an important role in regulating synaptic plasticity. Through this property, neurons can modulate synaptic strength in response to the intensity and duration of a synaptic input, which is thought to underlie the formation of memory [28]. However, the majority of studies in human *postmortem* brain tissue have shown reduced expression of AMPAR, NMDAR, and KAR at the end stages of the disease, as summarised in [29].

Even though glutamatergic neurotransmission is required for synaptic plasticity, excessive activation of excitatory receptors results in excitotoxicity, whereby an elevated influx of Ca^{2+} initiates a signalling cascade that promotes cell death [30]. Hence, glutamatergic stimuli and levels of intracellular Ca^{2+} need to be tightly regulated in order to promote synaptic plasticity and avoid neuronal damage. However, a range of evidence suggests there is significant disruption to Ca^{2+} homeostasis, hyperactive (increased activity of) glutamatergic signalling, and neuronal excitability in AD. (1) Studies using different transgenic animal models expressing mutant *APP*, *PSEN*, and/or *MAPT* genes to model AD pathogenesis have detected elevated basal and stimulus-evoked glutamate release in the entorhinal cortex, dentate gyrus, CA1, and CA3 regions of the hippocampus [31]. (2) Concurrently, various studies have found neuronal hyperactivity in animal and cell models of AD, as well as in patients in the early stages of AD [32]. (3) Clinical studies have also detected a hyperactive phenotype of cortical and hippocampal brain regions in individuals at risk of developing AD, while brain hypoactivity was observed at later stages of the disease [33,34]. (4) Similar abnormalities have been detected in animal and cell models of AD, in which neurons displayed elevated Ca^{2+} signals and increased excitability [35–38]. Cell studies using induced pluripotent stem cells (iPSCs) from AD patients comprise a smaller fraction of research on neuronal excitability changes in AD but represent a useful tool to assess human cells and the impact of FAD-causing mutations compared to isogenic control lines.

Altogether, the current literature has identified numerous abnormalities in Ca^{2+} signalling and glutamatergic transmission in AD. Here, we aimed to assess the early calcium phenotypes of FAD neurons to identify whether these may precede AD pathology. In particular, we aimed to quantify differences in the magnitude of Ca^{2+} responses to iGluR agonists in early neurons representing FAD patients. Thus, we treated iPSC-derived excitatory cortical neurons from FAD patients, isogenic controls, and healthy individuals with glutamate, NMDA, AMPA, and kainate and measured the associated Ca^{2+} responses to determine the impact of iGluR agonists on the calcium phenotype in FAD neurons. Our data shows that FAD neurons demonstrate increased calcium responses to AMPA, even in the absence of amyloid and tau phenotypes, highlighting an early phenotype in FAD neurons that may contribute to calcium dysregulation.

2. Materials and Methods

2.1. Cell Culture and Neuronal Differentiation: Induced Pluripotent Stem Cell Culture

The induced pluripotent stem cell lines used in this study are described in Table 1. All cell lines were checked for karyotype abnormalities, RNA and protein expression of pluripotency markers, ability to differentiate into 3 germ layers, STR profiling, and genomic DNA sequencing. Experiments using human-derived iPSCs were approved by the University of Wollongong Human Ethics Committee and conducted in accordance with HE13/299 (Wollongong, Australia). To establish iPSCs from frozen stocks, cells were thawed at 37 °C in a water bath and mixed with warm DMEM/F12 medium. The cells were centrifuged for 5 min at 300× *g*, the supernatant was aspirated, and the pellet was resuspended in mTeSR1 (Stem Cell Technologies; Vancouver, BC, Canada) medium. Colonies were then plated into a 60 mm tissue culture dish coated with 0.1 mg/mL Matrigel (Corning; Corning, NY, USA), fed with mTeSR1, and kept in a humidified incubator at 37 °C with 5% CO_2 and 3% O_2. The medium was changed every day, and spontaneously differentiated cells were manually removed using a sterile pipette tip. Colonies were passaged every 5–7 days with Mg^{2+}/Ca^{2+}-free Phosphate Buffered Saline (PBS) (Thermo Fisher Scientific; Waltham, MA, USA) with 0.5 mM Ethylenediaminetetraacetic acid (EDTA; Life Technologies; Carlsbad, CA, USA) (PBS/EDTA). After 3–5 min in PBS-EDTA, colonies were detached by flushing DMEM/F12 into the dish. The medium containing the cells was then transferred into a centrifuge tube and centrifuged for 5 min at 300× *g*. The supernatant was discarded, and colonies were resuspended in 1 mL of mTeSR1 before being plated on a new Matrigel-coated dish.

Table 1. Details of induced pluripotent cell lines.

iPSC Name	Disease Status	WT/Mutation	APOE Genotype	Age at Collection	Sex	iPSC Line Characterisation
FAD1	Familial AD	$PSEN1^{S290C}$	ε3/3	47	M	[39]
IC1	Isogenic control	$PSEN1^{WT}$	ε3/3	47	M	[39]
HC1	Healthy control	$PSEN1^{WT}$	ε2/4	57	F	[40]
FAD2	Familial AD	$PSEN1^{A246E}$	ε3/4	56	F	[41]
IC2	Isogenic control	$PSEN1^{WT}$	ε3/4	56	F	Anastacio et al., 2024 in revision
HC2	Healthy control	$PSEN1^{WT}$	ε2/3	75	F	[41]

Abbreviations: FAD—Familial Alzheimer's disease; HC—Healthy control; IC—Isogenic control; iPSC—induced pluripotent stem cell.

2.2. Neuronal Differentiation

To generate excitatory cortical neurons from induced pluripotent stem cells, a well-characterised protocol combines the use of a lentiviral vector to express the neurogenin-2 (NGN2) transcription factor and drive neuronal differentiation, as well as small molecules inhibiting TGF-β and BMP signalling (SB431542 and LDN193189) to drive neurons towards a forebrain identity [42]. Representative images of the cells and the differentiation timeline are shown in Supplementary Figure S1.

Differentiation of iPSCs to excitatory cortical neurons was performed according to Nehme et al. (2018) [42] with minor modifications. Throughout differentiation, cells were maintained in a humidified incubator at 37 °C with 5% CO_2 and 3% O_2. After being thawed from frozen stocks, iPSCs were passaged at least once before being differentiated into neurons. Once iPSCs achieved the desired confluence, they were dissociated to single cells with Accutase (Thermo Fischer Scientific) and counted with a Countess II Automated Cell Counter (Thermo Fischer Scientific) using Trypan Blue (Sigma-Aldrich; Burlington, MA, USA). Single cells were plated at the desired density depending on the assay with mTeSR1 (Stem Cell Technologies) and 10 µM Y-27632 (Focus Bioscience, St Lucia, QLD, Australia) in Matrigel (Corning) and Poly-D-lysine (10–50 µg/mL; Sigma-Aldrich) coated plates. After 24 h, the cells were transduced with lentivirus (1–2 µL virus/1 mL medium) containing rtTA and NGN2-GFP expression vectors. The following day (day 0), the medium was replaced with neuronal differentiation medium containing 1 μg/mL doxycycline (Sigma-Aldrich), 10 μM SB431542 (Focus Bioscience), and 0.1 μM LDN-193189 (Focus Bioscience). On day 1, the culture medium was fully changed to a neuronal differentiation medium containing 1 μg/mL doxycycline, 10 μM SB431542, 0.1 μM LDN-193189, and 0.5 μg/mL puromycin. The same procedure was repeated on days 2 and 3. On day 4, the medium was gradually transitioned to neuronal maturation medium by adding 25% neuronal maturation medium and 75% neuronal differentiation medium containing 10 ng/mL BDNF (Miltenyi Biotec; Gladbach, Germany). The medium was changed every day until neurons were in 100% neuronal maturation medium (day 7). After that, a half-medium change was carried out every other day with neuronal maturation medium containing 1 μg/mL doxycycline and 10 ng/mL BDNF. Doxycycline was maintained until day 10 of differentiation. Neurons were grown until day 35 before experiments were conducted.

2.3. Lentivirus Production

Lentiviruses were produced by co-transfecting HEK293T cells (ATCC; Manassas, VA, USA) with two packaging plasmids, one envelope plasmid, and one transfer plasmid (NGN2 or rtTA). Throughout the experiment, the cells were maintained in a humidified incubator at 37 °C with 5% CO_2 and 3% O_2. For each lentivirus produced (NGN2 and rtTA), one T-75 flask was plated with 4 × 106 HEK293T cells in DMEM/F12 with 1× Glutamax (Thermo Fischer Scientific) and 5% fetal bovine serum (FBS; Gibco; Grand Island, NY, USA). The following day, 1.5 mL of Opti-MEM was mixed with the plasmid DNAs. On a different falcon tube, 75 μL of polyethylenimine (PEI) transfection reagent (3:1 ratio with DNA) was mixed with 1.5 mL of Opti-MEM (Gibco) and incubated for 5 min. Then, the DNA/Opti-MEM mix was combined with the PEI/Opti-MEM solution and incubated for 20 min to allow the formation of DNA/lipid complexes. This solution was then gently added to the T-75 flasks containing the HEK293T cells. After 5–7 h, the medium was discarded and replaced with fresh DMEM/F12 medium (with 1× Glutamax and 5% FBS). The next day, the medium containing viral particles was collected and stored at 4 °C, and fresh medium was added to the cells. The same procedure was repeated for another two days. After 3 days of viral media collection, the media was combined and centrifuged at 200× *g* for 2 min to pellet and remove HEK293T cells. The viral supernatant was filtered through a 50 mL syringe fitted with a 0.45 μm syringe filter and moved to an ultracentrifuge tube. The tubes were weighted to ensure they were within 0.1 g of each other, and, if necessary, sterile H_2O was added. The virus was spun at 23,500 rpm for 2 h at 4 °C in an ultracentrifuge. Following centrifugation, the supernatant was discarded and PBS (200× initial volume) was added to the viral pellet, which was gently triturated for approximately 10 min. The viral particles were aliquoted and stored at −80 °C until further use.

2.4. Neuronal Characterization by Immunocytochemistry

Cells grown for immunocytochemistry were plated on glass coverslips coated with Matrigel (Corning) or Matrigel and Poly-D-lysine (10–50 μg/mL; Sigma-Aldrich) for iPSCs and neurons, respectively. Cells were washed with TBS and fixed with 4% (*w*/*v*) paraformaldehyde (PFA) in PBS for 20 min, and then washed 3 times with TBS. Cells were permeabilised in 0.3% Triton X-100 in TBS for 10 min at room temperature. Following permeabilisation, cells were blocked in 10% goat serum in TBS for 1 h at room temperature. Then, cells were incubated with primary antibodies (in 10% goat serum in TBS) at specified dilutions overnight at 4 °C. The next day, cells were washed three times in wash buffer (0.1% triton X-100 in TBS) and incubated at room temperature for 1 h with the secondary antibody (in 10% goat serum in TBS) at specified dilutions. After that, cells were washed 3 times in wash buffer and incubated with Hoechst 33342 (1:1000; Thermo Fischer Scientific) in TBS for 15 min at room temperature. Then, cells were washed in TBS, and coverslips were mounted on glass slides using ProLong Gold Antifade Mountant (Thermo Fischer Scientific). Cells were imaged on the Leica TCS SP8 confocal microscope (Leica

Microsystems) and analysed with the Leica Application Suite—Advanced Fluorescence (LAS-AF) 2.6.1–7314 software (Leica Microsystems, Wetzlar, Germany). For the no primary antibody control, no primary antibody was added, and both secondary antibodies were added using the same conditions used for MAP2 and GFAP.

Antibodies: Anti-GFAP Chicken, pAb 1:800 Sigma-Aldrich #ab5541; Anti-MAP2 Mouse, mAb 1:200 Sigma-Aldrich #MAB3418; Anti-Oct4 Mouse, mAb 1:1000 Stem Cell Technologies #01550; Anti-SSEA-4 Mouse, mAb 1:200 Abcam #ab16287; Goat anti-mouse IgG (H + L) Alexa Fluor 488 Goat 1:1000 Thermo Fischer Scientific # A-11001; Goat anti-mouse IgG (H + L) Alexa Fluor 555 Goat 1:1000 Thermo Fischer Scientific # A-21422; Goat anti-chicken IgY (H + L) Alexa Fluor 647 Goat 1:1000 Thermo Fischer Scientific # A-21449.

2.5. Amyloid-β Enzyme-Linked Immunosorbent Assay (ELISA)

Aβ40 and Aβ42 levels were measured using an enzyme-linked immunosorbent assay (ELISA) (Thermo Scientific). The conditioned medium from cultured neurons was collected with 1× cOmplete protease inhibitor cocktail (Roche, Switzerland) and 1× PhosphoSTOP phosphatase inhibitor (Roche). Assays were performed following the manufacturer's instructions. Absorbance at 450 nm was measured on a SpectraMax Plus Microplate Reader (Molecular Devices; San Jose, CA, USA). Each sample was measured in triplicate, and the final protein concentration was determined using a standard curve generated using the Aβ peptide standard provided in the assay kit.

The Aβ40 and Aβ42 enzyme-linked immunosorbent assays (ELISA) (Thermo Fischer Scientific) were performed as per the manufacturer's instructions.

2.6. Western Blotting for Tau: Protein Extraction

Cells were washed twice with 1× TBS and incubated with RIPA buffer (50 mM Tris HCl pH 7.4, 1% Sodium deoxycholate, 150 mM NaCl, 1 mM EDTA, 1% Triton-X and 0.1% SDS in MilliQ) containing 1× cOmplete protease inhibitor cocktail (Roche) and 1× PhosphoSTOP phosphatase inhibitor (Roche) for 20 min on ice. Cell lysate was vortexed for 10 sec and centrifuged at 10,000× g for 5 min. The supernatant was transferred into a new microcentrifuge tube and stored at −80 °C until use.

2.7. Detergent Compatible (DC) Protein Assay

The Detergent Compatible Protein Assay (Bio-Rad; Hercules, CA, USA) was used to measure total protein concentration according to the manufacturer's instructions. A total of 5 µL of sample was mixed with 25 µL of reagent A' (cell lysate) or A (tissue lysate) and 200 µL of reagent B and incubated for 10 min on an orbital plate rocker (Labnet; Edison, NJ, USA). Each sample was measured in triplicate, and the final protein concentration was determined using a standard curve generated using Bovine Serum Albumin (BSA; Sigma-Aldrich) diluted in either RIPA buffer or lysis buffer. Absorbance was read at 750 nm using a SpectraMax Plus Microplate Reader (Molecular Devices).

2.8. Western Blotting

Cell pellets were diluted in RIPA buffer. Samples were mixed with Laemmli buffer (Bio-Rad) containing 5% (v/v) 2-mercaptoethanol (Sigma-Aldrich). Samples were denatured at 70 °C for 10 min before being loaded into a 4–20% Criterion TGX Stain-Free gel (Bio-Rad). The Precision Plus Protein Dual Colour (Bio-Rad) molecular weight marker was also loaded into the gel. Samples were run in duplicates, and a pooled sample was run on all gels to normalise for variations between gels. Following separation by electrophoresis at 160 V for 50 min in SDS buffer, proteins were transferred to an Immobilon-P PVDF membrane (Merck Millipore; Burlington, MA, USA) at 100 V for 60 min in ice-cold transfer buffer. Total membrane protein was imaged with ChemiDoc MP (Bio-Rad). The membrane was then washed 3× in Tris-Buffered Saline with 0.1% (v/v) Tween (TBST) and blocked in 5% (w/v) skim milk or 10% goat serum in TBST for 1 h at room temperature. Incubation with primary antibodies in blocking solution was carried out overnight at 4 °C, and then the membrane was washed 3× for 5 min in TBST before being incubated with secondary antibodies in blocking solution for 1 h at room temperature. Following this step, the membrane was washed 3× for 5 min in TBST and then incubated for 5 min with SuperSignal™ West Pico PLUS Chemiluminescent Substrate (Thermo Scientific). Protein bands were detected by chemiluminescence with Amersham 600 RGB (GE Healthcare; Chicago, IL, USA), and relative densitometry values were quantified using ImageLab (Bio-Rad). Each protein band was normalised to its respective total protein and the mean pool total protein.

Antibodies: Anti-AMPAR1 Rabbit pAb 1:1000 Abcam #ab31232; Anti-AMPAR2 Rabbit mAb 1:2000 Abcam #ab133477; Anti-AMPAR3 Mouse mAb Merck Millipore #MAB5416; Anti-AMPAR4 Rabbit mAb Cell Signalling #8070; Anti-Tau-5 Mouse mAb Abcam #ab80579; Anti-Tau s404 Rabbit mAb 1:2500 Abcam #ab92676; Abbreviations: mAb—monoclonal antibody; pAb—polyclonal antibody; Goat anti-mouse IgG (H + L) HRP Goat 1:5000 Merck Millipore #AP308P; Goat anti-rabbit IgG (H + L) HRP Goat 1:5000 Merck Millipore #AP307P.

2.9. Flexstation to Measure Calcium Responses to Glutamate, NMDA, AMPA and Kainate

To study neuronal calcium response, iPSC-derived neurons were grown in a clear bottom 96 Well Black Polystyrene Microplate (Corning) for 35 days. A minimum of 6 wells per cell line were used for each drug treatment (glutamate, NMDA, AMPA, and kainate). A standard bathing solution (SBS; 160 mM NaCl, 2.5 mM KCl, 5 mM $CaCl_2$, 1 mM $MgCl_2$, 5 mM glucose, and 10 mM HEPES) was used for baseline perfusion of neurons. A 60 mM high potassium (High K^+, 102.5 mM NaCl, 2.5 mM KCl, 5 mM $CaCl_2$, 1 mM $MgCl_2$, 5 mM glucose, and 10 mM HEPES) buffer was used as a positive control due to its ability to depolarise neurons and induce a calcium response. To measure neuronal response to iGluR agonists, 100 μM glutamate (Tocris, UK), 100 μM NMDA (Tocris), 100 μM AMPA (Tocris), and 100 μM kainate (Tocris) were prepared in SBS. All buffers and drugs were made on the same day the assay was run and adjusted to have a pH of 7.4 and osmolarity between 310 and 320 mOsm, the same as the conditioned neuronal cultured media.

Neurons were washed with SBS and loaded with 2 μM Fura-2-AM buffer (Thermo Fischer Scientific) and Pluronic F-127 acid (20% *w/v* in DMSO; Biotium; Fremont, CA, USA) at 37 °C for 30 min. After incubation, cells were washed again in SBS for 30 min before recording was conducted at 37 °C using a FlexStation 3 Multi-Mode Microplate Reader (Molecular Devices). The fluorescence signal was measured per well, corresponding to the calcium response of the neuronal population grown on each well. Fura-2 fluorescence emission at 510 nm was recorded every four seconds following excitation at 340 and 380 nm. To each well, a total of three solutions (SBS, drug of interest, and High K^+) were added, and fluorescence was measured. Baseline recordings in SBS were taken for 20 s before the addition of the first compound. Each compound was recorded for 60 s before the addition of the following compound. The drugs glutamate, NMDA, AMPA, or kainate were added after SBS, followed by 60 mM High K^+. The maximum Ca^{2+} response was calculated from the maximum peak of the Fura-2 AM 340/380 nm excitation ratio normalised to baseline fluorescence (ΔF340/380). Wells that did not respond to high K^+ were excluded from the analysis. The SoftMax Pro 7 software (Molecular Devices) was used for data acquisition.

2.10. RT-qPCR for AMPAR Subunits: RNA Extraction and Purification

After being washed twice in PBS, cells were harvested in TRIsure (Bioline; Cincinnati, OH, USA) and left to incubate for 5 min at room temperature. To each 1 mL of cells in TRIsure, 200 μL of chloroform was added, and the sample was shaken vigorously for 15 s and incubated for 15 min at room temperature. Then, cells were centrifuged at 12,000× *g* for 15 min at 4 °C. The upper aqueous phase (RNA) was transferred to a fresh tube, and 500 μL of isopropanol was added. The samples were incubated for 10 min at room temperature and then centrifuged at 12,000× *g* for 10 min at 4 °C. The supernatant was discarded, and the RNA pellet was washed with 70% ethanol and centrifuged at 7500× *g* for 5 min at 4 °C. The supernatant was removed, and the RNA pellet was air-dried for approximately 30 min. Once dried, the pellet was resuspended in UltraPure DNase/RNase-Free Distilled Water

(Thermo Fischer Scientific). The NanoDrop 2000c (Thermo Fischer Scientific) was used to measure nucleic acid concentration and purity.

To remove genomic DNA, the TURBO DNA-free kit (Thermo Fischer Scientific) was used following the manufacturer's instructions. The RNA was diluted in 50 µL of nuclease-free water, and 1× TURBO DNase buffer and 1 µL TURBO DNase inactivation reagents were added. After the solution was incubated at 37 °C for 30 min, the DNase Inactivation Reagent was added and incubated for 5 min at room temperature. The samples were centrifuged at 10,000× *g* for 90 s, and the supernatant (RNA) was transferred to a fresh tube. RNA concentration was calculated with NanoDrop 2000c. The treated RNA was purified by ethanol precipitation. The RNA was diluted in 200 µL UltraPure DNase/RNase-Free Distilled Water and mixed with 220 µL isopropanol and 20 µL 3 M sodium acetate. After being vortexed, the samples were kept at −80 °C overnight and then centrifuged at 16,000× *g* for 20 min at 4 °C. The supernatant was removed, and the pellets were washed with 70% ethanol before being centrifuged at 16,000× *g* °C for another 2 min. The supernatant was discarded, and the pellet was air dried for approximately 30 min before being resuspended in UltraPure DNase/RNase-Free Distilled Water. The final concentration and purity of mRNA were measured using NanoDrop 2000c.

2.11. cDNA Synthesis

Complementary DNA (cDNA) was synthesised using the Tetro cDNA Synthesis Kit (Bioline). The protocol was followed according to the manufacturer's instructions: 2 µg RNA was mixed with 1 µL oligo (dT)18, 1 µL 10 mM dNTP mix, 1 µL tetro reverse transcriptase enzyme, 4 µL 5× RT buffer, 1 µL RNAse inhibitor, and enough RNAse-free water to complete the 20 µL solution. Samples were incubated for 30 min at 45 °C, followed by 5 min at 85 °C, in the Mastercycler Pro (Eppendorf, Germany).

2.12. qPCR

Primers designed for the coding regions of the genes of interest and primers for housekeeper genes B2M, GAPDH, and HPRT1 were used (Table 2). Each cDNA sample was run in triplicate using 30 ng of cDNA per reaction. No reverse transcriptase RNA control and no template control were run as negative controls. QuantStudio 5 (Thermo Fischer Scientific) was used with the following settings: denaturation at 95 °C for 2 min, followed by 40 cycles of denaturation at 95 °C for 10 s, annealing at a specific primer melting temperature for 10 s, and extension at 72 °C for 10 s. The melt curve protocol started at 95 °C for 5 s, followed by 65 °C for 1 min, ramping up to 97 °C. Primers are listed in Table 2.

Table 2. List of primers used for reverse transcription-quantitative polymerase chain reaction (RT-qPCR).

Target	Sequence	Tm
B2M	F: AAGGACTGGTCTTTCTATCTC R: GATCCCACTTAACTATCTTGG	55 °C
GAPDH	F: GAGCACAAGAGGAAGAGAGAGACCC R: GTTGAGCACAGGGTACTTTATTGATGGTACATG	58 °C
HPRT1	F: TGACACTGGCAAAACAATGCA R: GGTCCTTTTCACCAGCAAGCT	58 °C
GRIA1	F: CTAGAAGATCCTTATGTGATGC R: CTCCGTATTTTCCATCACTG	58 °C
GRIA2	F: GGAATCTCCGTATGTTATGATG R: TTGTACTTGAACCCACAATG	55 °C
GRIA3	F: TATTGTATCTGGGGCGTTAC R: TTGAGAACTCAAGAAGGGAG	55 °C
GRIA4	F: GGTACGATAAAGGTGAATGTG R: AAAAGGTCAGCTTCATTCTC	58 °C

Abbreviations: F—forward; R—reverse; Tm—primer melting temperature.

2.13. Data Analysis

Data analysis was performed in R v3.3.1 (https://www.r-project.org, accessed on 16 March 2023) using packages 'library(readxl)', 'library(ggplot2)', 'library(ggpubr), 'library(lme4)', and 'library(lmerTest)'. The normal distribution of the data was assessed with a Shapiro–Wilk test using the function 'shapiro.test()'. Data not normally distributed were normalised via logarithmic (log2 or log10), square root (sqrt), or reciprocal (1/x) transformation. For comparison between disease and control groups, the data were subjected to linear regression modelling using the 'lm()' function, with independent differentiations included as co-variates.

3. Results

3.1. Generation of iPSC-Derived Neurons from FAD and Control Lines

Our aim in this study was to specifically assess the early calcium phenotypes of AD, and for this reason, we used day 35 of neuronal maturation, in which cells express neuronal markers and neuronal ion channels but may not yet demonstrate AD pathology. In this well-characterised protocol to generate excitatory cortical neurons, Nehme et al. (2018) [42] demonstrated by RNAseq the expression of AMPAR subunits GluA1-4 over time and found no expression of genes for hindbrain, inhibitory, hypothalamic, diencephalic, hippocampal, dopaminergic, and spinal motor neurons. Single-cell RT-qPCR of day 21 neurons detected mainly markers of cortical upper layer neurons, which was also confirmed by immunostaining and no expression of inhibitory neurons or glia. Furthermore, patch clamping experiments after 28 days of differentiation demonstrated that approximately 75% of the cells recorded were able to fire multiple action potentials [42]. Application of the AMPAR and NMDAR antagonists, NBQX and D-AP5, respectively, showed each compound individually reduced spiking rates and the occurrence of network-wide bursts, confirming the functional expression of AMPAR and NMDARs [42]. On the other hand, the $GABA_A$ receptor antagonist, picrotoxin, caused no changes in firing rates or the occurrence of network-wide bursts [42].

To study AD phenotypes in a 2D cell culture model using iPSC-derived neurons, a total of 6 iPSC lines were used, including two familial Alzheimer's disease cell lines (FAD1 and FAD2) and their respective isogenic controls (IC1 and IC2) and two healthy controls (HC1 and HC2) (Table 1). Each FAD line contains a different mutation in the *PSEN1* gene, which causes early-onset familial Alzheimer's disease, while the isogenic controls have the same genome as their respective FAD lines, with the exception of the FAD mutation, which has been reverted back to WT. HC1 cells were obtained from a healthy donor with no known family association with FAD1, while the HC2 cell line donor is a healthy family relative of FAD2. Due to their genetic background and family associations, throughout this study, FAD1 neurons were compared against their isogenic control (IC1) and HC1 only, while FAD2 neurons were compared against IC2 and HC2.

To generate iPSC-derived neurons from AD patients, isogenic controls, and healthy individuals, a previously published protocol using small molecules and NGN2 lentivirus was used to generate cortical excitatory neurons [42]. The neurons were differentiated for a total of 35 days prior to analysis.

iPSC-Derived Neurons Express Neuronal Marker MAP2

To confirm the generation of iPSC-derived neurons after 35 days of differentiation, immunocytochemistry was used. Immunostaining against microtubule-associated protein 2 (MAP2), a neuron-specific cytoskeletal protein, confirmed extensive expression of this marker in all 6 cell lines (FAD1, IC1, HC1, FAD2, IC2, and HC2) (Figures 1 and 2). To check for the presence of glial cells, the cells were also stained against the glial fibrillary acidic protein (GFAP), a type III intermediate filament protein expressed in glial cells. However, no positive staining was observed for GFAP for any of the 6 cell lines (Figures 1 and 2). A non-primary antibody control showed no signal for the MAP2 and GFAP channels

(Supplementary Figures S2 and S3). These results confirm the generation of neurons and the absence of GFAP-positive cells after 35 days of differentiation.

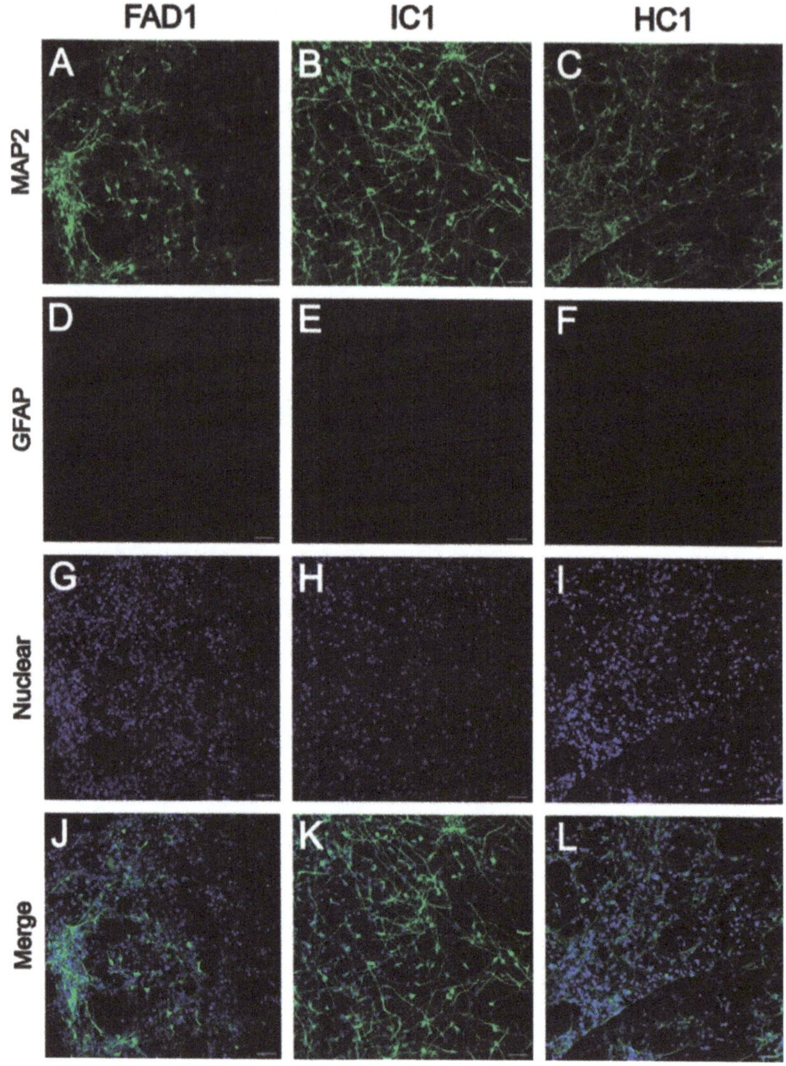

Figure 1. MAP2 and GFAP immunoreactivity in FAD1, IC1, and HC1 iPSC-derived neurons. Representative images showing MAP2 (**A–C**), GFAP (**D–F**), the nuclear marker Hoechst (**G–I**), and the merged image (**J–L**) from FAD1, IC1, and HC1. Scale bar = 50 μm. FAD—familial Alzheimer's disease; HC—healthy control; IC—isogenic control.

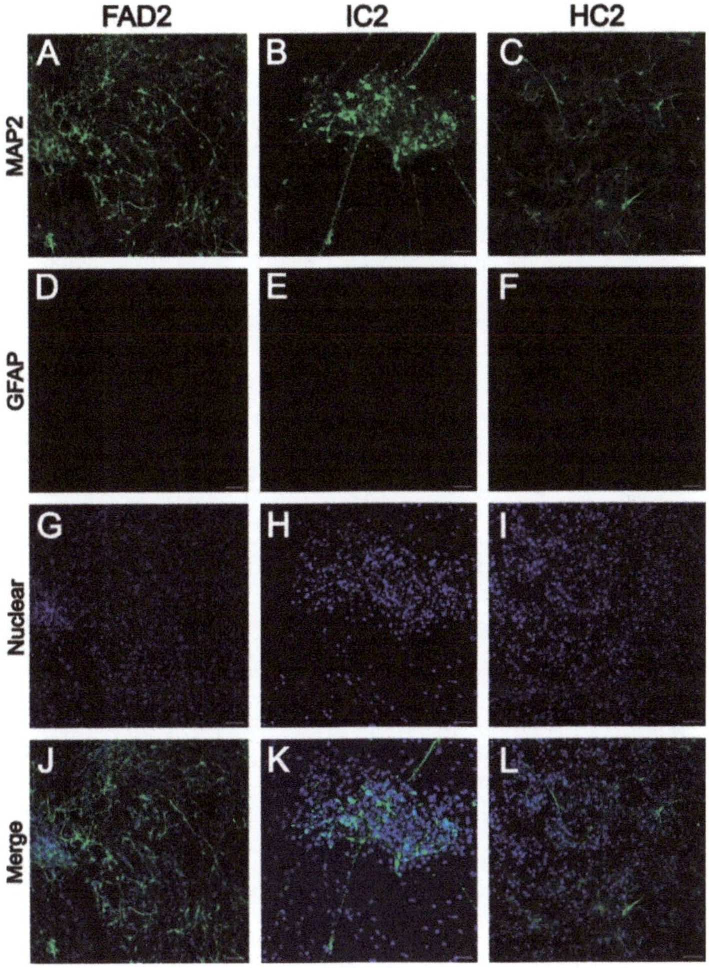

Figure 2. MAP2 and GFAP immunoreactivity in FAD2, IC2, and HC2 iPSC-derived neurons. Representative images showing MAP2 (**A–C**), GFAP (**D–F**), the nuclear marker Hoechst (**G–I**), and the merged image (**J–L**) from FAD2, IC2, and HC2. Scale bar = 50 μm. FAD—familial Alzheimer's disease; HC—healthy control; IC—isogenic control.

3.2. *FAD NGN2-Derived iPSC Neurons Did Not Show Evidence of Aβ Pathology at Day 35 of Maturation*

The decreased Aβ42/Aβ40 ratio in both CSF and plasma is an important biomarker of AD [43,44]. While Aβ40 is the most abundant Aβ peptide, Aβ42 is more prone to aggregation. Due to inter-individual differences in Aβ peptide production, the Aβ42/Aβ40 ratio is considered a more accurate measurement of Aβ pathology than the levels of Aβ42 alone [43,45,46].

In 2D in vitro cultures, since Aβ plaques cannot be formed due to the regular medium changes, the elevated production of Aβ42 results in a higher Aβ42/Aβ40 ratio in some long-term cell culture models [36,47–50]. For example, in day >50–60 neurons, both Aβ and tau pathology can be identified in AD neurons [47,48]. To measure the Aβ42/Aβ40 ratio in disease and control day 35 neurons in this study, an ELISA was performed using medium from neuronal cultures collected at day 35 of differentiation to quantify both Aβ40 and Aβ42 levels separately.

While no significant differences were detected in Aβ40 concentration between disease (FAD1—152.1 ± 38.64, FAD2—223.7 ± 51.23) and isogenic control (IC1—150.1 ± 45.28, p = 0.642; IC2—249.6 ± 68.31, p = 0.846) or healthy control (HC1—206 ± 77.18, p = 0.143; HC2—187.8 ± 52.06, p = 0.122) cell lines (Figure 3A), FAD2 (9.37 ± 1.859) neurons showed significantly higher levels of Aβ42 compared to HC2 (5.346 ± 0.994; p = 0.001), but no significant differences compared to IC2 (8.333 ± 1.168; p = 0.448) (Figure 3B). For the cell lines FAD1, IC1, and HC1, the concentration of Aβ42 in the medium could not be quantified as it was below the detection threshold of the assay, and therefore the Aβ42/Aβ40 ratio could not be calculated. No significant differences were found between the FAD2 (0.044 ± 0.005) and control cell lines (HC2—0.032 ± 0.005, p = 0.281; IC2—0.041 ± 0.01, p = 0.782) in the Aβ42/Aβ40 ratio (Figure 3C).

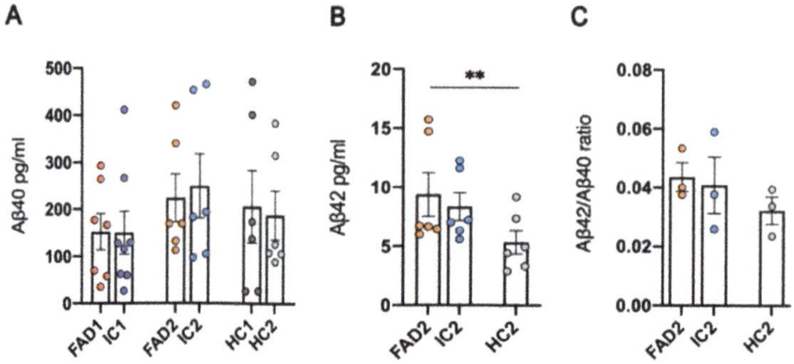

Figure 3. Aβ42/Aβ40 ratio in neuronal cultures. Aβ40 (**A**) and Aβ42 (**B**) levels were measured separately in pg/mL by ELISA and were used for calculation of the Aβ42/Aβ40 ratio (**C**). Data shown are day 35 neurons, n = 2 replicates from 3–4 independent differentiations, and error bars represent ± SEM. For the Aβ42/Aβ40 ratio, data = mean ± SEM, ** $p < 0.01$. Data are normally distributed (FAD2/HC2/IC2 Aβ42/Aβ40 ratio) or normalised to log2 (FAD1/HC1/IC1 Aβ40), sqrt (FAD2/HC2/IC2 Aβ42), or 1/x (FAD2/HC2/IC2 Aβ40) and analysed by linear regression modelling. Independent differentiations were included as co-variates. FAD—familial Alzheimer's disease; HC—healthy control; IC—isogenic control.

These results indicate Aβ40 levels can be detected in the conditioned medium of iPSC-derived neurons, but the levels of Aβ42 may be too low to allow detection by commercial assays. For the group of cell lines in which both Aβ42 and Aβ40 could be measured, FAD1 iPSC-derived day 35 neurons did not show AD-associated changes in Aβ42/Aβ40 ratios.

3.3. FAD NGN2 Derived iPSC Neurons Did Not Show Tau Pathology at Day 35 of Maturation

Intracellular accumulation of misfolded phosphorylated tau (p-Tau) protein is a molecular hallmark of AD that is associated with cognitive decline. Our previous work has shown that day >50–60 AD neurons demonstrate increased phosphorylated tau relative to total tau, consistent with AD-related tau pathology [48]. To assess whether day 35 neurons demonstrate tau phenotypes, we performed western blotting to measure total tau (Tau-5) and phosphorylated tau at serine 404 (p-Tau S404) protein levels between AD and control neurons after 35 days of differentiation (Figure 4A,B). Although tau phosphorylation occurs at different protein epitopes, phosphorylation at serine 404 is one of the earliest events in tau pathology [51], hence why p-Tau S404 was chosen as a marker of AD pathology for this neuronal model of AD. For both Tau-5 and p-Tau S404 western blots, a single band of approximately 50 kDa was observed.

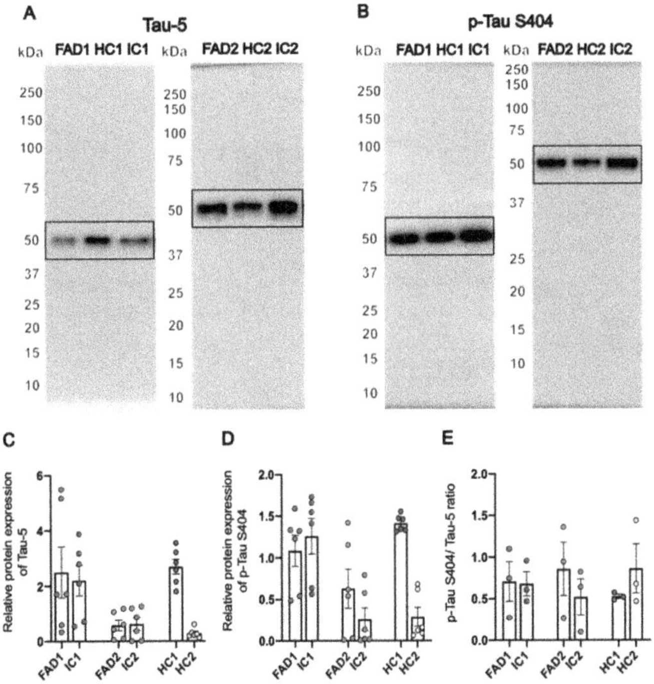

Figure 4. Expression of phosphorylated and total protein tau in neuronal cultures. Representative western blots of Tau-5 (**A**) and p-Tau S404 (**B**) and relative protein expression of Tau-5 (**C**) and p-Tau S404 (**D**) and ratio of p-Tau S404/Tau-5 relative protein expression (**E**). Data shown are day 35 neurons; n = 2 replicates from 3 independent differentiations, normalised to total protein; error bars represent ± SEM. For the p-Tau S404/Tau-5 ratio, data = mean ± SEM. Data are normally distributed (FAD1/HC1/IC1 and FAD2/HC2/IC2 p-Tau S404/Tau-5 ratio and FAD1/HC1/IC1 Tau-5 and p-Tau S404) or normalised to log10 (FAD2/HC2/IC2 p-Tau S404) or sqrt (FAD2/HC2/IC2 Tau-5) and analysed by linear regression modelling. Independent differentiations were included as co-variates; there were no significant differences between groups. FAD—familial Alzheimer's disease; HC—healthy control; IC—isogenic control; p-Tau—phosphorylated tau.

Regarding protein expression of Tau-5, there were no significant differences between FAD (FAD1—2.505 ± 0.923, FAD2—0.575 ± 0.194) and control cell lines (HC1—2.69 ± 0.265, $p = 0.84$; IC1—2.202 ± 0.551, $p = 0.742$; HC2—0.283 ± 0.069, $p = 0.299$; IC2—0.624 ± 0.231, $p = 0.919$) (Figure 4C and Supplementary Figures S4 and S5). Similar results were obtained after analysis of the relative protein amount of p-Tau S404. FAD neurons (FAD1—1.084 ± 0.187, FAD2—0.629 ± 0.234) showed no significant differences compared to their isogenic or healthy controls (HC1—1.41 ± 0.04, $p = 0.073$; IC1—1.259 ± 0.213, $p = 0.316$; HC2—0.29 ± 0.115, $p = 0.819$; IC2—0.259 ± 0.139, $p = 0.274$) (Figure 4D and Supplementary Figures S6 and S7). The ratio of phosphorylated S404 Tau to Tau-5 (total tau) normalised to total protein was calculated to compensate for variations in the production and clearance of tau between cell lines. Data analysis using linear regression modelling showed there were no significant differences between the FAD lines (FAD1—0.706 ± 0.239, FAD2—0.856 ± 0.319) and their respective isogenic (IC1—0.679 ± 0.145, $p = 0.91$; IC2—0.521 ± 0.217, $p = 0.43$) and healthy controls (HC1—0.527 ± 0.024, $p = 0.465$; HC2—0.861 ± 0.294, $p = 0.99$) (Figure 4E).

Overall, both Tau-5 and p-Tau S404 proteins were detected in all cell lines, with no significant differences between FAD and control neurons. These results show that iPSC-derived neurons from FAD donors at day 35 of differentiation did not show the AD-associated phenotype of abnormally elevated S404 tau phosphorylation.

3.4. FAD Neurons Demonstrated Increased AMPAR Ca^{2+} Signalling Compared to Isogenic Controls

Dysfunctional calcium signalling and abnormal glutamatergic transmission are common phenotypes of AD. To assess if these phenotypes were recapitulated in iPSC-derived FAD neurons even in the absence of Aβ and tau phenotypes, quantification of calcium responses was conducted in Alzheimer's disease, isogenic control, and healthy control day 35 neurons. Neurons were treated with 60 mM High K^+ (to mimic depolarisation), and Ca^{2+} responses were quantified. Ca^{2+} responses were also quantified following the application of 100 μM glutamate, NMDA, AMPA, or kainate in iPSC-derived neurons from FAD patients and controls. Live cell calcium imaging was performed by loading the cells with the ratiometric dye Fura-2 AM prior to treatments. The Flexstation microplate reader was used to measure Ca^{2+} signals from a population of neurons in the wells of a 96 well plate. The maximum Ca^{2+} response from each well was calculated from the maximum peak of the Fura-2 AM 340/380 nm excitation ratio normalised to baseline fluorescence ($\Delta F_{340/380}$).

3.5. The iPSC-Derived Neurons Displayed Ca^{2+} Responses to High K^+

To verify if day 35 neurons were able to depolarise and produce Ca^{2+} responses, the cells were treated with a high concentration of K^+ (High K^+, 60 mM). After loading the neurons with Fura-2 AM, a standard bathing solution (SBS) was added to the cells to allow the measurement of baseline fluorescence, then High K^+ was applied and Ca^{2+} signals were quantified. Neurons from all cell lines exhibited an increase in the 340/380 fluorescence ratio, indicative of Ca^{2+} responses (Figure 5A,B).

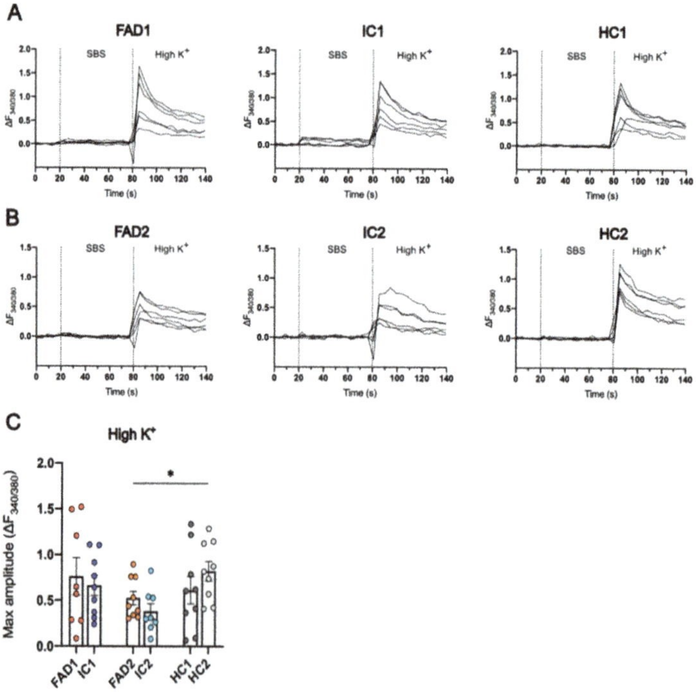

Figure 5. High K^+ elicits Ca^{2+} responses in iPSC-derived neurons. Representative traces of the change in fluorescence signal ratio over baseline ($\Delta F_{340/380}$) in response to 60 mM K^+ (High K^+) were recorded from FAD1/IC1/HC1 (**A**) and FAD2/IC2/HC2 neurons (**B**). The maximum Ca^{2+} response to High K^+ was calculated as the maximum amplitude of the Fura-2 AM fluorescence ratio change over baseline ($\Delta F_{340/380}$) (**C**). Day 35 neurons were loaded with the ratiometric dye Fura-2 AM, perfused with SBS, and treated with High K^+. Data shown are single wells of a 96-well plate containing day 35 neurons,

n = 8–15 replicates from three independent differentiations; error bars represent ± SEM; * $p < 0.05$ (FAD2 vs. HC2, but not IC2). Data are normally distributed and analysed using linear regression modelling. Independent differentiations were included as co-variates. FAD—familial Alzheimer's disease; HC—healthy control; IC—isogenic control.

FAD2 neurons (0.525 ± 0.074) had a significantly lower Ca^{2+} response than HC2 cells (0.821 ± 0.107, $p = 0.026$), which displayed the highest maximum amplitude of all cell lines. Although IC2 (0.38 ± 0.083, $p = 0.272$) neurons had the lowest maximum amplitude, this value was not significantly different from the FAD2 line. The peak $\Delta F_{340/380}$ of FAD1 (0.760 ± 0.202) cells had no statistically significant difference from IC1 (0.662 ± 0.111, $p = 0.442$) and HC1 (0.612 ± 0.149, $p = 0.249$) cells (Figure 5C).

In sum, iPSC-derived day 35 neurons depolarised and displayed Ca^{2+} signals after treatment with 60 mM K^+. There were no significant differences in these responses between the FAD and IC neurons.

3.6. FAD Neurons Displayed Increased Ca^{2+} Responses to Glutamate

Glutamate, as the main excitatory neurotransmitter in the CNS, is an important regulator of synaptic plasticity in the brain. However, numerous studies have reported aberrant neuronal excitability in AD patients and lab models of the disease [32]. To investigate how day 35 iPSC-derived neurons respond to glutamatergic stimuli, neurons were treated with 100 μM glutamate, and the Fura-2 AM 340/380 nm excitation ratio normalised to baseline fluorescence ($\Delta F_{340/380}$) was recorded over time.

Neurons from all FAD, IC, and HC cell lines displayed Ca^{2+} responses to glutamate (Figure 6A,B). To compare the maximum amplitude of Ca^{2+} responses to glutamate between cell lines, the maximum peak of the Fura-2 AM 340/380 nm fluorescence change ($\Delta F_{340/380}$) over baseline was calculated. This analysis confirmed that FAD1 (0.686 ± 0.111) neurons showed significantly higher Ca^{2+} responses compared to IC1 (0.446 ± 0.056, $p = 0.033$) and HC1 (0.377 ± 0.047, $p = 0.005$). Similarly, FAD2 (0.530 ± 0.095) neurons had significantly greater Ca^{2+} responses to 100 μM glutamate than neurons from the relevant isogenic control IC2 (0.345 ± 0.067, $p = 0.013$), but were not significantly different from neurons from the HC2 line (0.526 ± 0.040, $p = 0.570$) (Figure 6C). Thus, FAD iPSC-derived day 35 neurons showed elevated Ca^{2+} signals in response to treatment with 100 μM glutamate compared to their relevant isogenic controls.

3.7. FAD Neurons Displayed Increased Ca^{2+} Responses to AMPA but Not NMDA or Kainate Compared to Their Isogenic Control Lines

Glutamate can act on two subgroups of receptors in the post-synapse: mGluRs and iGluRs. The latter are divided into AMPAR, NMDAR, and KAR, and their activation mediates fast excitatory synaptic transmission and modulates synaptic plasticity. To investigate the contribution of specific iGluRs in the elevated Ca^{2+} responses of FAD neurons to glutamate, day 35 iPSC-derived neurons were treated with the AMPAR, NMDAR, and KAR agonists: AMPA, NMDA, and kainate, respectively.

Tracking of the 340/380 fluorescence change ($\Delta F_{340/380}$) over time showed all 6 cell lines responded to 100 μM AMPA (Figure 7A,B). Statistical analysis of the maximum amplitude of $\Delta F_{340/380}$ showed both FAD1 (0.709 ± 0.088) and FAD2 (0.673 ± 0.125) exhibited higher Ca^{2+} responses to 100 μM AMPA than their respective isogenic controls, IC1 (0.289 ± 0.048, $p = 1.53 \times 10^{-5}$) and IC2 (0.330 ± 0.086, $p = 0.010$). FAD1 also had increased Ca^{2+} responses to 100 μM AMPA compared to HC1 (0.432 ± 0.075, $p = 0.017$); however, FAD2 and HC2 (0.406 ± 0.067, $p = 0.077$) were not significantly different (Figure 7C).

Treatment with 100 μM NMDA showed no significant differences in Ca^{2+} responses from FAD1 (0.380 ± 0.085) and FAD2 (0.41 ± 0.134) relative to their isogenic controls, IC1 (0.353 ± 0.064, $p = 0.874$) and IC2 (0.337 ± 0.132, $p = 0.692$), respectively. However, FAD1 (0.38 ± 0.085) Ca^{2+} signals were significantly elevated compared to HC1 (0.139 ± 0.014, $p = 0.046$), while no significant differences were observed between FAD2 and HC2 (0.422 ± 0.1, $p = 0.307$) (Figure 7D).

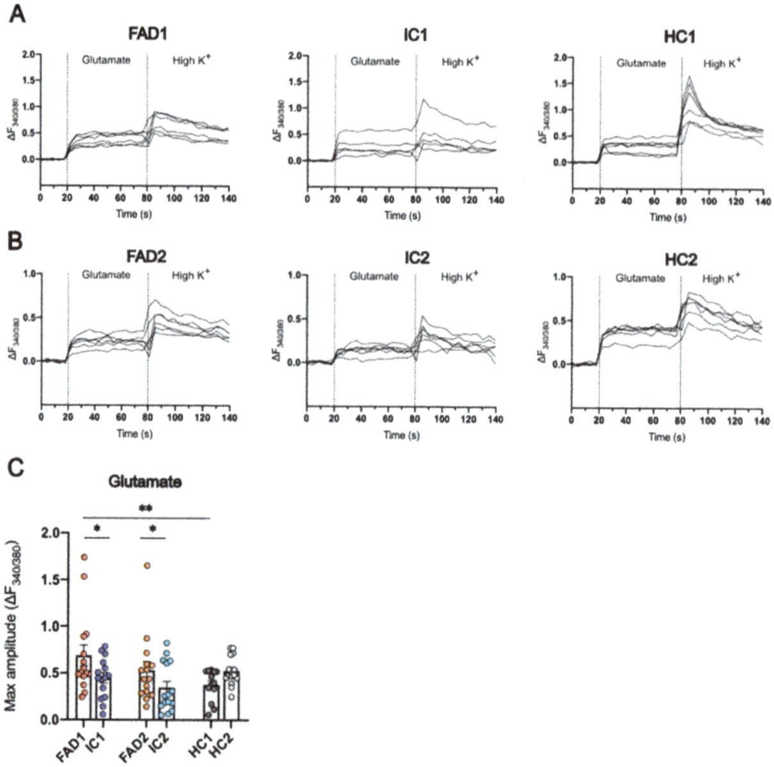

Figure 6. Ca^{2+} responses to glutamate are increased in FAD neurons. Representative traces of the change in fluorescence signal ratio over baseline ($\Delta F_{340/380}$) in response to 100 μM glutamate were recorded from FAD1/IC1/HC1 (**A**) and FAD2/IC2/HC2 neurons (**B**). The maximum Ca^{2+} response to 100 μM glutamate was calculated as the maximum amplitude of the Fura-2 AM fluorescence ratio change over baseline ($\Delta F_{340/380}$) (**C**). Day 35 neurons were loaded with the ratiometric dye Fura-2 AM, perfused with SBS, and treated with 100 μM glutamate, followed by 60 mM K^+ (High K^+). Data shown are single wells of a 96-well plate containing day 35 neurons; n = 8–15 replicates from three independent differentiations; error bars represent ± SEM; * $p < 0.05$, ** $p < 0.01$. Data were normalised to sqrt (FAD1/HC1/IC1) or log2 (FAD2/HC2/IC2) and analysed using linear regression modelling. Independent differentiations were included as co-variates. FAD—familial Alzheimer's disease; HC—healthy control; IC—isogenic control.

Analysis of Ca^{2+} signals following 100 μM kainate treatment revealed no significant differences between the FAD (FAD1 = 0.636 ± 0.111; FAD2 = 0.37 ± 0.037) lines and their isogenic controls (IC1 = 0.39 ± 0.069, $p = 0.128$; IC2 = 0.42 ± 0.051, $p = 0.446$). Additionally, FAD1 and HC1 (0.47 ± 0.067, $p = 0.604$) had no significant differences. Nonetheless, HC2 (0.601 ± 0.049, $p = 0.001$) neurons had greater Ca^{2+} responses to 100 μM kainate compared to FAD2 (0.37 ± 0.037) (Figure 7E).

Taken together, these data confirm that day 35 iPSC-derived neurons respond to excitatory stimulus and that FAD neurons display abnormally elevated Ca^{2+} signals when treated with glutamate and, more specifically, the AMPAR agonist AMPA, but not with NMDA or kainate. This suggests FAD neurons demonstrate a glutamatergic Ca^{2+} signalling phenotype, contributed (at least in part) by increased AMPAR signalling.

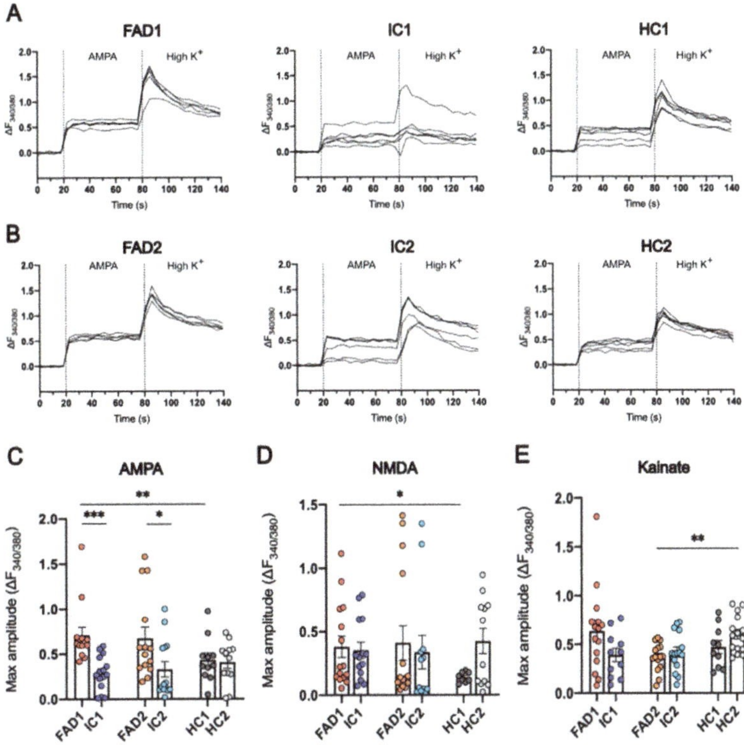

Figure 7. Ca^{2+} responses to AMPA are increased in FAD neurons. Representative traces of the change in fluorescence signal ratio over baseline ($\Delta F_{340/380}$) in response to 100 μM AMPA were recorded from FAD1/IC1/HC1 (**A**) and FAD2/IC2/HC2 neurons (**B**). The maximum Ca^{2+} response to 100 μM AMPA (**C**), 100 μM NMDA (**D**), and 100 μM kainate (**E**) was calculated as the maximum amplitude of the Fura-2 AM fluorescence ratio change over baseline ($\Delta F_{340/380}$) (**C**). Day 35 neurons were loaded with the ratiometric dye Fura-2 AM, perfused with SBS, and treated with either 100 μM AMPA, 100 μM NMDA, or 100 μM kainate, followed by 60 mM K^+ (High K^+). Individual data points are from individual wells of a 96-well plate containing day 35 neurons; n = 8–15 replicates from three independent differentiations; error bars represent ± SEM; *** $p< 0.001$ (AMPA: FAD1 vs. IC1); ** $p< 0.01$ (AMPA: FAD1 vs. HC1; Kainate: FAD2 vs. HC2); * $p< 0.05$ (AMPA: FAD2 vs. IC2; NMDA: FAD1 vs. HC1). Data are normally distributed (FAD2/HC2/IC2 kainate) or normalised to log2 (FAD1/HC1/IC1 kainate and NMDA and FAD2/HC2/IC2 NMDA) or sqrt (FAD1/HC1/IC1 and FAD2/HC2/IC2 AMPA) and analysed using linear regression modelling. Independent differentiations were included as co-variates. FAD—familial Alzheimer's disease; HC—healthy control; IC—isogenic control.

3.8. Regulation at the Level of mRNA or Protein of the AMPA Receptor Subunits Does Not Explain the Increased Calcium Responses to AMPA in FAD Neurons Compared to Isogenic Control Neurons

Since the FAD neurons exhibited significantly increased maximum Ca^{2+} responses to AMPA, we next investigated whether changes in mRNA or protein levels of the AMPAR subunits may be responsible for the increased signalling. To test this hypothesis, RT-qPCR was performed to investigate if differences observed in neuronal Ca^{2+} responses to AMPA between FAD and isogenic control neurons were due to differences in mRNA expression of AMPAR subunits 1–4 encoded by the genes *GRIA1*, *GRIA2*, *GRIA3*, and *GRIA4* (Figure 8A–D).

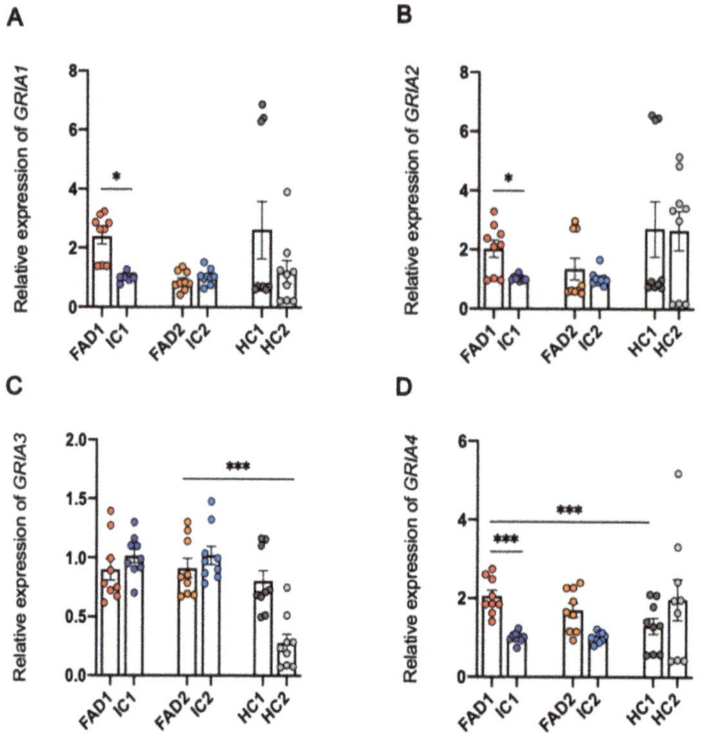

Figure 8. AMPAR subunits mRNA expression of FAD and control neurons. RT-qPCR analysis of GRIA1 (**A**), GRIA2 (**B**), GRIA3 (**C**), and GRIA4 (**D**) mRNA expression of day 35 neurons normalised to housekeeper genes GAPDH, HPRT1, and B2M. Data shown are replicates from one experiment using cDNA collected from three independent differentiations run in triplicate; error bars represent ± SEM; * $p < 0.05$, *** $p < 0.001$. Data are normally distributed (FAD1/HC1/IC1 GRIA3 and GRIA4 and FAD2/HC2/IC2 GRIA3) or normalised to log2 (FAD1/HC1/IC1 GRIA1 and FAD2/HC2/IC2 GRIA1, GRIA2 and GRIA4) or 1/x (FAD1/HC1/IC1 GRIA2) and analysed using linear regression modelling. Independent differentiations were included as co-variates. FAD—familial Alzheimer's disease; HC—healthy control; IC—isogenic control.

Expression of *GRIA1* mRNA was significantly higher in FAD1 (2.386 ± 0.257) than IC1 (1.007 ± 0.43, $p = 0.011$) in day 35 neurons, while there was no significant difference between FAD2 (0.888 ± 0.105) and IC2 (1.03 ± 0.087; $p = 0.527$). FAD1 and FAD2 *GRIA1* mRNA expression showed no significant differences from HC1 (2.622 ± 0.977; $p = 0.132$) and HC2 (1.197 ± 0.389, $p = 0.795$), respectively.

Similar results were obtained from the analysis of *GRIA2* gene expression. FAD1 (2.024 ± 0.284) had significantly greater mRNA levels of *GRIA2*, compared to IC1 (1.028 ± 0.028; $p = 0.027$), but was not significantly different from HC1 (2.712 ± 0.941; $p = 0.116$). Although FAD2 (1.353 ± 0.364) had a higher mean mRNA expression of *GRIA2* compared to IC2 (1.039 ± 0.084; $p = 0.975$), this difference did not reach statistical significance. FAD2 was also not significantly different from HC2 (2.649 ± 0.66, $p = 0.437$).

Regarding the relative expression of *GRIA3* mRNA, the only significant difference observed was between FAD2 (0.911 ± 0.082) and HC2 (0.277 ± 0.075, $p = 2.34 \times 10^{-6}$), with FAD2 neurons displaying higher expression of *GRIA3*. No significant differences were detected between FAD1 (0.899 ± 0.09) and IC1 (1.014 ± 0.058; $p = 0.24$) or HC1 (0.804 ± 0.089, $p = 0.325$) or between FAD2 and IC2 (1.021 ± 0.077, $p = 0.293$).

Lastly, RT-qPCR analysis of *GRIA4* showed FAD1 (2.056 ± 0.153) mRNA levels were significantly higher than both IC1 (1.009 ± 0.045; $p = 3.38 \times 10^{-7}$) and HC1 (1.303 ± 0.207, $p = 3.85 \times 10^{-5}$). For FAD2 (1.708 ± 0.182), even though there was a trend towards higher mRNA than IC2 (1.007 ± 0.043, $p = 0.065$), this difference was not statistically significant. There was no significant difference between FAD2 and HC2 (1.977 ± 0.522; $p = 0.563$).

Altogether, these results showed FAD1 iPSC-derived day 35 neurons had increased *GRIA1, GRIA2,* and *GRIA4* mRNA expression compared to IC1, but there was no difference in *GRIA3* expression. Furthermore, no significant differences between FAD2 and IC2 were observed in *GRIA1, GRIA2, GRIA3,* or *GRIA4 mRNA* expression, suggesting that transcriptional regulation of the GRIA genes cannot explain the increased calcium responses to AMPA.

Protein Expression of GluA1 and GluA2 Is Not Significantly Different between FAD and Control Neurons

Following the observed elevation in Ca^{2+} responses of FAD neurons compared to their IC neurons when treated with both glutamate and AMPA, it was hypothesised that these differences could be due to a difference in AMPAR subunit protein expression. We focused on GluA1 and GluA2 as the major AMPAR subunits in excitatory neurons. To investigate this hypothesis, total protein extraction from whole cell lysates of day 35 neurons was prepared for western blotting assays.

Western blots for all cell lines detected a strong band of approximately 100 kDa for both GluA1 and GluA2 (Figure 9A,B and Supplementary Figures S8–S11), present at their expected molecular weight. For some cell lines, a band of higher molecular weight, approximately 250 kDa, was also observed, which was not included in the quantification. Densitometry values of the bands at 100 kDa were normalised to respective total protein signals and a mean pooled sample, which were used to account for variability in protein loading and to normalise between separate blots, respectively. Linear regression analysis revealed no significant differences in GluA1 or GluA2 protein expression between FAD (FAD1—0.802 ± 0.148; FAD2—0.46 ± 0.066) and control cell lines IC (IC1—1.1 ± 0.35, $p = 0.515$; IC2—0.733 ± 0.226, $p = 0.25$) and HC (HC1—1.261 ± 0.36; $p = 0.4$; HC2—0.626 ± 0.152, $p = 0.479$) (Figure 9C,D).

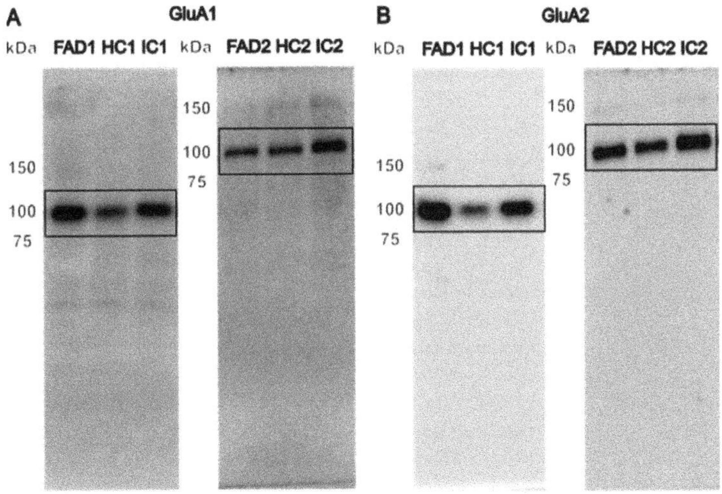

Figure 9. *Cont.*

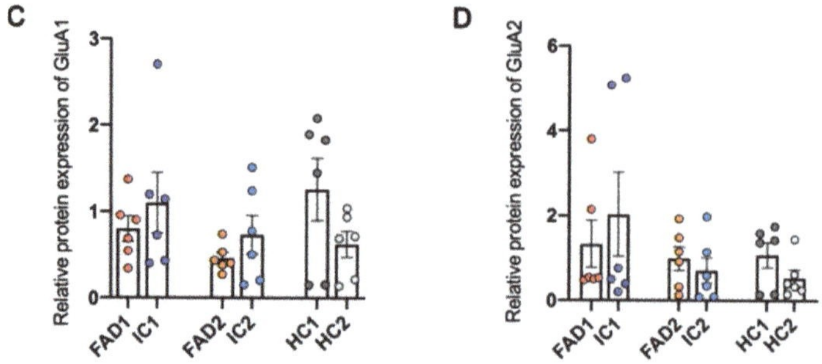

Figure 9. GluA1 and GluA2 protein expression of FAD and control neurons. Representative western blots (**A**,**B**) and relative protein expression of GluA1 (**C**) and GluA2 (**D**). Data shown are day 35 neurons; n = 2 replicates from 3 independent differentiations, normalised to total protein; error bars represent ± SEM. Data are normally distributed (FAD2/HC2/IC2 GluA1) or normalised to log2 (FAD1/HC1/IC1 and FAD2/HC2/IC2 GluA2) or sqrt (FAD1/HC1/IC1 GluA1) and analysed using linear regression modelling. Independent differentiations were included as co-variates; there were no significant differences between groups. FAD—familial Alzheimer's disease; HC—healthy control; IC—isogenic control.

These results indicate that the alterations in AMPAR Ca^{2+} signalling in FAD and control iPSC-derived neurons differentiated for 35 days were not due to differences in protein expression of GluA1 and GluA2 AMPAR subunits.

4. Discussion

Glutamate and calcium dyshomeostasis have been documented in Alzheimer's disease; however, it has been difficult to determine whether or not calcium and glutamate phenotypes precede amyloid and tau phenotypes in humans. To address this question in a 2D human cell model of the early stages of the disease, this study aimed to generate iPSC-derived neurons from FAD patients, isogenic controls, and healthy controls and measure Ca^{2+} responses following the application of glutamate, NMDA, AMPA, and kainate. The results obtained in this study showed that day 35 FAD iPSC-derived neurons did not demonstrate evidence of AD-associated Aβ and tau pathology and yet displayed altered Ca^{2+} responses. Neurons from FAD patients had elevated Ca^{2+} responses to both glutamate and AMPA but not to kainate and NMDA when compared to their isogenic controls. Together, the data suggest that mutations in *PSEN1* cause increased Ca^{2+} responses to AMPA as an early phenotype of AD.

4.1. FAD Neurons Lacking Aβ and Tau Pathology Show Elevated Ca^{2+} Responses to Glutamate and AMPA Compared to Isogenic Controls

In AD, cortical brain regions are severely affected by both Aβ and tau pathology and, eventually, neuronal loss. Hence, cortical glutamatergic neurons represent an appropriate model to study FAD neuronal calcium responses to excitatory stimuli.

The impact of FAD-causing mutations in the PSEN1 gene, $PSEN1^{S290C}$ and $PSEN1^{A246E}$, was investigated using iPSC-derived neurons from FAD patients as well as isogenic and healthy controls. Although the $PSEN1^{S290C}$ mutation results in the deletion of exon 9, preventing the endoproteolysis of PSEN1, $PSEN1^{S290C}$ is still generally considered to be functionally active, leading to an increased $A\beta_{42}/A\beta_{40}$ peptide ratio [52,53], similar to $PSEN1^{A246E}$ [52,54]. This has been suggested to occur due to the ability of $PSEN1^{S290C}$ to adopt a conformation similar to the heterodimeric active PSEN1 as opposed to the inactive PSEN1 holoprotein [55]. Furthermore, the pathologic function of $PSEN1^{S290C}$ in generating elevated $A\beta_{42}$ levels is believed to be independent of its inability to undergo endoproteolysis but rather due to the introduction of the point mutation (S290C), which occurs as a result of the exon 9 deletion [53]. Hence, although PSEN1^{S290C} is structurally different from PSEN1^{A246E}, it is not clear whether this results in different AD-associated phenotypes.

In this study, iPSC-derived neurons were differentiated for a total of 35 days. Changes in cell morphology from iPSCs into neurons and expression of the fluorescent reporter GFP (co-expressed with NGN2) could be observed from day 1 of neuronal differentiation. Immunocytochemistry results from this study confirmed the cell cultures generated expressed the neuronal marker MAP2, while no cells positive for the astrocytic marker GFAP were observed.

The presence of amyloid-β plaques and neurofibrillary tangles in human *postmortem* brain tissue provides the criteria for the neuropathologic diagnosis of AD [56]. Both plasma and cerebral spinal fluid levels of Aβ are decreased in AD and are inversely associated with plaque burden [43,44]. Since Aβ plaques cannot be formed in 2D neuronal cultures due to the frequent media changes, the levels of Aβ peptides released into the cell medium are commonly used as a measurement of Aβ pathology. Several long-term 2D cell culture models of AD have detected elevated Aβ42/Aβ40 ratios [36,47–50] or increased secreted Aβ42 in the media [47], while others have found no differences [36]. This discrepancy may be explained by the use of different protocols to generate neuronal subtypes or by the length of time that neurons are cultured/matured for. Another important contributing factor is the use of cell lines harbouring different FAD mutations or sporadic AD lines bearing different genetic risk factors and their comparison with either healthy or isogenic controls. Statistical analysis of Aβ42 levels for day 35 iPSC-derived neurons generated for this work showed FAD2 neurons had higher levels of Aβ42 than a healthy control line HC2 but, critically, was not statistically different from its isogenic control IC2. Since healthy controls have an entirely different genome from a familial disease cell line, differences found between these cells may be biased by factors other than the disease-associated mutation and, therefore, complicate the interpretation of the results. Isogenic controls, on the other hand, should have the same genome as their parental cell line, except for the disease-causing mutation that is reverted to WT, making it a more useful comparison to understand the contribution of specific mutations to disease phenotypes. Thus, we expect to observe differences in responses between unrelated lines. Nevertheless, the inclusion of additional healthy controls is helpful to assess variations in responses. Apart from the different types of controls used, the presence of glial cells in the culture can also affect the Aβ phenotype, as they are involved in the clearance of this peptide [57]. Although our iPSC-derived neuronal model did not show any GFAP+ cells after immunostaining analysis, it cannot be ruled out that GFAP-negative astrocytic or other glial populations may be present in the culture and impact Aβ processing.

The microtubule-associated protein tau is enriched in the axons of mature neurons; however, in pathological conditions such as AD, it is hyperphosphorylated and accumulates in the cell soma and dendrites, where it forms insoluble aggregates and neurofibrillary tangles. A total of 59 tau phosphorylation sites have been detected in human *postmortem* brain tissue of AD patients [58,59], and serine 404 (p-Tau S404) is considered one of the first epitopes to be phosphorylated in the disease [51]. Tau phosphorylation at different epitopes has been previously replicated in cell models of AD, including p-Tau S404 [48], hence its inclusion as a measurement for early stages of tau pathology. However, statistical analysis revealed no significant differences in the levels of p-Tau S404 between FAD and control cell lines in day 35 neurons. Although a combination of factors, such as lipid metabolism, endocytosis, and the immune response, contribute to pathologic tau accumulation [60], elevated Aβ levels are also believed to accelerate tau pathology [61]. Given that an Aβ pathology phenotype at this stage of neuronal maturation was not observed in the FAD neurons, this may explain the absence of a tau phenotype as well. Therefore, neither Aβ accumulation nor increased tau phosphorylation were detected in day 35 FAD iPSC-derived neurons in this study. Thus, this model permits the study of early FAD phenotypes associated with *PSEN1* mutations in human cells.

Another phenotype of AD is neuronal hyperexcitability (increased excitability), which is thought to precede neuronal hypoexcitability (reduced excitability) and cell death [32]. This elevated neuronal activity is thought to promote neurodegeneration in specific neurons [30] and correlate with cognitive decline. Numerous abnormalities have been suggested to contribute to the change in neuronal excitability, including calcium and glutamate dyshomeostasis [32]. To understand whether there is a calcium phenotype in early FAD neurons, it was first confirmed that all cell lines responded to high K^+. HC2 neurons demonstrated greater responses to high K^+ than FAD2 neurons, suggesting HC2 may have a greater density of K^+ channels compared to FAD2 cells. This could impact neuronal differentiation and excitability, key functions of K^+ channels [62,63]. Nevertheless, no significant differences were found between the isogenic control IC2 and FAD2 neurons, suggesting the differences observed between HC2 and FAD2 lines originate from genetic differences between the cell lines unrelated to the FAD mutation. Then, to investigate how neurons respond to excitatory stimuli, the experiments aimed to measure neuronal calcium signals after treatment with glutamate and the iGluR agonists, NMDA, AMPA, and kainate. Analysis of the maximum amplitude of the calcium response to each drug revealed FAD neurons had significantly higher calcium responses to both glutamate and AMPA compared to their isogenic controls. No significant differences were observed between the disease and isogenic controls following treatment with NMDA or kainate. Although this work focused on iGluR agonists, the differences in Ca^{2+} responses to glutamate could also stem from changes in mGluR, previously shown to have altered expression and/or activity in AD [64]. Hence, future work is crucial to elucidate the contribution of mGluR to Ca^{2+} dyshomeostasis in the model studied here.

Both Aβ and tau contribute to calcium dyshomeostasis and AMPAR trafficking dysfunction, resulting in a reduction in AMPAR expression and function and defective synaptic plasticity, as reviewed in [65,66]. Nevertheless, the data identified an aberrant calcium signalling phenotype that appears to occur independently of Aβ or tau phenotypes. Consequently, in the absence of evidence for Aβ and early phosphorylated tau S404 changes, alternative mechanisms need to be considered. Notch-1 expression is altered in neurons bearing these same *PSEN1* mutations [67]. Notch signalling is implicated in stem cell renewal, proliferation, and differentiation, as well as neuronal development [68]. Thus, reduced Notch-1 expression in FAD neurons could lead to differential regulation of AMPAR expression in FAD, control iPSC-derived neurons, and explain the phenotypes observed here.

4.2. *Aberrant Ca^{2+} Signalling of FAD Neurons Occurs Independently of Changes in GluA1 and GluA2 Protein Expression*

While the data failed to demonstrate alterations in GluA1 and GluA2 protein levels in FAD compared to isogenic control neurons, there were some increases in *GRIA1*, *GRIA2*, and *GRIA4* at the mRNA level. The transcriptional regulatory mechanisms were not assessed since the transcriptional up-regulation did not lead to concomitant increases in protein levels. AMPAR are tetrameric assemblies of GluA1-GluA4 subunits that permit Na^+ and Ca^{2+} influx into the cells. Ca^{2+}-permeable AMPAR is important for long-term potentiation (LTP) induction and long-term memory formation [69]. The permeability of AMPARs to Ca^{2+} depends on subunit composition, and GluA2 subunit permeability to Ca^{2+} is regulated by RNA editing. The post-transcriptional modification of *GRIA2* RNA from a codon encoding glutamine (Q) to a codon encoding arginine (R) renders this subunit impermeable to Ca^{2+}. Thus, AMPARs lacking GluA2 or containing the unedited version of GluA2 are Ca^{2+} permeable [70]. The unedited *GRIA2* mRNA has very low expression in the adult human brain, comprising less than 10% of *GRIA2* in the white matter and less than 1% in the grey matter [71]. In spite of its low expression, unedited GluA2 can contribute to synaptic plasticity and excitotoxic neuronal cell death, as reviewed in [72] Interestingly, lower RNA editing of GluA2 has been reported in AD brains [73–75]. Although RT-qPCR of *GRIA2* performed in this study is insufficient to discern edited from unedited *GRIA2*, previous studies have employed different strategies, such as restriction enzyme digestion and Sanger sequencing of PCR products [76,77], to quantify unedited *GRIA2* mRNA. Future work utilising these assays could inform whether changes in the amount of edited or unedited *GRIA2* occur in FAD neurons.

The expression of AMPAR subunits varies across regions of the mammalian brain. In murine hippocampal synapses, the majority of AMPARs are comprised of GluA1/2 subunits, followed by a smaller fraction of GluA2/3 heterodimers [78,79]. In the adult rat cortex, GluA1 is the predominant subunit (~45% of total AMPAR), followed by GluA2 (21%), GluA3 (27%), and very low levels of GluA4 (less than 6%) [13]. Several studies have shown GluA1 and GluA2 protein expression is decreased in AD *postmortem* brain tissue, as summarised in [29]. Lower expression of GluA1 has been reported in the frontal cortex [80–82] and cerebellum [83] of AD brains, while no differences were observed in the temporal cortex compared to healthy individuals [69], suggesting there is a brain region-specific phenotype. Regarding GluA2, this AMPAR subunit has been the most extensively studied in AD. A recent compilation of 38 proteomic studies of AD reported a total of 12 studies where GluA2 protein expression was reduced in the frontal, entorhinal, and parahippocampal cortex, as well as the hippocampus and precuneus brain regions of AD brains [29]. Most of the studies (7 out of 12) found these changes in the frontal cortex. Nonetheless, contrasting results have been reported by a couple of studies that detected an upregulation of GluA2 protein expression in the temporal cortex and hippocampus of AD patients [69,84]. The work conducted by Barbour and colleagues [69] is the only study to examine GluA2 expression in the temporal cortex, indicating GluA2 could be differentially regulated in specific brain regions; however, more studies analysing this area of the brain are required to confirm these results. While Stepler et al. [85] found decreased levels of GluA2 in tissue homogenates of the whole hippocampus, Yeung et al. [84] found GluA2 expression was elevated in a subregion of the hippocampus, the stratum moleculare layer of the dentate gyrus, but no significant changes were detected in the stratum granulosum and hilus areas of the dentate gyrus, or in any of the CA1, CA2, or CA3 regions of the hippocampus. Also, no changes were observed in the superior temporal gyrus, subiculum, and entorhinal cortex of AD versus control donors. Thus, a potential explanation for these discrepancies is that GluA2 expression is differentially altered in specific subregions of the hippocampus. Furthermore, although the *postmortem* brain tissue used by Yeung et al. [84] was obtained from a brain bank located in New Zealand, the ethnicity of their cohort was not specified. In contrast, Stepler et al. [85] analysed brains from African American and non-Hispanic White individuals and identified 185 proteins differentially expressed in these two groups. African American and Hispanic populations have a greater risk of developing AD and non-AD dementia than non-Hispanic white adults; however, these groups are commonly underrepresented in proteomic studies of AD. Even though GluA4 was not differentially expressed between the two ethnicity groups included in this study, these results highlight the need for more research, including ethnically diverse cohorts. This will allow for a better understanding of variabilities in disease-associated phenotypes among different populations so that future therapies can target various ethnic groups.

Overall, the work of Askenazi et al. (2023) [29], compiling studies on AD proteome alterations, postulates that expression of GluA1 and GluA2 is reduced in later stages of AD. Interestingly, this phenotype is not observed in studies of the early stages of the disease, including our study, in which western blot analysis showed no changes in total cell protein expression of GluA1 and GluA2. Since the AD model presented here represents an early stage of the disease, preceding Aβ and tau phenotypes, it can be speculated that abnormalities in AMPAR protein levels have not yet developed at this stage.

In this study, protein expression of GluA3 and GluA4 in iPSC-derived neurons was not analysed, but changes in their expression have been reported in AD *postmortem* brain tissue. GluA3 and GluA4 follow a similar pattern of protein expression as those observed for GluA1 and GluA2. Even though no changes were detected in the early stages of the disease, lower levels of GluA3 were found in the hippocampus, precuneus, frontal, and entorhinal cortex in the late stages of AD pathology [29]. Despite the low number of studies on GluA4, decreased expression of this subunit has been reported in the frontal and entorhinal cortex of advanced stages of AD [86,87]. Thus, future work measuring GluA3 and GluA4 protein levels in day 35 iPSC-derived neurons is required to fully characterise the protein expression of AMPAR subunits in this model. In summary, *postmortem* brain tissue of AD patients shows reduced AMPAR expression, but these changes are only observed in the late/end stages of the disease. Likewise, the results from this study found no differences in total protein levels of GluA1 and GluA2 subunits in FAD iPSC-derived neurons representing early stages of pathology compared to isogenic controls. This suggests the mechanism leading to changes in how FAD neurons respond to AMPA, compared to controls, is not explained by alterations in the total cellular protein expression of these two subunits. Nonetheless, analysis of total cellular protein expression may mask changes in protein expression in specific cellular compartments. The expression of AMPARs in the plasma membrane, specifically, is crucial for receptor function and can be regulated through post-translational modifications. Reversible post-translational modifications, such as phosphorylation, palmitoylation, and ubiquitination, affect receptor subunit exocytosis, endocytosis, degradation, and gating [88]. Phosphorylation of GluA1 regulates synaptic plasticity by potentiating AMPAR responses to glutamate [89] and increasing channel conductance [90] and open probability [91]. While phosphorylation of GluA1 at epitopes S818, S831, and S845 promotes synaptic insertion of AMPAR and dephosphorylation causes endocytosis, dephosphorylation of GluA1 at S567 increases receptor expression at the synapse [88]. GluA2 phosphorylation, on the other hand, is required for AMPAR internalisation in the CA1 region of the hippocampus both in vitro and in vivo [92,93].

Ubiquitination comprises the attachment of a single ubiquitin or polymeric ubiquitin chains to the lysine residues of a substrate protein [94]. All AMPAR subunits can undergo ubiquitination when treated with AMPA [95]. This post-translational modification is calcium-dependent [95,96] and only occurs in receptors present in the plasma membrane [95]. The ubiquitination of AMPAR signals for lysosomal [96,97] or proteasomal degradation [98] of this receptor, and most studies report it also regulates endocytosis [96,98,99]. Importantly, ubiquitination of AMPAR also modulates synaptic transmission [96,98–100]. In rat neuronal cultures, AMPA treatment increases the number of internalised GluA1. This phenotype is abolished in GluA1 mutants lacking ubiquitination sites. Instead, mutant GluA1 shows reduced GluA1 degradation and internalisation and increased GluA1 expression at the cell surface [96,98]. Finally, human *postmortem* brain tissue of AD patients showed increased expression of ubiquitinated GluA1 protein, suggesting ubiquitination of AMPAR subunits may play an important role in modulating plasma membrane expression and function. Proteins can also be modified by being covalently bound to lipids, such as fatty acids. This process is termed fatty acylation and includes the post-translational modification palmitoylation, which is the addition of the 16-carbon saturated fatty acid palmitate to one or more intracellular cysteine residues of target proteins. All four AMPAR subunits can be palmitoylated, and they are differentially regulated depending on the site of palmitoylation [101]. In HEK293T cells and primary cortical neurons, palmitoylation of cysteines in the transmembrane domain 2 of GluA1 or GluA2 accumulates these subunits in the Golgi and reduces their surface expression. C-terminal palmitoylation of AMPAR, on the other hand, does not influence steady-state surface expression of this receptor but increases its internalisation after NMDA or AMPA stimulation [101,102]. A knock-in GluA1 C811 palmitoylation-deficient mouse model (GluA1C811S) showed elevated expression of GluA1 in the cortex [103]. Altogether, these studies demonstrate that various post-translational changes to AMPARs influence their trafficking and consequent membrane expression and receptor function. Hence, future work needs to be conducted to measure the expression of all AMPAR subunits intracellularly and in the plasma membrane and to measure post-translational changes that may regulate their trafficking to and from the membrane.

Apart from AMPAR regulation via post-translational modifications, the composition of the lipid membrane, where AMPARs are embedded, provides another layer of ion channel modulation. Presynaptic and postsynaptic membranes are enriched in cholesterol, a sterol lipid that has been linked to AD. Although cholesterol does not affect AMPA binding to its receptor [104], it can still modulate AMPAR activity. Analysis of intracellular calcium levels in cultures of rat hippocampal neurons demonstrated that cholesterol depletion decreased AMPAR-mediated calcium influx [105]. In rat hippocampal slices, cholesterol depletion reduced both the amplitude of AMPAR-mediated excitatory postsynaptic currents (EPSCs) [106] and basal synaptic transmission [105]. Using the same model, it was shown that AMPA treatment potentiated basal synaptic transmission in both normal and cholesterol-depleted conditions. However, in the absence of cholesterol, this potentiation lasted for a shorter period, and the responses to AMPA were not fully abolished after washing out the drug, as was the case for control slices [105]. Nevertheless, contrasting results were found in rat cerebellar granule cells and mouse cortical neurons, where cholesterol depletion reduced NMDA-evoked currents but had no effect on AMPAR-mediated currents [107,108]. This divergence could possibly be explained by the different cell types analysed, the concentration and duration of the stimulus, or the AMPAR agonist utilised by each study. Hence, further investigation is required to understand the effect of neuronal membrane cholesterol content on AMPAR function.

The poly-unsaturated fatty acid arachidonic acid (AA) has also been implicated in neuronal excitability. Mouse brain slices expressing human APP (hAPP) showed higher surface expression of GluA1 and GluA2 subunits of AMPARs when treated with AA, which resulted in increased neuronal activity. AA is generated from the hydrolysis of phospholipids by phospholipase-A2 (PLA_2), an enzyme that was hyperphosphorylated in hAPP animals, suggesting elevated activation [109]. To confirm this hypothesis, the authors blocked PLA_2 activity, which prevented the increase in AMPAR protein expression as well as neuronal hyperactivity. PLA_2 further acts in synaptic transmission by increasing AMPA affinity, binding to its receptor [110], and modulating AMPAR phosphorylation [111]. In rat brain slices, inhibition of calcium-independent PLA_2 increased GluA1 phosphorylation at residue S831, while inhibition of calcium-dependent PLA_2 enhanced phosphorylation of GluA2/3 at S880/891 [111]. Although the authors did not investigate the impact of AMPAR phosphorylation on neuronal activity, it is possible that it could affect AMPAR trafficking and expression.

5. Future Directions

In this study, iPSC-derived neurons from FAD patients bearing *PSEN1* mutations showed greater calcium responses to glutamate and AMPA than isogenic control neurons. This occurred in the absence of overt Aβ42 and tau S404 phosphorylation phenotypes or changes in total cellular protein expression of GluA1 and GluA2 subunits of AMPAR. Future work should measure total cellular protein expression of GluA3 and GluA4 subunits, as well as expression of all AMPAR subunits in the neuronal membrane and intracellular compartments separately. Trafficking of AMPARs to the membrane can be regulated by post-translational modifications such as phosphorylation, ubiquitylation, and palmitoylation; thus, investigating levels of these post-translational modifications to AMPAR subunits and the enzymes involved in these pathways could demonstrate the mechanisms promoting early calcium signalling dysfunction in FAD neurons.

6. Conclusions

The iPSC-derived cortical excitatory neurons from FAD patients display elevated calcium responses to glutamate and AMPA compared to their isogenic controls. This early calcium phenotype was observed in the absence of an Aβ or tau phenotype. Although mRNA levels of the AMPAR subunits *GRIA1*, *GRIA2*, and *GRIA4* were increased in FAD1 neurons compared to IC1 neurons, this difference was not significant between FAD2 and IC2 neurons. Furthermore, protein expression of the AMPAR subunits GluA1 and GluA2 was not significantly different between FAD and isogenic control neurons. Hence, the difference in how FAD neurons respond to AMPA cannot be explained by changes in total cellular protein expression of AMPAR subunits GluA1 or GluA2.

Supplementary Materials: The following supporting information can be downloaded at: https://www.mdpi.com/article/10.3390/life14050625/s1, Figure S1: Representative images and timeline of differentiation, Figure S2: Representative images of no primary antibody control for immunostaining from FAD1, IC1 and HC1, Figure S3: Representative images of no primary antibody control for immunostaining from FAD2, IC2 and HC2, Figure S4: Sample description and images of full membranes for Tau-5 antibody and total protein for FAD1/HC1/IC1, Figure S5: Sample description and images of full membranes for Tau-5 antibody and total protein for FAD2/HC2/IC2, Figure S6: Sample description and images of full membranes for p-Tau-s404 antibody and total protein for FAD1/HC1/IC1, Figure S7: Sample description and images of full membranes for p-Tau-s404 antibody and total protein for FAD2/HC2/IC2, Figure S8: Sample description and images of full membranes for GluA1 antibody and total protein for FAD1/HC1/IC1, Figure S9: Sample description and images of full membranes for GluA1 antibody and total protein for FAD2/HC2/IC2, Figure S10: Sample description and images of full membranes for GluA2 antibody and total protein for FAD1/HC1/IC1, Figure S11: Sample description and images of full membranes for GluA2 antibody and total protein for FAD2/HC2/IC2.

Author Contributions: Conceptualization, H.T.D.A. and L.O.; methodology, H.T.D.A. and N.M.; formal analysis, H.T.D.A.; resources, L.O.; data curation, H.T.D.A.; writing—original draft preparation, H.T.D.A.; writing—review and editing, H.T.D.A., N.M., and L.O.; supervision, N.M. and L.O.; project administration, L.O.; funding acquisition, L.O. All authors have read and agreed to the published version of the manuscript.

Funding: This research was funded by a National Health and Medical Research Council of Australia (NHMRC) Boosting Dementia Research Leadership Fellowship awarded to L.O., grant number APP1135720.

Institutional Review Board Statement: The study was conducted in accordance with the Declaration of Helsinki and approved by the University of Wollongong Human Research Ethics Committee (protocol code 13/299, 2013).

Informed Consent Statement: Informed consent was obtained from all subjects involved in the study.

Data Availability Statement: Data are available upon request.

Acknowledgments: The authors wish to acknowledge the participants, without whom this research would not be possible.

Conflicts of Interest: The authors declare no conflicts of interest.

References

1. Bird, T.D. Alzheimer Disease Overview. In *GeneReviews® [Internet]*; University of Washington: Seattle, WA, USA, 1998.
2. Blacker, D.; Tanzi, R.E. The Genetics of Alzheimer Disease: Current Status and Future Prospects. *Arch. Neurol.* **1998**, *55*, 294–296. [CrossRef] [PubMed]
3. Glenner, G.G.; Wong, C.W. Alzheimer's Disease: Initial Report of the Purification and Characterization of a Novel Cerebrovascular Amyloid Protein. *Biochem. Biophys. Res. Commun.* **1984**, *120*, 885–890. [CrossRef] [PubMed]
4. Brion, J.P.; Couck, A.M.; Passareiro, E.; Flament-Durand, J. Neurofibrillary Tangles of Alzheimer's Disease: An Immunohistochemical Study. *J. Submicrosc. Cytol.* **1985**, *17*, 89–96. [PubMed]
5. Ge, M.; Zhang, J.; Chen, S.; Huang, Y.; Chen, W.; He, L.; Zhang, Y. Role of Calcium Homeostasis in Alzheimer's Disease. *Neuropsychiatr. Dis. Treat.* **2022**, *18*, 487–498. [CrossRef]
6. Bezprozvanny, I.; Mattson, M.P. Neuronal Calcium Mishandling and the Pathogenesis of Alzheimer's Disease. *Trends Neurosci.* **2008**, *31*, 454–463. [CrossRef] [PubMed]
7. Zhang, H.; Sun, S.; Herreman, A.; De Strooper, B.; Bezprozvanny, I. Role of Presenilins in Neuronal Calcium Homeostasis. *J. Neurosci.* **2010**, *30*, 8566–8580. [CrossRef] [PubMed]
8. Tu, H.; Nelson, O.; Bezprozvanny, A.; Wang, Z.; Lee, S.F.; Hao, Y.H.; Serneels, L.; De Strooper, B.; Yu, G.; Bezprozvanny, I. Presenilins form ER Ca^{2+} Leak Channels, a Function Disrupted by Familial Alzheimer's Disease-Linked Mutations. *Cell* **2006**, *126*, 981–993. [CrossRef]
9. Green, K.N.; Demuro, A.; Akbari, Y.; Hitt, B.D.; Smith, I.F.; Parker, I.; LaFerla, F.M. SERCA Pump Activity Is Physiologically Regulated by Presenilin and Regulates Amyloid β Production. *J. Cell Biol.* **2008**, *181*, 1107–1116. [CrossRef]

10. Cheung, K.-H.; Shineman, D.; Müller, M.; Cárdenas, C.; Mei, L.; Yang, J.; Tomita, T.; Iwatsubo, T.; Lee, V.M.-Y.; Foskett, J.K. Supplemental Data—Mechanism of Ca^{2+} Disruption in Alzheimer's Disease by Presenilin Regulation of InsP3 Receptor Channel Gating. *Neuron* **2008**, *58*, 871–883. [CrossRef]
11. Lee, J.H.; McBrayer, M.K.; Wolfe, D.M.; Haslett, L.J.; Kumar, A.; Sato, Y.; Lie, P.P.Y.; Mohan, P.; Coffey, E.E.; Kompella, U.; et al. Presenilin 1 Maintains Lysosomal Ca^{2+} Homeostasis by Regulating VATPase-Mediated Lysosome Acidification. *Cell Rep.* **2015**, *12*, 1430. [CrossRef]
12. Wisden, W.; Seeburg, P.H. Mammalian Ionotropic Glutamate Receptors. *Curr. Opin. Neurobiol.* **1993**, *3*, 291–298. [CrossRef]
13. Schwenk, J.; Baehrens, D.; Haupt, A.; Bildl, W.; Boudkkazi, S.; Roeper, J.; Fakler, B.; Schulte, U. Regional Diversity and Developmental Dynamics of the AMPA-Receptor Proteome in the Mammalian Brain. *Neuron* **2014**, *84*, 41–54. [CrossRef] [PubMed]
14. Hsieh, H.; Boehm, J.; Sato, C.; Iwatsubo, T.; Tomita, T.; Sisodia, S.; Malinow, R. AMPAR Removal Underlies Aβ-Induced Synaptic Depression and Dendritic Spine Loss. *Neuron* **2006**, *52*, 831–843. [CrossRef]
15. Whitcomb, D.J.; Hogg, E.L.; Regan, P.; Piers, T.; Narayan, P.; Whitehead, G.; Winters, B.L.; Kim, D.H.; Kim, E.; St George-Hyslop, P.; et al. Intracellular Oligomeric Amyloid-Beta Rapidly Regulates GluA1 Subunit of AMPA Receptor in the Hippocampus. *Sci. Rep.* **2015**, *5*, 10934. [CrossRef]
16. Hoover, B.R.; Reed, M.N.; Su, J.; Penrod, R.D.; Kotilinek, L.A.; Grant, M.K.; Pitstick, R.; Carlson, G.A.; Lanier, L.M.; Yuan, L.L.; et al. Tau Mislocalization to Dendritic Spines Mediates Synaptic Dysfunction Independently of Neurodegeneration. *Neuron* **2010**, *68*, 1067–1081. [CrossRef] [PubMed]
17. Paoletti, P. Molecular Basis of NMDA Receptor Functional Diversity. *Eur. J. Neurosci.* **2011**, *33*, 1351–1365. [CrossRef]
18. Sun, X.-Y.; Tuo, Q.-Z.; Liuyang, Z.-Y.; Xie, A.-J.; Feng, X.-L.; Yan, X.; Qiu, M.; Li, S.; Wang, X.-L.; Cao, F.-Y.; et al. Extrasynaptic NMDA Receptor-Induced Tau Overexpression Mediates Neuronal Death through Suppressing Survival Signaling ERK Phosphorylation. *Cell Death Dis.* **2016**, *7*, e2449. [CrossRef] [PubMed]
19. Lesné, S.; Ali, C.; Gabriel, C.; Croci, N.; MacKenzie, E.T.; Glabe, C.G.; Plotkine, M.; Marchand-Verrecchia, C.; Vivien, D.; Buisson, A. NMDA Receptor Activation Inhibits α-Secretase and Promotes Neuronal Amyloid-β Production. *J. Neurosci.* **2005**, *25*, 9367–9377. [CrossRef]
20. Bordji, K.; Becerril-Ortega, J.; Nicole, O.; Buisson, A. Activation of Extrasynaptic, But Not Synaptic, NMDA Receptors Modifies Amyloid Precursor Protein Expression Pattern and Increases Amyloid-β Production. *J. Neurosci.* **2010**, *30*, 15927–15942. [CrossRef]
21. Li, S.; Hong, S.; Shepardson, N.E.; Walsh, D.M.; Shankar, G.M.; Selkoe, D. Soluble Oligomers of Amyloid β Protein Facilitate Hippocampal Long-Term Depression by Disrupting Neuronal Glutamate Uptake. *Neuron* **2009**, *62*, 788–801. [CrossRef]
22. Arias, C.; Arrieta, I.; Tapia, R. B-Amyloid Peptide Fragment 25–35 Potentiates the Calcium-dependent Release of Excitatory Amino Acids from Depolarized Hippocampal Slices. *J. Neurosci. Res.* **1995**, *41*, 561–566. [CrossRef]
23. Parpura-Gill, A.; Beitz, D.; Uemura, E. The Inhibitory Effects of β-Amyloid on Glutamate and Glucose Uptakes by Cultured Astrocytes. *Brain Res.* **1997**, *754*, 65–71. [CrossRef] [PubMed]
24. Snyder, E.M.; Nong, Y.; Almeida, C.G.; Paul, S.; Moran, T.; Choi, E.Y.; Nairn, A.C.; Salter, M.W.; Lombroso, P.J.; Gouras, G.K.; et al. Regulation of NMDA Receptor Trafficking by Amyloid-β. *Nat. Neurosci.* **2005**, *8*, 1051–1058. [CrossRef]
25. Winslow, B.T.; Onysko, M.K.; Stob, C.M.; Hazlewood, K.A. Treatment of Alzheimer Disease. *Am. Fam. Physician* **2011**, *83*, 1403–1412. [CrossRef] [PubMed]
26. Falcón-Moya, R.; Sihra, T.S.; Rodríguez-Moreno, A. Kainate Receptors: Role in Epilepsy. *Front. Mol. Neurosci.* **2018**, *11*, 217. [CrossRef]
27. Barthet, G.; Moreira-De-Sá, A.; Zhang, P.; Deforges, S.; Castanheira, J.; Gorlewicz, A.; Mulle, C. Presenilin and APP Regulate Synaptic Kainate Receptors. *J. Neurosci.* **2022**, *42*, 9253–9262. [CrossRef]
28. Malenka, R.C.; Bear, M.F. LTP and LTD: An Embarrassment of Riches. *Neuron* **2004**, *44*, 5–21. [CrossRef]
29. Askenazi, M.; Kavanagh, T.; Pires, G.; Ueberheide, B.; Wisniewski, T.; Drummond, E. Compilation of All Known Protein Changes in the Human Alzheimer's Disease Brain. *bioRxiv* **2023**. [CrossRef]
30. Choi, D.W. Glutamate Neurotoxicity and Diseases of the Nervous System. *Neuron* **1988**, *1*, 623–634. [CrossRef] [PubMed]
31. Cox, M.F.; Hascup, E.R.; Bartke, A.; Hascup, K.N. Friend or Foe? Defining the Role of Glutamate in Aging and Alzheimer's Disease. *Front. Aging* **2022**, *3*, 65. [CrossRef]
32. Targa Dias Anastacio, H.; Matosin, N.; Ooi, L. Neuronal Hyperexcitability in Alzheimer's Disease: What Are the Drivers behind This Aberrant Phenotype? *Transl. Psychiatry* **2022**, *12*, 257. [CrossRef] [PubMed]
33. Celone, K.A.; Calhoun, V.D.; Dickerson, B.C.; Atri, A.; Chua, E.F.; Miller, S.L.; DePeau, K.; Rentz, D.M.; Selkoe, D.J.; Blacker, D.; et al. Alterations in Memory Networks in Mild Cognitive Impairment and Alzheimer's Disease: An Independent Component Analysis. *J. Neurosci.* **2006**, *26*, 10222–10231. [CrossRef] [PubMed]
34. Dickerson, B.C.; Salat, D.H.; Greve, D.N.; Chua, E.F.; Rand-Giovannetti, E.; Rentz, D.M.; Bertram, L.; Mullin, K.; Tanzi, R.E.; Blacker, D.; et al. Increased Hippocampal Activation in Mild Cognitive Impairment Compared to Normal Aging and AD. *Neurology* **2005**, *65*, 404–411. [CrossRef] [PubMed]
35. Balez, R.; Ooi, L. Getting to NO Alzheimer's Disease: Neuroprotection versus Neurotoxicity Mediated by Nitric Oxide. *Oxid. Med. Cell. Longev.* **2016**, *2016*, 3806157. [CrossRef]

36. Ghatak, S.; Dolatabadi, N.; Trudler, D.; Zhang, X.; Wu, Y.; Mohata, M.; Ambasudhan, R.; Talantova, M.; Lipton, S.A. Mechanisms of Hyperexcitability in Alzheimer's Disease HiPSC-Derived Neurons and Cerebral Organoids vs. Isogenic Control. *eLife* **2019**, *8*, e50333. [CrossRef] [PubMed]
37. Šišková, Z.; Justus, D.; Kaneko, H.; Friedrichs, D.; Henneberg, N.; Beutel, T.; Pitsch, J.; Schoch, S.; Becker, A.; VonderKammer, H.; et al. Dendritic Structural Degeneration Is Functionally Linked to Cellular Hyperexcitability in a Mouse Model of Alzheimer's Disease. *Neuron* **2014**, *84*, 1023–1033. [CrossRef]
38. Lerdkrai, C.; Asavapanumas, N.; Brawek, B.; Kovalchuk, Y.; Mojtahedi, N.; Del Moral, M.O.; Garaschuk, O. Intracellular Ca^{2+} Stores Control in Vivo Neuronal Hyperactivity in a Mouse Model of Alzheimer's Disease. *Proc. Natl. Acad. Sci. USA* **2018**, *115*, E1279–E1288. [CrossRef] [PubMed]
39. Maksour, S.; Finol-Urdaneta, R.K.; Hulme, A.J.; Castro Cabral-da-Silva, M.; Targa Dias Anastacio, H.; Balez, R.; Berg, T.; Turner, C.; Sanz, S.; Engel, M.; et al. Alzheimer's Disease Induced Neurons Bearing PSEN1 Mutations Exhibit Reduced Excitability. *bioRxiv* **2024**. [CrossRef]
40. Engel, M.; Balez, R.; Muñoz, S.S.; Cabral-da-Silva, M.C.; Stevens, C.H.; Bax, M.; Do-Ha, D.; Sidhu, K.; Sachdev, P.; Ooi, L. Viral-Free Generation and Characterization of a Human Induced Pluripotent Stem Cell Line from Dermal Fibroblasts. *Stem Cell Res.* **2018**, *32*, 135–138. [CrossRef]
41. Muñoz, S.S.; Balez, R.; Castro Cabral-da-Silva, M.e.; Berg, T.; Engel, M.; Bax, M.; Do-Ha, D.; Stevens, C.H.; Greenough, M.; Bush, A.; et al. Generation and Characterization of Human Induced Pluripotent Stem Cell Lines from a Familial Alzheimer's Disease PSEN1 A246E Patient and a Non-Demented Family Member Bearing Wild-Type PSEN1. *Stem Cell Res.* **2018**, *31*, 227–230. [CrossRef]
42. Nehme, R.; Zuccaro, E.; Ghosh, S.D.; Li, C.; Sherwood, J.L.; Pietilainen, O.; Barrett, L.E.; Limone, F.; Worringer, K.A.; Kommineni, S.; et al. Combining NGN2 Programming with Developmental Patterning Generates Human Excitatory Neurons with NMDAR-Mediated Synaptic Transmission. *Cell Rep.* **2018**, *23*, 2509–2523. [CrossRef] [PubMed]
43. Hansson, O.; Lehmann, S.; Otto, M.; Zetterberg, H.; Lewczuk, P. Advantages and Disadvantages of the Use of the CSF Amyloid β (Aβ) 42/40 Ratio in the Diagnosis of Alzheimer's Disease. *Alzheimer's Res. Ther.* **2019**, *11*, 34. [CrossRef]
44. Schindler, S.E.; Bollinger, J.G.; Ovod, V.; Mawuenyega, K.G.; Li, Y.; Gordon, B.A.; Holtzman, D.M.; Morris, J.C.; Benzinger, T.L.S.; Xiong, C.; et al. High-Precision Plasma β-Amyloid 42/40 Predicts Current and Future Brain Amyloidosis. *Neurology* **2019**, *93*, e1647. [CrossRef] [PubMed]
45. Amft, M.; Ortner, M.; Eichenlaub, U.; Goldhardt, O.; Diehl-Schmid, J.; Hedderich, D.M.; Yakushev, I.; Grimmer, T. The Cerebrospinal Fluid Biomarker Ratio Aβ42/40 Identifies Amyloid Positron Emission Tomography Positivity Better than Aβ42 Alone in a Heterogeneous Memory Clinic Cohort. *Alzheimer's Res. Ther.* **2022**, *14*, 60. [CrossRef] [PubMed]
46. Hansson, O.; Zetterberg, H.; Buchhave, P.; Andreasson, U.; Londos, E.; Minthon, L.; Blennow, K. Prediction of Alzheimer's Disease Using the CSF Aβ42/Aβ40 Ratio in Patients with Mild Cognitive Impairment. *Dement. Geriatr. Cogn. Disord.* **2007**, *23*, 316–320. [CrossRef] [PubMed]
47. Balez, R.; Steiner, N.; Engel, M.; Muñoz, S.S.; Lum, J.S.; Wu, Y.; Wang, D.; Vallotton, P.; Sachdev, P.; O'Connor, M.; et al. Neuroprotective Effects of Apigenin against Inflammation, Neuronal Excitability and Apoptosis in an Induced Pluripotent Stem Cell Model of Alzheimer's Disease. *Sci. Rep.* **2016**, *6*, 31450. [CrossRef] [PubMed]
48. Balez, R.; Stevens, C.H.; Lenk, K.; Sidhu, K.; Sutherland, G.; Ooi, L. Increased Neuronal Nitric Oxide Synthase in Alzheimer's Disease Mediates Spontaneous Calcium Signalling and Divergent Glutamatergic Calcium Responses. *Antioxid. Redox Signal.* **2024**, 1–23. [CrossRef]
49. Lin, Y.T.; Seo, J.; Gao, F.; Feldman, H.M.; Wen, H.L.; Penney, J.; Cam, H.P.; Gjoneska, E.; Raja, W.K.; Cheng, J.; et al. APOE4 Causes Widespread Molecular and Cellular Alterations Associated with Alzheimer's Disease Phenotypes in Human IPSC-Derived Brain Cell Types. *Neuron* **2018**, *98*, 1141–1154.e7. [CrossRef]
50. Yang, J.; Zhao, H.; Ma, Y.; Shi, G.; Song, J.; Tang, Y.; Li, S.; Li, T.; Liu, N.; Tang, F.; et al. Early Pathogenic Event of Alzheimer's Disease Documented in IPSCs from Patients with PSEN1 Mutations. *Oncotarget* **2017**, *8*, 7900–7913. [CrossRef]
51. Mondragón-Rodríguez, S.; Perry, G.; Luna-Muñoz, J.; Acevedo-Aquino, M.C.; Williams, S. Phosphorylation of Tau Protein at Sites Ser396-404 Is One of the Earliest Events in Alzheimer's Disease and Down Syndrome. *Neuropathol. Appl. Neurobiol.* **2014**, *40*, 121–135. [CrossRef]
52. Borchelt, D.R.; Thinakaran, G.; Eckman, C.B.; Lee, M.K.; Davenport, F.; Ratovitsky, T.; Prada, C.M.; Kim, G.; Seekins, S.; Yager, D.; et al. Familial Alzheimer's Disease-Linked Presenilin I Variants Elevate Aβ1- 42/1-40 Ratio in Vitro and in Vivo. *Neuron* **1996**, *17*, 1005–1013. [CrossRef]
53. Steiner, H.; Romig, H.; Grim, M.G.; Philipp, U.; Pesold, B.; Citron, M.; Baumeister, R.; Haass, C. The Biological and Pathological Function of the Presenilin-1 Δexon 9 Mutation Is Independent of Its Defect to Undergo Proteolytic Processing. *J. Biol. Chem.* **1999**, *274*, 7615–7618. [CrossRef] [PubMed]
54. Murayama, O.; Tomita, T.; Nihonmatsu, N.; Murayama, M.; Sun, X.; Honda, T.; Iwatsubo, T.; Takashima, A. Enhancement of Amyloid β 42 Secretion by 28 Different Presenilin 1 Mutations of Familial Alzheimer's Disease. *Neurosci. Lett.* **1999**, *265*, 61–63. [CrossRef] [PubMed]
55. Berezovska, O.; Lleo, A.; Herl, L.D.; Frosch, M.P.; Stern, E.A.; Bacskai, B.J.; Hyman, B.T. Familial Alzheimer's Disease Presenilin 1 Mutations Cause Alterations in the Conformation of Presenilin and Interactions with Amyloid Precursor Protein. *J. Neurosci.* **2005**, *25*, 3009–3017. [CrossRef]

56. Hyman, B.T.; Phelps, C.H.; Beach, T.G.; Bigio, E.H.; Cairns, N.J.; Carrillo, M.C.; Dickson, D.W.; Duyckaerts, C.; Frosch, M.P.; Masliah, E.; et al. National Institute on Aging–Alzheimer's Association Guidelines for the Neuropathologic Assessment of Alzheimer's Disease. *Alzheimers. Dement.* **2012**, *8*, 1. [CrossRef]
57. Ries, M.; Sastre, M. Mechanisms of Aβ clearance and degradation by glial cells. *Front. Aging Neurosci.* **2016**, *8*, 160. [CrossRef]
58. Xia, Y.; Prokop, S.; Giasson, B.I. "Don't Phos Over Tau": Recent Developments in Clinical Biomarkers and Therapies Targeting Tau Phosphorylation in Alzheimer's Disease and Other Tauopathies. *Mol. Neurodegener.* **2021**, *16*, 37. [CrossRef]
59. Wesseling, H.; Mair, W.; Kumar, M.; Schlaffner, C.N.; Tang, S.; Beerepoot, P.; Fatou, B.; Guise, A.J.; Cheng, L.; Takeda, S.; et al. Tau PTM Profiles Identify Patient Heterogeneity and Stages of Alzheimer's Disease. *Cell* **2020**, *183*, 1699–1713.e13. [CrossRef]
60. van der Kant, R.; Goldstein, L.S.B.; Ossenkoppele, R. Amyloid-β-Independent Regulators of Tau Pathology in Alzheimer Disease. *Nat. Rev. Neurosci.* **2020**, *21*, 21–35. [CrossRef] [PubMed]
61. Kwak, S.S.; Washicosky, K.J.; Brand, E.; von Maydell, D.; Aronson, J.; Kim, S.; Capen, D.E.; Cetinbas, M.; Sadreyev, R.; Ning, S.; et al. Amyloid-B42/40 Ratio Drives Tau Pathology in 3D Human Neural Cell Culture Models of Alzheimer's Disease. *Nat. Commun.* **2020**, *11*, 1377. [CrossRef]
62. Spitzer, N.C.; Kingston, P.A.; Manning, T.J.; Conklin, M.W. Outside and in: Development of Neuronal Excitability. *Curr. Opin. Neurobiol.* **2002**, *12*, 315–323. [CrossRef] [PubMed]
63. Yasuda, T.; Bartlett, P.F.; Adams, D.J. Kir and Kv Channels Regulate Electrical Properties and Proliferation of Adult Neural Precursor Cells. *Mol. Cell. Neurosci.* **2008**, *37*, 284–297. [CrossRef] [PubMed]
64. Srivastava, A.; Das, B.; Yao, A.Y.; Yan, R. Metabotropic Glutamate Receptors in Alzheimer's Disease Synaptic Dysfunction: Therapeutic Opportunities and Hope for the Future. *J. Alzheimer's Dis.* **2020**, *78*, 1345–1361. [CrossRef] [PubMed]
65. Jurado, S. AMPA Receptor Trafficking in Natural and Pathological Aging. *Front. Mol. Neurosci.* **2018**, *10*, 446. [CrossRef]
66. Guntupalli, S.; Widagdo, J.; Anggono, V. Amyloid- β -Induced Dysregulation of AMPA Receptor Trafficking. *Neural Plast.* **2016**, *2016*, 3204519. [CrossRef] [PubMed]
67. Greenough, M.A.; Lane, D.J.R.; Balez, R.; Anastacio, H.T.D.; Zeng, Z.; Ganio, K.; McDevitt, C.A.; Acevedo, K.; Belaidi, A.A.; Koistinaho, J.; et al. Selective Ferroptosis Vulnerability Due to Familial Alzheimer's Disease Presenilin Mutations. *Cell Death Differ.* **2022**, *29*, 2123–2136. [CrossRef] [PubMed]
68. Lasky, J.L.; Wu, H. Notch Signaling, Brain Development, and Human Disease. *Pediatr. Res.* **2005**, *57*, 104–109. [CrossRef] [PubMed]
69. Barbour, A.; Gourmaud, S.; Li, X.; Stewart, D.; Irwin, D.; Talos, D.; Jensen, F. Seizures Exacerbate Excitatory: Inhibitory Imbalance in Alzheimer's Disease with Attenuation after Rapamycin Treatment in 5XFAD Mice. *bioRxiv* **2023**. [CrossRef]
70. Guo, C.; Ma, Y.Y. Calcium Permeable-AMPA Receptors and Excitotoxicity in Neurological Disorders. *Front. Neural Circuits* **2021**, *15*, 711564. [CrossRef]
71. Kawahara, Y.; Ito, K.; Sun, H.; Kanazawa, I.; Kwak, S. Low Editing Efficiency of GluR2 MRNA Is Associated with a Low Relative Abundance of ADAR2 MRNA in White Matter of Normal Human Brain. *Eur. J. Neurosci.* **2003**, *18*, 23–33. [CrossRef]
72. Wright, A.; Vissel, B. The Essential Role of AMPA Receptor GluA2 Subunit RNA Editing in the Normal and Diseased Brain. *Front. Mol. Neurosci.* **2012**, *5*, 34. [CrossRef] [PubMed]
73. Gaisler-Salomon, I.; Kravitz, E.; Feiler, Y.; Safran, M.; Biegon, A.; Amariglio, N.; Rechavi, G. Hippocampus-Specific Deficiency in RNA Editing of GluA2 in Alzheimer's Disease. *Neurobiol. Aging* **2014**, *35*, 1785–1791. [CrossRef] [PubMed]
74. Khermesh, K.; D'Erchia, A.M.; Barak, M.; Annese, A.; Wachtel, C.; Levanon, E.Y.; Picardi, E.; Eisenberg, E. Reduced Levels of Protein Recoding by A-to-I RNA Editing in Alzheimer's Disease. *RNA* **2016**, *22*, 290–302. [CrossRef] [PubMed]
75. Akbarian, S.; Smith, M.A.; Jones, E.G. Editing for an AMPA Receptor Subunit RNA in Prefrontal Cortex and Striatum in Alzheimer's Disease, Huntington's Disease and Schizophrenia. *Brain Res.* **1995**, *699*, 297–304. [CrossRef] [PubMed]
76. Konen, L.M.; Wright, A.L.; Royle, G.A.; Morris, G.P.; Lau, B.K.; Seow, P.W.; Zinn, R.; Milham, L.T.; Vaughan, C.W.; Vissel, B. A New Mouse Line with Reduced GluA2 Q/R Site RNA Editing Exhibits Loss of Dendritic Spines, Hippocampal CA1-Neuron Loss, Learning and Memory Impairments and NMDA Receptor-Independent Seizure Vulnerability. *Mol. Brain* **2020**, *13*, 27. [CrossRef] [PubMed]
77. Pachernegg, S.; Münster, Y.; Muth-Köhne, E.; Fuhrmann, G.; Hollmann, M. GluA2 Is Rapidly Edited at the Q/R Site during Neural Differentiation in Vitro. *Front. Cell. Neurosci.* **2015**, *9*, 69. [CrossRef] [PubMed]
78. Lu, W.; Shi, Y.; Jackson, A.C.; Bjorgan, K.; During, M.J.; Sprengel, R.; Seeburg, P.H.; Nicoll, R.A. Subunit Composition of Synaptic AMPA Receptors Revealed by a Single-Cell Genetic Approach. *Neuron* **2009**, *62*, 254–268. [CrossRef] [PubMed]
79. Wenthold, R.J.; Petralia, R.S.; Blahos, J.; Niedzielski, A.S. Evidence for Multiple AMPA Receptor Complexes in Hippocampal CA1/CA2 Neurons. *J. Neurosci.* **1996**, *16*, 1982–1989. [CrossRef]
80. Zhang, Q.; Ma, C.; Gearing, M.; Wang, P.G.; Chin, L.S.; Li, L. Integrated Proteomics and Network Analysis Identifies Protein Hubs and Network Alterations in Alzheimer's Disease. *Acta Neuropathol. Commun.* **2018**, *6*, 19. [CrossRef]
81. Mendonça, C.F.; Kuras, M.; Nogueira, F.C.S.; Plá, I.; Hortobágyi, T.; Csiba, L.; Palkovits, M.; Renner, É.; Döme, P.; Marko-Varga, G.; et al. Proteomic Signatures of Brain Regions Affected by Tau Pathology in Early and Late Stages of Alzheimer's Disease. *Neurobiol. Dis.* **2019**, *130*, 104509. [CrossRef]
82. Johnson, E.C.B.; Dammer, E.B.; Duong, D.M.; Yin, L.; Thambisetty, M.; Troncoso, J.C.; Lah, J.J.; Levey, A.I.; Seyfried, N.T. Deep Proteomic Network Analysis of Alzheimer's Disease Brain Reveals Alterations in RNA Binding Proteins and RNA Splicing Associated with Disease. *Mol. Neurodegener.* **2018**, *13*, 52. [CrossRef] [PubMed]

83. Xu, J.; Patassini, S.; Rustogi, N.; Riba-Garcia, I.; Hale, B.D.; Phillips, A.M.; Waldvogel, H.; Haines, R.; Bradbury, P.; Stevens, A.; et al. Regional Protein Expression in Human Alzheimer's Brain Correlates with Disease Severity. *Commun. Biol.* **2019**, *2*, 43. [CrossRef] [PubMed]
84. Yeung, J.H.Y.; Walby, J.L.; Palpagama, T.H.; Turner, C.; Waldvogel, H.J.; Faull, R.L.M.; Kwakowsky, A. Glutamatergic Receptor Expression Changes in the Alzheimer's Disease Hippocampus and Entorhinal Cortex. *Brain Pathol.* **2021**, *31*, e13005. [CrossRef] [PubMed]
85. Stepler, K.E.; Mahoney, E.R.; Kofler, J.; Hohman, T.J.; Lopez, O.L.; Robinson, R.A.S. Inclusion of African American/Black Adults in a Pilot Brain Proteomics Study of Alzheimer's Disease. *Neurobiol. Dis.* **2020**, *146*, 105129. [CrossRef]
86. Sweet, R.A.; MacDonald, M.L.; Kirkwood, C.M.; Ding, Y.; Schempf, T.; Jones-Laughner, J.; Kofler, J.; Ikonomovic, M.D.; Lopez, O.L.; Garver, M.E.; et al. Apolipoprotein E*4 (APOE*4) Genotype Is Associated with Altered Levels of Glutamate Signaling Proteins and Synaptic Coexpression Networks in the Prefrontal Cortex in Mild to Moderate Alzheimer Disease. *Mol. Cell. Proteom.* **2016**, *15*, 2252–2262. [CrossRef] [PubMed]
87. Johnson, E.C.B.; Carter, E.K.; Dammer, E.B.; Duong, D.M.; Gerasimov, E.S.; Liu, Y.; Liu, J.; Betarbet, R.; Ping, L.; Yin, L.; et al. Large-Scale Deep Multi-Layer Analysis of Alzheimer's Disease Brain Reveals Strong Proteomic Disease-Related Changes Not Observed at the RNA Level. *Nat. Neurosci.* **2022**, *25*, 213–225. [CrossRef] [PubMed]
88. Lu, W.; Roche, K.W. Posttranslational Regulation of AMPA Receptor Trafficking and Function. *Curr. Opin. Neurobiol.* **2012**, *22*, 470–479. [CrossRef] [PubMed]
89. Roche, K.W.; O'Brien, R.J.; Mammen, A.L.; Bernhardt, J.; Huganir, R.L. Characterization of Multiple Phosphorylation Sites on the AMPA Receptor GluR1 Subunit. *Neuron* **1996**, *16*, 1179–1188. [CrossRef]
90. Derkach, V.; Barria, A.; Soderling, T.R. Ca^{2+}/Calmodulin-Kinase II Enhances Channel Conductance of α-Amino-3-Hydroxy-5-Methyl-4-Isoxazolepropionate Type Glutamate Receptors. *Proc. Natl. Acad. Sci. USA* **1999**, *96*, 3269–3274. [CrossRef]
91. Banke, T.G.; Bowie, D.; Lee, H.K.; Huganir, R.L.; Schousboe, A.; Traynelis, S.F. Control of GluR1 AMPA Receptor Function by CAMP-Dependent Protein Kinase. *J. Neurosci.* **2000**, *20*, 89–102. [CrossRef]
92. Fox, C.J.; Russell, K.; Titterness, A.K.; Yu, T.W.; Christie, B.R. Tyrosine Phosphorylation of the GluR2 Subunit Is Required for Long-Term Depression of Synaptic Efficacy in Young Animals in Vivo. *Hippocampus* **2007**, *17*, 600–605. [CrossRef]
93. Ahmadian, G.; Ju, W.; Liu, L.; Wyszynski, M.; Lee, S.H.; Dunah, A.W.; Taghibiglou, C.; Wang, Y.; Lu, J.; Wong, T.P.; et al. Tyrosine Phosphorylation of GluR2 Is Required for Insulin-Stimulated AMPA Receptor Endocytosis and LTD. *EMBO J.* **2004**, *23*, 1040–1050. [CrossRef] [PubMed]
94. Hershko, A.; Ciechanover, A. The Ubiquitin System. *Annu. Rev. Biochem.* **1998**, *67*, 425–479. [CrossRef]
95. Widagdo, J.; Chai, Y.J.; Ridder, M.C.; Chau, Y.Q.; Johnson, R.C.; Sah, P.; Huganir, R.L.; Anggono, V. Activity-Dependent Ubiquitination of GluA1 and GluA2 Regulates AMPA Receptor Intracellular Sorting and Degradation. *Cell Rep.* **2015**, *10*, 783–795. [CrossRef]
96. Schwarz, L.A.; Hall, B.J.; Patrick, G.N. Activity-Dependent Ubiquitination of GluA1 Mediates a Distinct AMPA Receptor Endocytosis and Sorting Pathway. *J. Neurosci.* **2010**, *30*, 16718–16729. [CrossRef] [PubMed]
97. Widagdo, J.; Guntupalli, S.; Jang, S.E.; Anggono, V. Regulation of AMPA Receptor Trafficking by Protein Ubiquitination. *Front. Mol. Neurosci.* **2017**, *10*, 347. [CrossRef]
98. Lin, A.; Hou, Q.; Jarzylo, L.; Amato, S.; Gilbert, J.; Shang, F.; Man, H.Y. Nedd4-Mediated AMPA Receptor Ubiquitination Regulates Receptor Turnover and Trafficking. *J. Neurochem.* **2011**, *119*, 27–39. [CrossRef]
99. Huo, Y.; Khatri, N.; Hou, Q.; Gilbert, J.; Wang, G.; Man, H.Y. The Deubiquitinating Enzyme USP46 Regulates AMPA Receptor Ubiquitination and Trafficking. *J. Neurochem.* **2015**, *134*, 1067–1080. [CrossRef] [PubMed]
100. Lussier, M.P.; Herring, B.E.; Nasu-Nishimura, Y.; Neutzner, A.; Karbowski, M.; Youle, R.J.; Nicoll, R.A.; Roche, K.W. Ubiquitin Ligase RNF167 Regulates AMPA Receptor-Mediated Synaptic Transmission. *Proc. Natl. Acad. Sci. USA* **2012**, *109*, 19426–19431. [CrossRef]
101. Hayashi, T.; Rumbaugh, G.; Huganir, R.L. Differential Regulation of AMPA Receptor Subunit Trafficking by Palmitoylation of Two Distinct Sites. *Neuron* **2005**, *47*, 709–723. [CrossRef]
102. Lin, D.T.; Makino, Y.; Sharma, K.; Hayashi, T.; Neve, R.; Takamiya, K.; Huganir, R.L. Regulation of AMPA Receptor Extrasynaptic Insertion by 4.1N, Phosphorylation and Palmitoylation. *Nat. Neurosci.* **2009**, *12*, 879–887. [CrossRef] [PubMed]
103. Itoh, M.; Yamashita, M.; Kaneko, M.; Okuno, H.; Abe, M.; Yamazaki, M.; Natsume, R.; Yamada, D.; Kaizuka, T.; Suwa, R.; et al. Deficiency of AMPAR–Palmitoylation Aggravates Seizure Susceptibility. *J. Neurosci.* **2018**, *38*, 10220–10235. [CrossRef] [PubMed]
104. Baudry, M.; Massicotte, G.; Hauge, S. Phosphatidylserine Increases the Affinity of the AMPA/Quisqualate Receptor in Rat Brain Membranes. *Behav. Neural Biol.* **1991**, *55*, 137–140. [CrossRef] [PubMed]
105. Frank, C.; Rufini, S.; Tancredi, V.; Forcina, R.; Grossi, D.; D'Arcangelo, G. Cholesterol Depletion Inhibits Synaptic Transmission and Synaptic Plasticity in Rat Hippocampus. *Exp. Neurol.* **2008**, *212*, 407–414. [CrossRef] [PubMed]
106. Korinek, M.; Gonzalez-Gonzalez, I.M.; Smejkalova, T.; Hajdukovic, D.; Skrenkova, K.; Krusek, J.; Horak, M.; Vyklicky, L. Cholesterol Modulates Presynaptic and Postsynaptic Properties of Excitatory Synaptic Transmission. *Sci. Rep.* **2020**, *10*, 12651. [CrossRef]
107. Korinek, M.; Vyklicky, V.; Borovska, J.; Lichnerova, K.; Kaniakova, M.; Krausova, B.; Krusek, J.; Balik, A.; Smejkalova, T.; Horak, M.; et al. Cholesterol Modulates Open Probability and Desensitization of NMDA Receptors. *J. Physiol.* **2015**, *593*, 2279–2293. [CrossRef]

108. Antonini, A.; Caioli, S.; Saba, L.; Vindigni, G.; Biocca, S.; Canu, N.; Zona, C. Membrane Cholesterol Depletion in Cortical Neurons Highlights Altered NMDA Receptor Functionality in a Mouse Model of Amyotrophic Lateral Sclerosis. *Biochim. Biophys. Acta—Mol. Basis Dis.* **2018**, *1864*, 509–519. [CrossRef]
109. Sanchez-Mejia, R.O.; Newman, J.W.; Toh, S.; Yu, G.-Q.; Zhou, Y.; Halabisky, B.; Cissé, M.; Scearce-Levie, K.; Cheng, I.H.; Gan, L.; et al. Phospholipase A2 Reduction Ameliorates Cognitive Deficits in a Mouse Model of Alzheimer's Disease. *Nat. Neurosci.* **2008**, *11*, 1311–1318. [CrossRef]
110. Bernard, J.; Chabot, C.; Gagné, J.; Baudry, M.; Massicotte, G. Melittin Increases AMPA Receptor Affinity in Rat Brain Synaptoneurosomes. *Brain Res.* **1995**, *671*, 195–200. [CrossRef]
111. Ménard, C.; Patenaude, C.; Massicotte, G. Phosphorylation of AMPA Receptor Subunits Is Differentially Regulated by Phospholipase A2 Inhibitors. *Neurosci. Lett.* **2005**, *389*, 51–56. [CrossRef]

life

Review

"Dirty Dancing" of Calcium and Autophagy in Alzheimer's Disease

Hua Zhang [1] and Ilya Bezprozvanny [1,2,*]

[1] Department of Physiology, UT Southwestern Medical Center, Dallas, TX 75390, USA; hua.zhang@utsouthwestern.edu
[2] Laboratory of Molecular Neurodegeneration, Peter the Great St. Petersburg State Polytechnical University, St. Petersburg 195251, Russia
* Correspondence: ilya.bezprozvanny@utsouthwestern.edu

Abstract: Alzheimer's disease (AD) is the most common cause of dementia. There is a growing body of evidence that dysregulation in neuronal calcium (Ca^{2+}) signaling plays a major role in the initiation of AD pathogenesis. In particular, it is well established that Ryanodine receptor (RyanR) expression levels are increased in AD neurons and Ca^{2+} release via RyanRs is augmented in AD neurons. Autophagy is important for removing unnecessary or dysfunctional components and long-lived protein aggregates, and autophagy impairment in AD neurons has been extensively reported. In this review we discuss recent results that suggest a causal link between intracellular Ca^{2+} signaling and lysosomal/autophagic dysregulation. These new results offer novel mechanistic insight into AD pathogenesis and may potentially lead to identification of novel therapeutic targets for treating AD and possibly other neurodegenerative disorders.

Keywords: ryanodine receptor; autophagy; calcium signaling; calcineurin; Alzheimer's disease

Citation: Zhang, H.; Bezprozvanny, I. "Dirty Dancing" of Calcium and Autophagy in Alzheimer's Disease. *Life* **2023**, *13*, 1187. https://doi.org/10.3390/life13051187

Academic Editor: Carlo Musio

Received: 24 March 2023
Revised: 9 May 2023
Accepted: 11 May 2023
Published: 15 May 2023

1. Introduction

Alzheimer's disease (AD) is an age-related brain disorder that causes progressive neurodegeneration predominantly in the cortical and hippocampal brain regions. Major hallmarks of AD are the progressive impairment of memory storage and accumulation of fibrillary amyloid plaques in patient's brains. Despite decades of research and effort, there is still no effective disease-modifying treatment for AD. Although amyloid pathology is a hallmark and defining feature of AD, targeting amyloid pathways has been very challenging due to low efficacy and serious side effects. Therefore, it is important to explore alternative approaches for treating memory loss in AD [1]. Increasing studies suggest that Ca^{2+} dysregulation in AD plays an important role in AD pathology and is associated with other AD abnormalities, such as excessive inflammation, increased ROS, impaired autophagy, neurodegeneration, and synapse and cognitive dysfunction [2–5]. Autophagy is important for removing unnecessary or dysfunctional components and long-lived protein aggregates, and a substantial amount of evidence in both AD patients and AD animal models indicates autophagy dysregulation plays an important role in AD pathogenesis [6–15]. Autophagy can also be regulated by intracellular Ca^{2+} signals arising from different organelles, including ER, mitochondria and lysosomes [16–18]. The majority of Ca^{2+} release from the ER is mediated by two families of Ca^{2+}-permeable channels, inositol 1,4,5-trisphosphate receptors ($InsP_3Rs$) and ryanodine receptors (RyanRs). The role of $InsP_3Rs$ in regulation of autophagy had been intensively studied in non-excitable cells, and $InsP_3R$-mediated Ca^{2+} signals have been suggested to be involved in both inhibitory and stimulatory effects on autophagy [17,19–23]. Recently, the role of RyanR in regulating lysosomal acidification [24] and autophagy [25–27] is receiving greater attention. On another hand, some recent studies suggested that changes in lysosomal acidification may affect intracellular Ca^{2+} signaling by modulating activity of TRPML1 lysosomal Ca^{2+}

channels [28,29]. In this review, we will focus on emerging interplay between dysregulated Ca^{2+} signaling and dysfunction of the lysosomal/autophagic system in AD neurons.

2. Intracellular Calcium Signaling Dysregulation in AD

There are two types of intracellular Ca^{2+} release channels in neurons—$InsP_3Rs$ and RyanRs. There are three isoforms of $InsP_3Rs$, the predominant one in neurons being $InsP_3R$ type 1. It is known that expression of FAD PS1 mutants in *Xenopus* oocytes potentiates $InsP_3$-mediated Ca^{2+} release [30], and that $InsP_3R$ activity can be potentiated in PS1(M146V) knock-in mice [31]. Similar potentiation was also reported in human lymphoblasts expressing FAD mutant PS1-M146L [32]. The $InsP_3Rs$ are known to be enriched in MAMs, and up-regulated Ca^{2+} release from ER through $InsP_3R$ can overload mitochondria, cause openings of the mitochondrial permeability transition pore (PTP), cause a reduction in the inner mitochondrial membrane potential, result in drop in NADH and ATP levels, and potentially lead to cell death [33]. Computational modeling of single $InsP_3R1$ channel activity showed that significantly lower $InsP_3$ levels are needed to induce the same level of $InsP_3R1$ channel activity in the presence of FAD-PS1 mutants [34]. Besides FAD-PS1, Aβ can also affect $InsP_3R$ function, and it was reported that Aβ42 can induce a cytosolic Ca^{2+} increase in both an $InsP_3R$-dependent and $InsP_3R$-independent manner [35]. Genetic reduction $InsP_3R1$ by 50% normalized exaggerated Ca^{2+} signaling in PS1M146V knock-in mice and restored hippocampal long-term potentiation (LTP). In 3xTg mice, reduced $InsP_3R1$ expression reduced amyloid β accumulation and tau hyperphosphorylation and restored hippocampal LTP and memory deficits [36]. Proper $InsP_3Rs$ function is important to maintain spine morphology, synaptic plasticity and memory consolidation [37,38], and limited inhibition of $InsP_3R$ may be considered as a potential therapeutic approach for AD.

$InsP_3R1$ plays an important role in Ca^{2+} signaling in cerebellar Purkinje cells, but in most other neurons RyanRs play a predominant role in control of intracellular Ca^{2+} levels. There are three structurally similar mammalian isoforms of RynaRs—RyanR1, RyanR2 and RyanR3. RyanR1 was initially found in skeletal muscles but also expressed in cerebellar Purkinje cells. RyanR2 is expressed in cardiac muscle cells and in most neurons. RyanR3 is also expressed in neurons, but RyanR2 is the most dominant neuronal isoform with the exception of cerebellar Purkinje cells [39–42]. Neuronal RyanRs are activated by Ca^{2+}-induced Ca^{2+} release (CICR) mechanisms in response to initial Ca^{2+} influx via voltage-gated Ca^{2+} channels or NMDARs. RyanRs can also open in response to endoplasmic reticulum (ER) Ca^{2+} overloads in resting neurons [43]. These spontaneous Ca^{2+} release events ("Ca^{2+} sparks") act to control ER Ca^{2+} levels and also contribute to setting basal levels of cytosolic Ca^{2+} in resting cells.

There is extensive evidence that activity and expression of RyanR2 is elevated in AD neurons. The expression and function of RyanR2 is increased in animal models with familial AD (FAD) and in early stages of sporadic AD in patients [42,44–50]. Aging is the most important risk factor for sporadic AD, and enhanced activity of RyanR may play a critical role in aging-related cognitive impairments [51]. Microarray analysis shows that RyanR2 expression continues to increase in the hippocampus from 6 months of age and onwards in mice, while the level of neuronal FKBP1b protein which binds to RyanR2 and stabilizes their opening is high at early developmental stages and begins to decrease at approximately 3 months of age [52]. This down-regulation continues throughout the aging process, leading to minimal amounts of FKBP1b being detected at 23 months of age, likely leading to further RyanR2 overactivation [52]. Consistent with the importance of RyanR overactivation for AD pathogenesis, pharmacological inhibitors of RyanR such as dantrolene demonstrated beneficial effects in a variety of AD cellular and animal models [45,48,53,54]. In our previous studies we used a genetic strategy and evaluated effects of RyanR3 knockout in AD mouse models, and we found a dual role for RyanR3 in AD pathology depending on the different stages of the disease [42]. Beneficial effects of a gating mutation of RyanR2 in AD mouse models have been previously demonstrated [55–57]. However, the mechanisms linking RyanR overactivation with AD pathology are less clear.

Some investigators suggested a mechanism that involves the restoration of neuronal hyperexcitability in AD mice [42,55–57], some suggest that blocking RyanR-mediated Ca^{2+} leakage may lead to decreased endoplasmic reticulum (ER) stress [58], some focused on the role of RyanR-mediated Ca^{2+} changes in synaptic functions [54], and some suggested dantrolene can decrease β- and γ-secretase activities and APP phosphorylation by affecting Cdk5 and GSk3β kinase activities [45].

3. Dysregulation of Autophagy in AD

Autophagy is a process that maintains healthy cells, organelles, proteins, and nutrient homeostasis in living organisms. Three types of autophagy are observed in mammalian cells depending on the mode of substrate delivery: macroautophagy, chaperone-mediated autophagy, and microautophagy [59]. In macroautophagy, a double-membrane vesicle known as an autophagosome engulfs its targets by isolating a portion of the cytoplasm. The autophagosomal membrane fuses with lysosomes to form autophagic vesicles, the contents of which are degraded by lysosomal proteases. In microautophagy, substrate proteins are internalized into lysosome lumen via membrane invagination. In chaperone-mediated autophagy (CMA), substrate proteins are translocated across the lysosomal membrane by chaperone proteins [60]. In most cases, term "autophagy" refers to macroautophagy.

Autophagy is a conserved cellular process that removes damaged or nonfunctional cellular organelles to recycle the materials and maintain the cell function and efficiency. It also destroys pathogens that invade the cells in order to protect the cells, and plays an important role in removing long-lived proteins and aggregates that help to maintain cell homeostasis [61]. Neurons are post-mitotic and long-lived cells; misfolded proteins and damaged organelles cannot be diluted through cell division and thus they are more easily affected by protein homeostasis impairment. Since $Atg5^{-/-}$ and $Atg7^{-/-}$ mice die soon after birth, researchers made neural-cell-specific Atg5 or Atg7 knockout mice. Both mice show dramatic abnormal intracellular protein accumulation and form intraneuronal aggregates and inclusions that increased in size and number with aging. They also show progressive neurodegeneration and cell death, which suggests autophagy is essential for normal neuronal function and survival [62,63]. Neurons are highly polarized cells with axon and synaptic terminals far away from the cell body, which is the primary site for protein synthesis and degradation. Recently, researchers found autophagosomes are constitutively formed at synaptic sites in the distal axon, with new autophagosomes engulfing soluble and aggregated proteins as well as mitochondrial [64], ER [65] and damaged synaptic vesicles [66]. They rapidly fuse with a late endosome or lysosome, and subsequently retrograde transport along the axon to the cell body. This conserved mechanism plays an important role in the maintenance of synaptic homeostasis [67,68]. Specific deletion of Atg7 in Purkinje cells initially causes cell-autonomous progressive degeneration of the axon terminals with little sign of dendritic or spine atrophy, suggesting that axon terminals are much more vulnerable to autophagy impairment than dendrites [69]. Autophagy can regulate synaptic plasticity by modulating synaptic transmission through removal of SVs and their associated proteins [70] or through controlling the axonal ER [65]. interestingly, it was reported that autophagy protein LC3B is also an RNA-binding protein that can trigger rapid mRNA degradation to control local protein synthesis [71]. Even though the main sites of autophagosome biogenesis are at distal axons, autophagy in postsynaptic sites were also recently reported to play important role. It was shown that NMDAR-dependent LTD induction triggers a profound reorganization of PSD-95, which requires the autophagy mechanism to remove the T19-phosphorylated form of PSD-95 from synapses [72]. It was also shown that during NMDAR-LTD, key postsynaptic proteins are sequestered for autophagic degradation, pharmacological inhibition of AV biogenesis, or knockdown of atg5. Specifically, postsynaptic pyramidal neurons in the CA1 area abolish LTD. These data suggest that local autophagy of postsynaptic proteins in dendrites involve LTD expression [73]. Synaptic plasticity is the foundation of learning and memory, which is impaired in AD, so autophagy dysregulation will also play important role in AD [7].

Autophagy dysregulation plays an important role in AD pathogenesis according to the studies of both AD patients and animal models. Using immunogold staining with compartment-specific markers and electron microscopy of AD patients' brain samples, it was demonstrated that autophagysome, multivesicular bodies, multilamellar bodies, and cathepsin-containing autophagosomes were accumulated in dystrophic neurites and synaptic terminals in AD brains [10]. APP/PS1 double transgenic mice also showed that autophagosomes and late autophagic vacuoles (AVs) accumulate markedly in dystrophic dendrites, implying an impaired maturation of AVs to lysosomes. In the hippocampus of young (4- to 6-month-old) PS1(M146L)/APP(751SL) mice models, many autophagic vesicles accumulated in the dystrophic neurites and presynaptic terminals surrounding amyloid plaques, and the autophagosome marker LC3 was also increased around plaques [13]. Interestingly, silver-enhanced immunogold labeling revealed that APP preferentially localized to the AVs within plaque-associated dystrophic neurites [13].

Autophagy is the main mechanism for regulating the processing of APP and intracellular Aβ peptide. It was discovered that AVs enriched Aβ, βCTF, and also the components of the γ-secretase complex, and that Aβ production increased after macroautophagy was acutely stimulated, implying that the cleavage of APP occurs in the AVs [12]. Interestingly, autophagosomes can also transport BACE1 and regulate BACE1 trafficking and degradation. Autophagic vacuole-associated BACE1 is accumulated in the distal axon of Alzheimer's disease-related mutant human APP transgenic neurons and mouse brains which exacerbates the AD pathological changes [74]. Therefore, it is possible that accumulated β-secretase, γ-secretase complex and APP in autophagic vacuoles enhances the β and γ processing, resulting in Aβ overproduction. Autophagy also possibly participates in the secretion of Aβ [75]. When transgenic mice overexpressing an amyloid precursor protein (APP) were crossed with the mice lacking autophagy in excitatory forebrain neurons, the amyloid plaques were dramatically reduced, while intraneuronal Aβ accumulated in the perinuclear region and caused neuron degeneration and cognitive dysfunction [76]. Moreover, it was shown that these changes were due to reduced Aβ secretion, suggesting that autophagy is important for Aβ secretion [76].

Autophagy also plays an important role in the degradation of both soluble and insoluble Tau, another signaling component of AD pathology. Wang et al. demonstrated in an inducible neuronal cell model of tauopathy that the autophagy-lysosomal system contributes to both Tau fragmentation into pro-aggregating forms and to clearance of Tau aggregates. Inhibition of macroautophagy enhances Tau aggregation and cytotoxicity. Unlike the proteasome, autophagy degrades tau regardless of phosphorylation. Tau phosphorylated at the KXGS motifs of the repeat domain, which cannot be degraded by the UPS, is efficiently degraded by autophagy [77]. However, they also found that N terminus truncated species are preferentially trapped by CMA, and C terminal truncation occurred. The C terminal-truncated tau promoted tau aggregation. Even though soluble tau, aggregated tau, and C terminal-truncated tau mainly degraded through the autophagy lysosome system, full-length tau is preferentially digested through proteasomes [78]. Recently, it was reported that a large fraction of neuronal tau is degraded by CMA, whereas upon acetylation, tau is preferentially degraded by macroautophagy and endosomal microautophagy [79]. Consistent with these findings, trehalose and rapamycin have been shown to result in a significant reduction in cortical tau tangles, less tau hyperphosphorylation, and lowered levels of insoluble tau in the forebrain in P301S mutant tau transgenic mice by stimulation of autophagy [80,81]. Secreted soluble tau species spread trans-cellularly were reported in AD, and data have shown that autophagy inducers can promote tau secretion and knockdown. Beclin1 or autophagy inhibitors can inhibit tau secretion, and researchers have further identified that six isoforms of tau protein are secreted in an autophagy-dependent manner [82,83]. Accumulated data showed that secreted Tau contributes to synaptic impairment in AD, especially in GABAergic transmissions [84,85].

4. Mechanisms Underlying Autophagy Impairment in AD

Despite extensive evidence regarding dysregulation of autophagy in AD, there is no clear understanding of mechanisms that cause this impairment. The build-up of AVs in neurodegenerative diseases may reflect enhanced autophagy induction, impaired later lysosomal degradation steps in the autophagic pathway, or a lower rate of autophagy initiation combined with insufficient lysosome fusion and digestion [7]. Several reports suggest that autophagy induction is impaired in AD neurons. Beclin1, a protein with a key role in autophagy initiation, was decreased in affected brain regions of AD patients early in the disease process [86] and the expression of p62, an autophagic cargo receptor, was reported decreased in AD brains relative to age-matched controls [87]. There is a report that autophagy is transcriptionally down-regulated during normal aging in the human brain, and in contrast to normal aging, they observe transcriptional up-regulation of autophagy in the brains of AD patients, suggesting that there might be a compensatory regulation of autophagy [88]. A critical role of lysosomal proteolytic failure in AD neurons has been suggested by other investigators [89,90]. Presenilin1 (PS1) mutations are the most common cause of early-onset familial AD (FAD), in addition to its role as a catalytic subunit of the γ-secretase complex, PS1 is also essential for v-ATPase targeting to lysosomes, lysosome acidification, and proteolysis during autophagy. Fibroblasts from patients with FAD caused by PS1 mutations also exhibit markedly defective lysosome acidification and autolysosome maturation, a similar mechanism to that seen in PS1-null cells [91]. However, there are also reports that endo-lysosomal dysfunction in PSEN-deficient cells is due to lysosomal calcium homeostasis defects, not proton pump defects [92,93] (please see discussion in [94]). An APP-dependent compromise of lysosomal acidification was also reported in multiple AD mouse models in which FAD-mutant APP alone is expressed [95]. The mechanism for this lysosome failure is not clear, although it has been proposed that decreased expression of the motor proteins kinesin and dynein can induce axonal transport impairments and further impair lysosome trafficking, maturation and function [96]. Defects of lysosomal acidification observed in AD models suggest that stimulation of lysosomal acidification should elicit beneficial effects in AD. This can be achieved, for example, by stimulating the activity of lysosomal v-ATPase by disrupting its association with STK11IP, a recently identified lysosome-specific substrate of mTORC1 that regulates lysosomal acidification [97].

5. Dysregulated Ca^{2+} Signaling and Autophagy Defects in AD

Recent reports suggest that dysregulation of Ca^{2+} signaling and impaired autophagy in AD neurons may have a causal relationship to each other. It has been suggested that lysosomal Ca^{2+} contributes to autophagy and is important for lysosomal degradation. Thus, intracellular Ca^{2+} distribution may affect lysosomal acidification (Figure 1A) [24]. Indeed, it has been reported that $InsP_3R$ preferentially associate with ER-lysosome contact sites and selectively deliver Ca^{2+} to lysosomes [98]. Ca^{2+} uptake by lysosomes may regulate lysosomal pH due to the activation of a lysosomal Ca^{2+}/H^+ exchanger (Figure 1A). For example, in hepatic cell HepG2, Ca^{2+} release from the ER caused a disruption of the lysosomal acidity and impaired protein degradation [99]. Recently increased Ca^{2+} release via RyanRs was reported to be associated with reduced expression of the lysosome proton pump vacuolar-ATPase (vATPase) subunits (V1B2 and V0a1), which cause lysosome deacidification and disrupt proteolytic activity. These findings are reported in both AD mouse models and human-induced neurons (HiN) [24] (Figure 1A). Normalizing AD-associated aberrant RyanR Ca^{2+} signaling with the negative allosteric modulator, dantrolene (Ryanodex), restored vATPase levels, lysosomal acidification and proteolytic activity, and autophagic clearance of intracellular protein aggregates in AD neurons [24]. These results directly implicate intracellular Ca^{2+} dysregulation in altering lysosomal vATPase expression levels and reinforces the important role of inter-organelle Ca^{2+} communication for vATPase trafficking to lysosomes, assembly of both domains, and lysosomal acidification (Figure 1A).

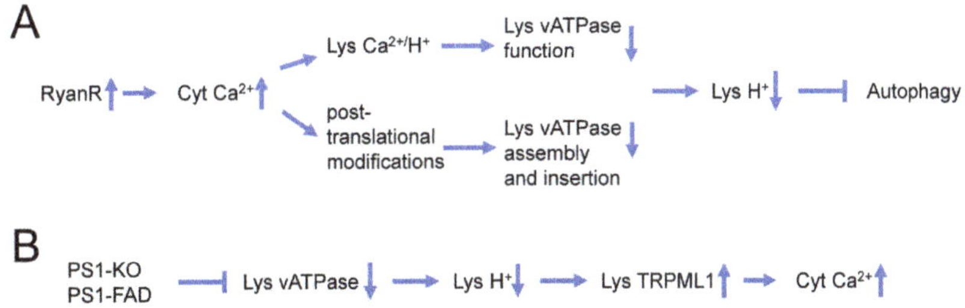

Figure 1. Causal relationship between dysregulation of Ca^{2+} signaling and impaired autophagy in AD neurons. (**A**) Supranormal RyanR-mediated Ca^{2+} release in AD neurons may decrease lysosomal vATPase expression levels through post-translational modifications that affect its assembly and insertion to the lysosome or disrupts vATPase function through stimulation of lysosomal Ca^{2+}/H^+ exchangers. Based on [24,99]. (**B**) Presenilin 1 (PS1) knockout or FAD linked mutations in PS1 lead to impaired glycosylation and instability of vATPase V0a1 subunit, further induce deficient lysosomal vATPase assembly and impaired its function and elevated lysosomal pH. Increased lysosomal pH induces abnormal Ca^{2+} efflux from lysosomes to cytoplasm by enhancing activity of lysosomal TRPML1 channels. Based on [28,29,91].

In addition to the effects of intracellular Ca^{2+} signaling on lysosomal acidification, there are also reports that suggest that changes in lysosomal acidification may also affect intracellular Ca^{2+} signaling (Figure 1B) [29]. Defects in V-ATPase targeting to lysosomes and lysosomal acidification were reported for PS1 knockout and FAD mutant cells [91]. It has been further reasoned that in addition to ER, lysosomes may also act as Ca^{2+} signaling organelles and that TRPML channels may mediate Ca^{2+} release from lysosomal compartments. It has been reported that lysosomal Ca^{2+} efflux through TRPML1 can trigger membrane fusion/fission events and regulate membrane trafficking [100], and that impairment in the TRPML1 function leads to various lysosomal storage diseases [101]. Moreover, it has been reported that impaired lysosomal acidification in presenilin knockout neurons causes an overactivation of lysosomal TRPMl1 channels [28,29], suggesting that impaired lysosomal acidification may contribute to dysregulated Ca^{2+} signaling in PS1 knockout and FAD mutant neurons (Figure 1B).

Some recent findings also suggest that dysregulated Ca^{2+} signaling may play a more direct role in control of neuronal autophagy, independently from impaired lysosomal acidification (Figure 2A). In experiments with hippocampal neural stem (HCN) cells, it was shown that RyanR agonist caffeine significantly promoted the autophagic death of insulin-deficient HCN cells, and treatment with the RyanR inhibitor dantrolene prevented the induction of autophagy following insulin withdrawal [102]. Moreover, CRISPR/Cas9-mediated knockout of the RyanR3 gene in HCN cells abolished autophagic cell death [102]. Neferine, a natural alkaloid from Nelumbo nucifera, was reported to induce autophagy through Ulk-1-PERK and AMPK-mTOR signaling pathways in cancer cells which involved RyanRs activation [103]. Similarly, high doses of propofol (a commonly used intravenous anesthetic) can induce cytotoxicity in cortical progenitor cells, and data showed that blocking both InsP3R and RyanR can reduce autophagy and increase cell viability, suggesting that RyanR mediated excessive autophagy plays an important role in propofol induced toxicity [104]. All these studies suggest that excessive Ca^{2+} release via RyanR stimulates autophagy, most likely by stimulating the CaMKK2-AMPK-mTOR signaling pathway [102,103] (Figure 2A).

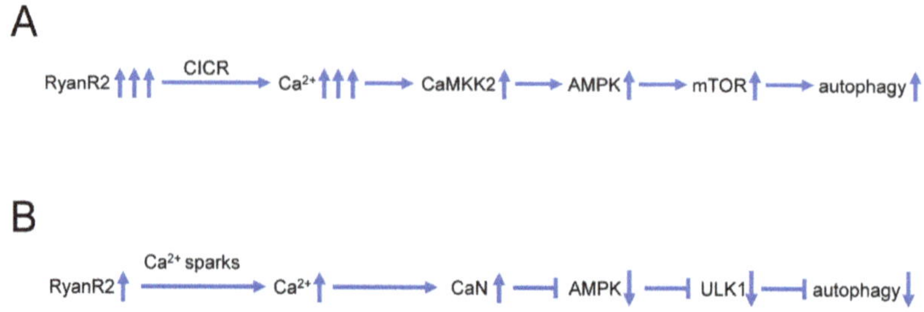

Figure 2. Biphasic effects of RyanR2-mediated Ca^{2+} signals on neuronal autophagy. (**A**) Evoked or stimulated Ca^{2+} release via RyanR2 leads to significant elevation of cytosolic Ca^{2+} levels, resulting in activation of CAMKK2-AMPK-mTOR pathway that promotes autophagy. Based on [102,103]. (**B**) Ca^{2+} sparks resulting from spontaneous basal activity of RyanR2 lead to modest elevation of cytosolic Ca^{2+} levels, sufficient to stimulate CaN and inhibit AMPK/ULK1 pathway, resulting in inhibition of autophagy. Based on [25,27].

However, some reports also suggest that basal RyanR activity may actually inhibit autophagic flux (Figure 2B). Indeed, pharmacological inhibition of RyanR augmented autophagic flux in ectopic RyanR-expressing models such as HEK293 cells transfected with RyanR constructs or C2C12 myoblasts [25]. These studies have been performed using pharmacological modulators of RyanR activity that may exert off-target effects. In our recent studies, we took advantage of RyanR2-E4872Q knock-in mouse models (EQ) to test the importance of RyanR2 in control of neuronal autophagy in wild type and AD neurons. In EQ mice, the basal RyanR2-mediated Ca^{2+} influx is reduced due to shortened open channel time [105]. In our studies, we discovered augmented autophagic flux in primary hippocampal neuron cultures in a RyanR2-E4872Q knock-in mouse model (EQ) [27] (Figure 2B). Strikingly, we discovered thar EQ mutation is able to repair autophagic defects in two different AD mouse models (APPKI and APPPS1) [27]. Based on a series of additional pharmacological experiments, we demonstrated that overactivation of RyanR2 in AD neurons leads to persistent overactivation of cytosolic calcineurin (CaN), which is able to suppress autophagy by inhibiting AMPK/ULK1 pathways [27] (Figure 2B). Indeed, we have been able to achieve similar restoration of autophagic flux in AD neurons by inhibiting RyanR or by directly inhibiting CaN [27].

To reconcile these results, we propose that low levels of basal Ca^{2+} determined by spontaneous activity of RyanR2 primarily stimulates CaN-AMPK-ULK1 pathways to inhibit autophagy as we discovered [27] (Figure 2B), but high levels of Ca^{2+} during evoked or stimulated Ca^{2+} release led to the activation of the CAMKK2-AMPK-mTOR pathway and promoted autophagy as has been reported [102,103] (Figure 2A). Both CaN and CaMKK2 are regulated by Ca^{2+}/calmodulin, and differential modulation of CaN and CaMKII by low and high Ca^{2+} elevations has already been described in the context of synaptic plasticity [106,107]. The model that we propose (Figure 2) has important implications for AD pathogenesis. Reduced autophagy impairs clearance of APP and APP proteolytic fragments, eventually leading to the accumulation of soluble Aβ42 oligomers, amyloid plaques and impaired synaptic plasticity. Indeed, in our studies, we demonstrated that the genetic cross of EQ mice with AD mouse models (APPKI and APPPS1) was able to reduce levels of accumulated Aβ, reduce amyloid plaque loads and repair LTP defects in these mice [27]. These results suggest that the beneficial effects of RyanR and CaN inhibitors in AD may at least in part be explained by their ability to repair autophagic defects.

In addition to RyanR2-mediated pathways (Figure 2), there are also other Ca^{2+} signaling pathways that are connected with autophagy that may also play important roles in AD pathogenesis (Figure 3). It was shown that Ca^{2+} release through lysosomal TRMPL1 channels can activate CaN, leading to the dephosphorylation of TFEB transcription factors and resulting in increased expression of lysosomal and autophagic genes [108]. TRPML1 activity can also modulate autophagy through the activation of the CaMKKβ/AMPK/ULK1/VPS34 pathway [109]. Recently, functional defects in the TRPML1 Ca^{2+} channels were identified in LOAD patients' brain samples [110]. In the same study it was shown that decreased TRPML1-mediated lysosomal Ca^{2+} released in a neuronal apoE4 iPSC model can be reduced by treatment with ML-SA1, a small-molecule TRPML1 agonist [110]. It was also shown that TFEB mediates mutant tau releases in iPSC-derived neurons in a TRPML1-dependent manner [111]. All of these results suggest that TRPML1 may also be considered as a potential therapeutic treatment for AD (Figure 3).

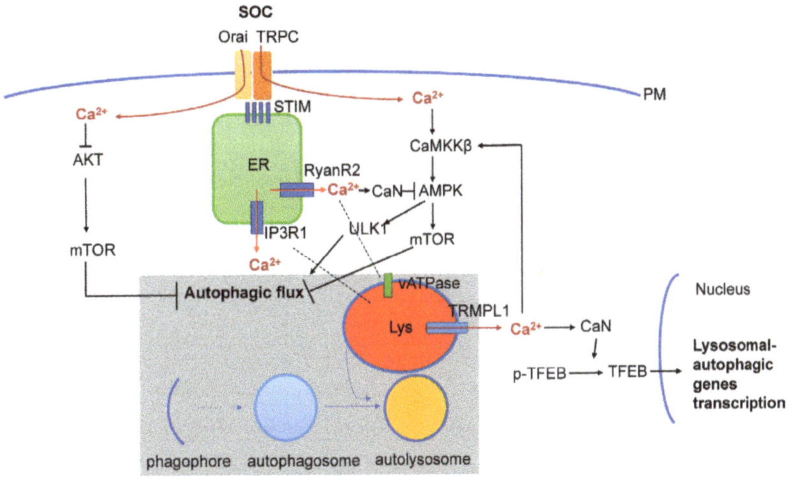

Figure 3. Interplay between Ca^{2+} signaling and neuronal autophagy in AD. Store-operated Ca^{2+} influx (SOC), Ca^{2+} release from ER via $InsP_3R1$ and RyanR2 channels and Ca^{2+} release form lysosomal compartment via TRPM1 channel set basal levels of cytosolic Ca^{2+}. Cytosolic Ca^{2+} controls autophagic pathways by directly acting on AKT, by modulating activity of CaMKKβ or CaN/AMPK/ULK1 and by modulating expression levels of genes involved in lysosomal and autophagic function via TFEB transcription factor. Dysregulation of Ca^{2+} signaling in AD neurons is closely linked with dysregulation of lysosomal function and autophagic pathways.

Store-operated Ca^{2+} entry (SOCE) is a major calcium-entry pathway in non-excitable cells, but it is also an important Ca signaling pathway in neurons and plays an important role in AD pathogenesis [112,113] (Figure 3). ER Ca^{2+} depletion induces ER calcium sensors STIM1 or STIM2 to translocate to the plasma membrane, where they activate Orai and/or TRPC channels, causing Ca^{2+} entry. In non-neuronal cell types, STIM1/ORAI1/TRPC have been shown to regulate autophagy and apoptosis [17], and recently, STIM1 mediated SOCE was reported to induce autophagy through AKT/mTOR pathways in hippocampal neurons under hypoxic conditions [114]. Dexmedetomidine (DEX) was reported to exert neuroprotective effects in PC12 cells through STIM1/Orai1 signaling pathways by regulating autophagy and apoptosis [115]. Previously we reported that STIM2/ORAI2/TRPC6 was impaired in AD [116], and TRPC6 was reported to induce autophagy through CaMKKβ-AMPK-mTOR pathways [117]. Thus, it will be interesting to explore the role of STIM2/ORAI2/TRPC6 impaired function in AD autophagy dysfunction. Moreover, it has been reported that ER stress can degrade STIM2 through autophagy, represses SOCE and disrupts dendrite arbor in primary neuronal cultures [118]. In the same paper it was also proposed that autophagy represses SOCE by degrading STIM proteins, leading to synapse loss in AD [119], providing additional potential feedback between Ca^{2+} signaling and autophagy in AD neurons.

6. Conclusions

Dysregulation of intracellular neuronal Ca^{2+} signaling and impairment of neuronal autophagy are two well-established phenomenon observed in AD neurons. Until recently studies of these two signaling pathways proceeded largely independently of each other. However, several recent papers point to a causal connection between them. Some results suggest that defects in lysosomal acidification may contribute to Ca^{2+} signaling defects by causing an overactivation of lysosomal TRPML1 channels [29]. Some data suggest that enhanced activity of RyanR may lead to defects in lysosomal acidification by affecting the activity of lysosomal V-ATPase [24]. Our recent findings suggest that basal Ca^{2+} released by RyanR2 may control steady-state levels of autophagy via CaN-AMPK-ULK1 pathways and that overactivation of RyanR2 in AD may lead to the overstimulation of CaN and the inhibition of autophagic flux [27]. Further studies will be needed to dissect relationships between intracellular Ca^{2+} dysregulation and autophagy in AD, but obtained results already offer some novel mechanistic insight and may potentially lead to the identification of novel therapeutic methods for treating AD and potentially other neurodegenerative disorders.

Author Contributions: Conceptualization, H.Z. and I.B., writing—original draft preparation, H.Z.; writing—review and editing, I.B.; visualization, H.Z. and I.B.; supervision, I.B.; project administration, I.B.; funding acquisition, I.B. All authors have read and agreed to the published version of the manuscript.

Funding: This research was funded by the Russian Science Foundation Grant 22-15-00049 (I.B.) and by the National Institutes of Health grant R01AG071310 (I.B.) I.B. holds the Carl J. and Hortense M. Thomsen Chair in Alzheimer's Disease Research.

Institutional Review Board Statement: Not applicable.

Informed Consent Statement: Not applicable.

Data Availability Statement: Not applicable.

Acknowledgments: We are thankful to members of Bezprozvanny laboratory for useful discussions.

Conflicts of Interest: The authors declare no conflict of interest.

References

1. Bezprozvanny, I. Alzheimer's disease—Where do we go from here? *Biochem. Biophys. Res. Commun.* **2022**, *633*, 72–76. [CrossRef] [PubMed]
2. Briggs, C.A.; Chakroborty, S.; Stutzmann, G.E. Emerging pathways driving early synaptic pathology in Alzheimer's disease. *Biochem. Biophys. Res. Commun.* **2017**, *483*, 988–997. [CrossRef]
3. Bezprozvanny, I.; Mattson, M.P. Neuronal calcium mishandling and the pathogenesis of Alzheimer's disease. *Trends Neurosci.* **2008**, *31*, 454–463. [CrossRef] [PubMed]
4. Popugaeva, E.; Vlasova, O.L.; Bezprozvanny, I. Restoring calcium homeostasis to treat Alzheimer's disease: A future perspective. *Neurodegener. Dis. Manag.* **2015**, *5*, 395–398. [CrossRef] [PubMed]
5. Popugaeva, E.; Chernyuk, D.; Bezprozvanny, I. Reversal of Calcium Dysregulation as Potential Approach for Treating Alzheimer's Disease. *Curr. Alzheimer Res.* **2020**, *17*, 344–354. [CrossRef] [PubMed]
6. Zhang, Z.G.; Yang, X.F.; Song, Y.Q.; Tu, J. Autophagy in Alzheimer's disease pathogenesis: Therapeutic potential and future perspectives. *Ageing Res. Rev.* **2021**, *72*, 101464. [CrossRef] [PubMed]
7. Liu, J.; Li, L. Targeting Autophagy for the Treatment of Alzheimer's Disease: Challenges and Opportunities. *Front. Mol. Neurosci.* **2019**, *12*, 203. [CrossRef]
8. Kuang, H.; Tan, C.Y.; Tian, H.Z.; Liu, L.H.; Yang, M.W.; Hong, F.F.; Yang, S.L. Exploring the bi-directional relationship between autophagy and Alzheimer's disease. *Cns Neurosci. Ther.* **2020**, *26*, 155–166. [CrossRef] [PubMed]
9. Chen, J.; He, H.J.; Ye, Q.Q.; Feng, F.F.; Wang, W.W.; Gu, Y.Y.; Han, R.Y.; Xie, C.L. Defective Autophagy and Mitophagy in Alzheimer's Disease: Mechanisms and Translational Implications. *Mol. Neurobiol.* **2021**, *58*, 5289–5302. [CrossRef]
10. Nixon, R.A.; Wegiel, J.; Kumar, A.; Yu, W.H.; Peterhoff, C.; Cataldo, A.; Cuervo, A.M. Extensive involvement of autophagy in Alzheimer disease: An immuno-electron microscopy study. *J. Neuropathol. Exp. Neurol.* **2005**, *64*, 113–122. [CrossRef]
11. Boland, B.; Kumar, A.; Lee, S.; Platt, F.M.; Wegiel, J.; Yu, W.H.; Nixon, R.A. Autophagy induction and autophagosome clearance in neurons: Relationship to autophagic pathology in Alzheimer's disease. *J. Neurosci.* **2008**, *28*, 6926–6937. [CrossRef]

12. Yu, W.H.; Cuervo, A.M.; Kumar, A.; Peterhoff, C.M.; Schmidt, S.D.; Lee, J.H.; Mohan, P.S.; Mercken, M.; Farmery, M.R.; Tjernberg, L.O.; et al. Macroautophagy—A novel Beta-amyloid peptide-generating pathway activated in Alzheimer's disease. *J. Cell Biol.* **2005**, *171*, 87–98. [CrossRef] [PubMed]
13. Sanchez-Varo, R.; Trujillo-Estrada, L.; Sanchez-Mejias, E.; Torres, M.; Baglietto-Vargas, D.; Moreno-Gonzalez, I.; De Castro, V.; Jimenez, S.; Ruano, D.; Vizuete, M.; et al. Abnormal accumulation of autophagic vesicles correlates with axonal and synaptic pathology in young Alzheimer's mice hippocampus. *Acta Neuropathol.* **2012**, *123*, 53–70. [CrossRef] [PubMed]
14. Cataldo, A.M.; Peterhoff, C.M.; Schmidt, S.D.; Terio, N.B.; Duff, K.; Beard, M.; Mathews, P.M.; Nixon, R.A. Presenilin mutations in familial Alzheimer disease and transgenic mouse models accelerate neuronal lysosomal pathology. *J. Neuropathol. Exp. Neurol.* **2004**, *63*, 821–830. [CrossRef] [PubMed]
15. Yang, D.S.; Stavrides, P.; Mohan, P.S.; Kaushik, S.; Kumar, A.; Ohno, M.; Schmidt, S.D.; Wesson, D.; Bandyopadhyay, U.; Jiang, Y.; et al. Reversal of autophagy dysfunction in the TgCRND8 mouse model of Alzheimer's disease ameliorates amyloid pathologies and memory deficits. *Brain* **2011**, *134*, 258–277. [CrossRef] [PubMed]
16. Medina, D.L. Lysosomal calcium and autophagy. *Int. Rev. Cell Mol. Biol.* **2021**, *362*, 141–170. [PubMed]
17. Sukumaran, P.; Da Conceicao, V.N.; Sun, Y.Y.; Ahamad, N.; Saraiva, L.R.; Selvaraj, S.; Singh, B.B. Calcium Signaling Regulates Autophagy and Apoptosis. *Cells* **2021**, *10*, 2125. [CrossRef]
18. La Rovere, R.M.L.; Roest, G.; Bultynck, G.; Parys, J.B. Intracellular Ca^{2+} signaling and Ca^{2+} microdomains in the control of cell survival, apoptosis and autophagy. *Cell Calcium* **2016**, *60*, 74–87. [CrossRef]
19. Decuypere, J.P.; Bultynck, G.; Parys, J.B. A dual role for Ca^{2+} in autophagy regulation. *Cell Calcium* **2011**, *50*, 242–250. [CrossRef] [PubMed]
20. Cardenas, C.; Miller, R.A.; Smith, I.; Bui, T.; Molgo, J.; Muller, M.; Vais, H.; Cheung, K.H.; Yang, J.; Parker, I.; et al. Essential regulation of cell bioenergetics by constitutive InsP3 receptor Ca^{2+} transfer to mitochondria. *Cell* **2010**, *142*, 270–283. [CrossRef]
21. Valladares, D.; Utreras-Mendoza, Y.; Campos, C.; Morales, C.; Diaz-Vegas, A.; Contreras-Ferrat, A.; Westermeier, F.; Jaimovich, E.; Marchi, S.; Pinton, P.; et al. IP3 receptor blockade restores autophagy and mitochondrial function in skeletal muscle fibers of dystrophic mice. *Biochim. Biophys. Acta Mol. Basis Dis.* **2018**, *1864*, 3685–3695. [CrossRef]
22. Lam, D.; Kosta, A.; Luciani, M.F.; Golstein, P. The inositol 1,4,5-trisphosphate receptor is required to signal autophagic cell death. *Mol. Biol. Cell* **2008**, *19*, 691–700. [CrossRef]
23. Khan, M.T.; Joseph, S.K. Role of inositol trisphosphate receptors in autophagy in DT40 cells. *J. Biol. Chem.* **2010**, *285*, 16912–16920. [CrossRef] [PubMed]
24. Mustaly-Kalimi, S.; Gallegos, W.; Marr, R.A.; Gilman-Sachs, A.; Peterson, D.A.; Sekler, I.; Stutzmann, G.E. Protein mishandling and impaired lysosomal proteolysis generated through calcium dysregulation in Alzheimer's disease. *Proc. Natl. Acad. Sci. USA* **2022**, *119*, e2211999119. [CrossRef]
25. Vervliet, T.; Pintelon, I.; Welkenhuyzen, K.; Bootman, M.D.; Bannai, H.; Mikoshiba, K.; Martinet, W.; Kasri, N.N.; Parys, J.B.; Bultynck, G. Basal ryanodine receptor activity suppresses autophagic flux. *Biochem. Pharmacol.* **2017**, *132*, 133–142. [CrossRef] [PubMed]
26. Vervliet, T. Ryanodine Receptors in Autophagy: Implications for Neurodegenerative Diseases? *Front. Cell. Neurosci.* **2018**, *12*, 89. [CrossRef] [PubMed]
27. Zhang, H.; Knight, C.; Chen, S.R.W.; Bezprozvanny, I. A gating mutation in ryanodine receptor type 2 rescues phenotypes of Alzheimer's disease mouse models by upregulating neuronal autophagy. *J. Neurosci.* **2023**, *43*, 1441–1454. [CrossRef]
28. Lee, J.H.; McBrayer, M.K.; Wolfe, D.M.; Haslett, L.J.; Kumar, A.; Sato, Y.; Lie, P.P.; Mohan, P.; Coffey, E.E.; Kompella, U.; et al. Presenilin 1 Maintains Lysosomal Ca^{2+} Homeostasis via TRPML1 by Regulating vATPase-Mediated Lysosome Acidification. *Cell Rep.* **2015**, *12*, 1344–1430. [CrossRef]
29. Lie, P.P.Y.; Yoo, L.; Goulbourne, C.N.; Berg, M.J.; Stavrides, P.; Huo, C.; Lee, J.H.; Nixon, R.A. Axonal transport of late endosomes and amphisomes is selectively modulated by local Ca^{2+} efflux and disrupted by PSEN1 loss of function. *Sci. Adv.* **2022**, *8*, eabj5716. [CrossRef]
30. Leissring, M.A.; Paul, B.A.; Parker, I.; Cotman, C.W.; LaFerla, F.M. Alzheimer's Alzheimer's presenilin-1 mutation potentiates inositol 1,4,5-trisphosphate-mediated calcium signaling in Xenopus oocytes. *J. Neurochem.* **1999**, *72*, 1061–1068. [CrossRef]
31. Stutzmann, G.E.; Caccamo, A.; LaFerla, F.M.; Parker, I. Dysregulated IP3 signaling in cortical neurons of knock-in mice expressing an Alzheimer's-linked mutation in presenilin1 results in exaggerated Ca^{2+} signals and altered membrane excitability. *J Neurosci.* **2004**, *24*, 508–513. [CrossRef]
32. Toglia, P.; Ullah, G. The gain-of-function enhancement of IP3-receptor channel gating by familial Alzheimer's disease-linked presenilin mutants increases the open probability of mitochondrial permeability transition pore. *Cell Calcium.* **2016**, *60*, 13–24. [CrossRef]
33. Toglia, P.; Cheung, K.H.; Mak, D.O.; Ullah, G. Impaired mitochondrial function due to familial Alzheimer's disease-causing presenilins mutants via Ca^{2+} disruptions. *Cell Calcium* **2016**, *59*, 240–250. [CrossRef] [PubMed]
34. Mak, D.O.; Cheung, K.H.; Toglia, P.; Foskett, J.K.; Ullah, G. Analyzing and Quantifying the Gain-of-Function Enhancement of IP3 Receptor Gating by Familial Alzheimer's Disease-Causing Mutants in Presenilins. *PLoS Comput. Biol.* **2015**, *11*, e1004529. [CrossRef] [PubMed]

35. Jensen, L.E.; Bultynck, G.; Luyten, T.; Amijee, H.; Bootman, M.D.; Roderick, H.L. Alzheimer's Alzheimer's disease-associated peptide Abeta42 mobilizes ER Ca^{2+} via InsP3R-dependent and -independent mechanisms. *Front Mol. Neurosci.* **2013**, *6*, 36. [CrossRef] [PubMed]
36. Shilling, D.; Muller, M.; Takano, H.; Mak, D.O.; Abel, T.; Coulter, D.A.; Foskett, J.K. Suppression of InsP3 receptor-mediated Ca^{2+} signaling alleviates mutant presenilin-linked familial Alzheimer's disease pathogenesis. *J Neurosci.* **2014**, *34*, 6910–6923. [CrossRef] [PubMed]
37. Baker, K.D.; Edwards, T.M.; Rickard, N.S. The role of intracellular calcium stores in synaptic plasticity and memory consolidation. *Neurosci. Biobehav. Rev.* **2013**, *37*, 1211–1239. [CrossRef]
38. Sugawara, T.; Hisatsune, C.; Le, T.D.; Hashikawa, T.; Hirono, M.; Hattori, M.; Nagao, S.; Mikoshiba, K. Type 1 inositol trisphosphate receptor regulates cerebellar circuits by maintaining the spine morphology of purkinje cells in adult mice. *J. Neurosci.* **2013**, *33*, 12186–12196. [CrossRef]
39. Lai, F.A.; Dent, M.; Wickenden, C.; Xu, L.; Kumari, G.; Misra, M.; Lee, H.B.; Sar, M.; Meissner, G. Expression of a Cardiac Ca-2+-Release Channel Isoform in Mammalian Brain. *Biochem. J.* **1992**, *288*, 553–564. [CrossRef]
40. Furuichi, T.; Furutama, D.; Hakamata, Y.; Nakai, J.; Takeshima, H.; Mikoshiba, K. Multiple Types of Ryanodine Receptor Ca^{2+} Release Channels Are Differentially Expressed in Rabbit Brain. *J. Neurosci.* **1994**, *14*, 4794–4805. [CrossRef]
41. Hertle, D.N.; Yeckel, M.F. Distribution of inositol-1,4,5-trisphosphate receptor isotypes and ryanodine receptor isotypes during maturation of the rat hippocampus. *Neuroscience* **2007**, *150*, 625–638. [CrossRef] [PubMed]
42. Liu, J.; Supnet, C.; Sun, S.; Zhang, H.; Good, L.; Popugaeva, E.; Bezprozvanny, I. The role of ryanodine receptor type 3 in a mouse model of Alzheimer disease. *Channels* **2014**, *8*, 230–242. [CrossRef]
43. Zima, A.V.; Mazurek, S.R. Functional Impact of Ryanodine Receptor Oxidation on Intracellular Calcium Regulation in the Heart. *Rev. Physiol. Biochem. Pharmacol.* **2016**, *171*, 39–62. [PubMed]
44. Smith, I.F.; Hitt, B.; Green, K.N.; Oddo, S.; LaFerla, F.M. Enhanced caffeine-induced Ca^{2+} release in the 3×Tg-AD mouse model of Alzheimer's disease. *J. Neurochem.* **2005**, *94*, 1711–1718. [CrossRef]
45. Oules, B.; Del Prete, D.; Greco, B.; Zhang, X.; Lauritzen, I.; Sevalle, J.; Moreno, S.; Paterlini-Brechot, P.; Trebak, M.; Checler, F.; et al. Ryanodine receptor blockade reduces amyloid-beta load and memory impairments in Tg2576 mouse model of Alzheimer disease. *J. Neurosci.* **2012**, *32*, 11820–11834. [CrossRef]
46. Kelliher, M.; Fastbom, J.; Cowburn, R.F.; Bonkale, W.; Ohm, T.G.; Ravid, R.; Sorrentino, V.; O'Neill, C. Alterations in the ryanodine receptor calcium release channel correlate with Alzheimer's disease neurofibrillary and beta-amyloid pathologies. *Neuroscience* **1999**, *92*, 499–513. [CrossRef] [PubMed]
47. Chakroborty, S.; Goussakov, I.; Miller, M.B.; Stutzmann, G.E. Deviant ryanodine receptor-mediated calcium release resets synaptic homeostasis in presymptomatic 3xTg-AD mice. *J. Neurosci.* **2009**, *29*, 9458–9470. [CrossRef]
48. Lacampagne, A.; Liu, X.P.; Reiken, S.; Bussiere, R.; Meli, A.C.; Lauritzen, I.; Teich, A.F.; Zalk, R.; Saint, N.; Arancio, O.; et al. Post-translational remodeling of ryanodine receptor induces calcium leak leading to Alzheimer's disease-like pathologies and cognitive deficits. *Acta Neuropathol.* **2017**, *134*, 749–767. [CrossRef]
49. Bruno, A.M.; Huang, J.Y.; Bennett, D.A.; Marr, R.A.; Hastings, M.L.; Stutzmann, G.E. Altered ryanodine receptor expression in mild cognitive impairment and Alzheimer's disease. *Neurobiol Aging.* **2012**, *33*, 1001.e1–1001.e6. [CrossRef]
50. Zhang, H.; Sun, S.; Herreman, A.; De Strooper, B.; Bezprozvanny, I. Role of presenilins in neuronal calcium homeostasis. *J. Neurosci.* **2010**, *30*, 8566–8580. [CrossRef]
51. Gant, J.C.; Sama, M.M.; Landfield, P.W.; Thibault, O. Early and simultaneous emergence of multiple hippocampal biomarkers of aging is mediated by Ca^{2+}-induced Ca^{2+} release. *J. Neurosci.* **2006**, *26*, 3482–3490. [CrossRef] [PubMed]
52. Gant, J.C.; Blalock, E.M.; Chen, K.C.; Kadish, I.; Porter, N.M.; Norris, C.M.; Thibault, O.; Landfield, P.W. FK506-binding protein 1b/12.6: A key to aging-related hippocampal Ca^{2+} dysregulation? *Eur. J. Pharmacol.* **2014**, *739*, 74–82. [CrossRef] [PubMed]
53. Peng, J.; Liang, G.; Inan, S.; Wu, Z.; Joseph, D.J.; Meng, Q.C.; Peng, Y.; Eckenhoff, M.F.; Wei, H.F. Dantrolene ameliorates cognitive decline and neuropathology in Alzheimer triple transgenic mice. *Neurosci. Lett.* **2012**, *516*, 274–279. [CrossRef] [PubMed]
54. Chakroborty, S.; Briggs, C.; Miller, M.B.; Goussakov, I.; Schneider, C.; Kim, J.; Wicks, J.; Richardson, J.C.; Conklin, V.; Cameransi, B.G.; et al. Stabilizing ER Ca^{2+} Channel Function as an Early Preventative Strategy for Alzheimer's Disease. *PLoS ONE* **2012**, *7*, e52056. [CrossRef] [PubMed]
55. Liu, Y.; Yao, J.; Song, Z.; Guo, W.; Sun, B.; Wei, J.; Estillore, J.P.; Back, T.G.; Chen, S.R.W. Limiting RyR2 open time prevents Alzheimer's disease-related deficits in the 3xTG-AD mouse model. *J Neurosci Res.* **2021**, *99*, 2906–2921. [CrossRef]
56. Sun, B.; Yao, J.; Chen, A.W.; Estillore, J.P.; Wang, R.; Back, T.G.; Chen, S.R.W. Genetically and pharmacologically limiting RyR2 open time prevents neuronal hyperactivity of hippocampal CA1 neurons in brain slices of 5xFAD mice. *Neurosci. Lett.* **2021**, *758*, 136011. [CrossRef]
57. Yao, J.; Sun, B.; Institoris, A.; Zhan, X.; Guo, W.; Song, Z.; Liu, Y.; Hiess, F.; Boyce, A.K.J.; Ni, M.; et al. Limiting RyR2 Open Time Prevents Alzheimer's Disease-Related Neuronal Hyperactivity and Memory Loss but Not beta-Amyloid Accumulation. *Cell Rep.* **2020**, *32*, 108169. [CrossRef]
58. Nakamura, Y.; Yamamoto, T.; Xu, X.J.; Kobayashi, S.; Tanaka, S.; Tamitani, M.; Saito, T.; Saido, T.C.; Yano, M. Enhancing calmodulin binding to ryanodine receptor is crucial to limit neuronal cell loss in Alzheimer disease. *Sci. Rep.* **2021**, *11*, 7289. [CrossRef]
59. Mizushima, N.; Levine, B. Autophagy in Human Diseases. *N. Engl. J. Med.* **2020**, *383*, 1564–1576. [CrossRef]

60. Fleming, A.; Bourdenx, M.; Fujimaki, M.; Karabiyik, C.; Krause, G.J.; Lopez, A.; Puri, C.; Scrivo, A.; Skidmore, J.; Son, S.M.; et al. The different autophagy degradation pathways and neurodegeneration. *Neuron* **2022**, *110*, 935–966. [CrossRef]
61. Metaxakis, A.; Ploumi, C.; Tavernarakis, N. Autophagy in Age-Associated Neurodegeneration. *Cells* **2018**, *7*, 37. [CrossRef]
62. Hara, T.; Nakamura, K.; Matsui, M.; Yamamoto, A.; Nakahara, Y.; Suzuki-Migishima, R.; Yokoyama, M.; Mishima, K.; Saito, I.; Okano, H.; et al. Suppression of basal autophagy in neural cells causes neurodegenerative disease in mice. *Nature* **2006**, *441*, 885–889. [CrossRef] [PubMed]
63. Komatsu, M.; Waguri, S.; Chiba, T.; Murata, S.; Iwata, J.; Tanida, I.; Ueno, T.; Koike, M.; Uchiyama, Y.; Kominami, E.; et al. Loss of autophagy in the central nervous system causes neurodegeneration in mice. *Nature* **2006**, *441*, 880–884. [CrossRef] [PubMed]
64. Goldsmith, J.; Ordureau, A.; Harper, J.W.; Holzbaur, E.L.F. Brain-derived autophagosome profiling reveals the engulfment of nucleoid-enriched mitochondrial fragments by basal autophagy in neurons. *Neuron* **2022**, *110*, 967–976.e8. [CrossRef]
65. Kuijpers, M.; Kochlamazashvili, G.; Stumpf, A.; Puchkov, D.; Swaminathan, A.; Lucht, M.T.; Krause, E.; Maritzen, T.; Schmitz, D.; Haucke, V. Neuronal Autophagy Regulates Presynaptic Neurotransmission by Controlling the Axonal Endoplasmic Reticulum. *Neuron* **2021**, *109*, 299–313.e9. [CrossRef] [PubMed]
66. Binotti, B.; Pavlos, N.J.; Riedel, D.; Wenzel, D.; Vorbruggen, G.; Schalk, A.M.; Kuhnel, K.; Boyken, J.; Erck, C.; Martens, H.; et al. The GTPase Rab26 links synaptic vesicles to the autophagy pathway. *eLife* **2015**, *4*, e05597. [CrossRef] [PubMed]
67. Stavoe, A.K.; Holzbaur, E.L. Axonal autophagy: Mini-review for autophagy in the CNS. *Neurosci. Lett.* **2019**, *697*, 17–23. [CrossRef]
68. Kuijpers, M. Keeping synapses in shape: Degradation pathways in the healthy and aging brain. *Neuronal Signal* **2022**, *6*, NS20210063. [CrossRef]
69. Komatsu, M.; Wang, Q.J.; Holstein, G.R.; Friedrich, V.L., Jr.; Iwata, J.; Kominami, E.; Chait, B.T.; Tanaka, K.; Yue, Z. Essential role for autophagy protein Atg7 in the maintenance of axonal homeostasis and the prevention of axonal degeneration. *Proc. Natl. Acad. Sci. USA* **2007**, *104*, 14489–14494. [CrossRef]
70. Hernandez, D.; Torres, C.A.; Setlik, W.; Cebrian, C.; Mosharov, E.V.; Tang, G.M.; Cheng, H.C.; Kholodilov, N.; Yarygina, O.; Burke, R.E.; et al. Regulation of Presynaptic Neurotransmission by Macroautophagy. *Neuron* **2012**, *74*, 277–284. [CrossRef]
71. Hwang, H.J.; Ha, H.; Lee, B.S.; Kim, B.H.; Song, H.K.; Kim, Y.K. LC3B is an RNA-binding protein to trigger rapid mRNA degradation during autophagy. *Nat. Commun.* **2022**, *13*, 1436. [CrossRef] [PubMed]
72. Compans, B.; Camus, C.; Kallergi, E.; Sposini, S.; Martineau, M.; Butler, C.; Kechkar, A.; Klaassen, R.V.; Retailleau, N.; Sejnowski, T.J.; et al. NMDAR-dependent long-term depression is associated with increased short term plasticity through autophagy mediated loss of PSD-95. *Nat. Commun.* **2021**, *12*, 2849. [CrossRef] [PubMed]
73. Kallergi, E.; Daskalaki, A.D.; Kolaxi, A.; Camus, C.; Ioannou, E.; Mercaldo, V.; Haberkant, P.; Stein, F.; Sidiropoulou, K.; Dalezios, Y.; et al. Dendritic autophagy degrades postsynaptic proteins and is required for long-term synaptic depression in mice. *Nat. Commun.* **2022**, *13*, 680. [CrossRef] [PubMed]
74. Feng, T.C.; Tammineni, P.; Agrawal, C.; Jeong, Y.Y.; Cai, Q. Autophagy-mediated Regulation of BACE1 Protein Trafficking and Degradation. *J. Biol. Chem.* **2017**, *292*, 1679–1690. [CrossRef]
75. Nilsson, P.; Saido, T.C. Dual roles for autophagy: Degradation and secretion of Alzheimer's disease Abeta peptide. *Bioessays* **2014**, *36*, 570–578. [CrossRef]
76. Nilsson, P.; Loganathan, K.; Sekiguchi, M.; Matsuba, Y.; Hui, K.; Tsubuki, S.; Tanaka, M.; Iwata, N.; Saito, T.; Saido, T.C. Abeta secretion and plaque formation depend on autophagy. *Cell Rep.* **2013**, *5*, 61–69. [CrossRef]
77. Wang, Y.P.; Martinez-Vicente, M.; Krüger, U.; Kaushik, S.; Wong, E.; Mandelkow, E.-M.; Cuervo, A.M.; Mandelkow, E. Tau fragmentation, aggregation and clearance: The dual role of lysosomal processing. *Hum. Mol. Genet.* **2009**, *18*, 4153–4170. [CrossRef]
78. Dolan, P.J.; Johnson, G.V. A Caspase Cleaved Form of Tau Is Preferentially Degraded through the Autophagy Pathway. *J. Biol. Chem.* **2010**, *285*, 21978–21987. [CrossRef]
79. Caballero, B.; Bourdenx, M.; Luengo, E.; Diaz, A.; Sohn, P.D.; Chen, X.; Wang, C.; Juste, Y.R.; Wegmann, S.; Patel, B.; et al. Acetylated tau inhibits chaperone-mediated autophagy and promotes tau pathology propagation in mice. *Nat. Commun.* **2021**, *12*, 2238. [CrossRef]
80. Ozcelik, S.; Fraser, G.; Castets, P.; Schaeffer, V.; Skachokova, Z.; Breu, K.; Clavaguera, F.; Sinnreich, M.; Kappos, L.; Goedert, M.; et al. Rapamycin Attenuates the Progression of Tau Pathology in P301S Tau Transgenic Mice. *PLoS ONE* **2013**, *8*, e62459. [CrossRef]
81. Schaeffer, V.; Lavenir, I.; Ozcelik, S.; Tolnay, M.; Winkler, D.T.; Goedert, M. Stimulation of autophagy reduces neurodegeneration in a mouse model of human tauopathy. *Brain* **2012**, *135*, 2169–2177. [CrossRef] [PubMed]
82. Katsinelos, T.; Zeitler, M.; Dimou, E.; Karakatsani, A.; Muller, H.M.; Nachman, E.; Steringer, J.P.; de Almodovar, C.R.; Nickel, W.; Jahn, T.R. Unconventional Secretion Mediates the Trans-cellular Spreading of Tau. *Cell Rep.* **2018**, *23*, 2039–2055. [CrossRef] [PubMed]
83. Kang, S.; Son, S.M.; Baik, S.H.; Yang, J.; Mook-Jung, I. Autophagy-mediated secretory pathway is responsible for both normal and pathological tau in neurons. *J. Alzheimers Dis.* **2019**, *70*, 667–680. [CrossRef]
84. Ruan, Z.; Pathak, D.; Kalavai, S.V.; Yoshii-Kitahara, A.; Muraoka, S.; Bhatt, N.; Takamatsu-Yukawa, K.; Hu, J.; Wang, Y.; Hersh, S.; et al. Alzheimer's disease brain-derived extracellular vesicles spread tau pathology in interneurons. *Brain* **2021**, *144*, 288. [CrossRef]

85. Sebastian-Serrano, A.; de Diego-Garcia, L.; Diaz-Hernandez, M. The Neurotoxic Role of Extracellular Tau Protein. *Int. J. Mol. Sci.* **2018**, *19*, 998. [CrossRef] [PubMed]
86. Pickford, F.; Masliah, E.; Britschgi, M.; Lucin, K.; Narasimhan, R.; Jaeger, P.A.; Small, S.; Spencer, B.; Rockenstein, E.; Levine, B.; et al. The autophagy-related protein beclin 1 shows reduced expression in early Alzheimer disease and regulates amyloid beta accumulation in mice. *J. Clin. Invest* **2008**, *118*, 2190–2199. [PubMed]
87. Du, Y.; Wooten, M.C.; Gearing, M.; Wooten, M.W. Age-associated oxidative damage to the p62 promoter: Implications for Alzheimer disease. *Free Radic. Biol. Med.* **2009**, *46*, 492–501. [CrossRef]
88. Lipinski, M.M.; Zheng, B.; Lu, T.; Yan, Z.; Py, B.F.; Ng, A.; Xavier, R.J.; Li, C.; Yankner, B.A.; Scherzer, C.R.; et al. Genome-wide analysis reveals mechanisms modulating autophagy in normal brain aging and in Alzheimer's disease. *Proc. Natl. Acad. Sci. USA* **2010**, *107*, 14164–14169. [CrossRef]
89. Colacurcio, D.J.; Pensalfini, A.; Jiang, Y.; Nixon, R.A. Dysfunction of autophagy and endosomal-lysosomal pathways: Roles in pathogenesis of Down syndrome and Alzheimer's Disease. *Free Radic. Biol. Med.* **2018**, *114*, 40–51. [CrossRef]
90. Lee, J.H.; Yang, D.S.; Goulbourne, C.N.; Im, E.; Stavrides, P.; Pensalfini, A.; Chan, H.; Bouchet-Marquis, C.; Bleiwas, C.; Berg, M.J.; et al. Faulty autolysosome acidification in Alzheimer's disease mouse models induces autophagic build-up of Abeta in neurons, yielding senile plaques. *Nat. Neurosci.* **2022**, *25*, 688–701. [CrossRef]
91. Lee, J.H.; Yu, W.H.; Kumar, A.; Lee, S.; Mohan, P.S.; Peterhoff, C.M.; Wolfe, D.M.; Martinez-Vicente, M.; Massey, A.C.; Sovak, G.; et al. Lysosomal Proteolysis and Autophagy Require Presenilin 1 and Are Disrupted by Alzheimer-Related PS1 Mutations. *Cell* **2010**, *141*, 1146–1158. [CrossRef]
92. Coen, K.; Flannagan, R.S.; Baron, S.; Carraro-Lacroix, L.R.; Wang, D.; Vermeire, W.; Michiels, C.; Munck, S.; Baert, V.; Sugita, S.; et al. Lysosomal calcium homeostasis defects, not proton pump defects, cause endo-lysosomal dysfunction in PSEN-deficient cells. *J. Cell Biol.* **2012**, *198*, 23–35. [CrossRef]
93. Zhang, X.; Garbett, K.; Veeraraghavalu, K.; Wilburn, B.; Gilmore, R.; Mirnics, K.; Sisodia, S.S. A role for presenilins in autophagy revisited: Normal acidification of lysosomes in cells lacking PSEN1 and PSEN2. *J. Neurosci.* **2012**, *32*, 8633–8648. [CrossRef] [PubMed]
94. Bezprozvanny, I. Presenilins: A novel link between intracellular calcium signaling and lysosomal function? *J. Cell Biol.* **2012**, *198*, 7–10. [CrossRef] [PubMed]
95. Colacurcio, D.J.; Nixon, R.A. Disorders of lysosomal acidification-The emerging role of v-ATPase in aging and neurodegenerative disease. *Ageing Res. Rev.* **2016**, *32*, 75–88. [CrossRef]
96. Torres, M.; Jimenez, S.; Sanchez-Varo, R.; Navarro, V.; Trujillo-Estrada, L.; Sanchez-Mejias, E.; Carmona, I.; Davila, J.C.; Vizuete, M.; Gutierrez, A.; et al. Defective lysosomal proteolysis and axonal transport are early pathogenic events that worsen with age leading to increased APP metabolism and synaptic Abeta in transgenic APP/PS1 hippocampus. *Mol. Neurodegener.* **2012**, *7*, 59. [CrossRef] [PubMed]
97. Zi, Z.; Zhang, Z.; Feng, Q.; Kim, C.; Wang, X.D.; Scherer, P.E.; Gao, J.; Levine, B.; Yu, Y. Quantitative phosphoproteomic analyses identify STK11IP as a lysosome-specific substrate of mTORC1 that regulates lysosomal acidification. *Nat. Commun.* **2022**, *13*, 1760. [CrossRef]
98. Atakpa, P.; Thillaiappan, N.B.; Mataragka, S.; Prole, D.L.; Taylor, C.W. IP_3 Receptors Preferentially Associate with ER-Lysosome Contact Sites and Selectively Deliver Ca^{2+} to Lysosomes. *Cell Rep.* **2018**, *25*, 3180–3193.e7. [CrossRef]
99. Kong, A.; Zhang, Y.; Ning, B.; Li, K.; Ren, Z.; Dai, S.; Chen, D.; Zhou, Y.; Gu, J.; Shi, H. Cadmium induces triglyceride levels via microsomal triglyceride transfer protein (MTTP) accumulation caused by lysosomal deacidification regulated by endoplasmic reticulum (ER) Ca^{2+} homeostasis. *Chem. Biol. Interact.* **2021**, *348*, 109649. [CrossRef]
100. Dong, X.P.; Shen, D.; Wang, X.; Dawson, T.; Li, X.; Zhang, Q.; Cheng, X.; Zhang, Y.; Weisman, L.S.; Delling, M.; et al. PI(3,5)P_2 controls membrane trafficking by direct activation of mucolipin Ca^{2+} release channels in the endolysosome. *Nat. Commun.* **2010**, *1*, 38. [CrossRef]
101. Krogsaeter, E.; Rosato, A.S.; Grimm, C. TRPMLs and TPCs: Targets for lysosomal storage and neurodegenerative disease therapy? *Cell Calcium* **2022**, *103*, 102553. [CrossRef] [PubMed]
102. Chung, K.M.; Jeong, E.J.; Park, H.; An, H.K.; Yu, S.W. Mediation of Autophagic Cell Death by Type 3 Ryanodine Receptor (RyR3) in Adult Hippocampal Neural Stem Cells. *Front. Cell. Neurosci.* **2016**, *10*, 116. [CrossRef] [PubMed]
103. Law, B.Y.K.; Michelangeli, F.; Qu, Y.Q.; Xu, S.W.; Han, Y.; Mok, S.W.F.A.; Dias, I.R.D.R.; Javed, M.U.; Chan, W.K.; Xue, W.W.; et al. Neferine induces autophagy-dependent cell death in apoptosis-resistant cancers via ryanodine receptor and Ca^{2+}-dependent mechanism. *Sci. Rep.* **2019**, *9*, 20034. [CrossRef] [PubMed]
104. Qiao, H.; Li, Y.; Xu, Z.D.; Li, W.X.; Fu, Z.J.; Wang, Y.Z.; King, A.; Wei, H.F. Propofol Affects Neurodegeneration and Neurogenesis by Regulation of Autophagy via Effects on Intracellular Calcium Homeostasis. *Anesthesiology* **2017**, *127*, 490–501. [CrossRef]
105. Chen, W.Q.; Wang, R.W.; Chen, B.Y.; Zhong, X.W.; Kong, H.H.; Bai, Y.L.; Zhou, Q.; Xie, C.H.; Zhang, J.Q.; Guo, A.; et al. The ryanodine receptor store-sensing gate controls Ca^{2+} waves and Ca^{2+}-triggered arrhythmias. *Nat. Med.* **2014**, *20*, 184–192. [CrossRef] [PubMed]
106. Saucerman, J.J.; Bers, D.M. Calmodulin Mediates Differential Sensitivity of CaMKII and Calcineurin to Local Ca^{2+} in Cardiac Myocytes. *Biophys. J.* **2008**, *95*, 4597–4612. [CrossRef]
107. Stefan, M.I.; Edelstein, S.J.; Le Novere, N. An allosteric model of calmodulin explains differential activation of PP2B and CaMKII. *Proc. Natl. Acad. Sci. USA* **2008**, *105*, 10768–10773. [CrossRef]

108. Medina, D.L.; Di Paola, S.; Peluso, I.; Armani, A.; De Stefani, D.; Venditti, R.; Montefusco, S.; Scotto-Rosato, A.; Prezioso, C.; Forrester, A.; et al. Lysosomal calcium signalling regulates autophagy through calcineurin and TFEB. *Nat. Cell Biol.* **2015**, *17*, 288–299. [CrossRef]
109. Rosato, A.S.; Montefusco, S.; Soldati, C.; Di Paola, S.; Capuozzo, A.; Monfregola, J.; Polishchuk, E.; Amabile, A.; Grimm, C.; Lombardo, A.; et al. TRPML1 links lysosomal calcium to autophagosome biogenesis through the activation of the CaMKK beta/VPS34 pathway. *Nat. Commun.* **2019**, *10*, 5630. [CrossRef]
110. Somogyi, A.; Kirkham, E.D.; Lloyd-Evans, E.; Winston, J.; Allen, N.D.; Mackrill, J.J.; Anderson, K.E.; Hawkins, P.T.; Gardiner, S.E.; Waller-Evans, H.; et al. The synthetic TRPML1 agonist ML-SA1 rescues Alzheimer-related alterations of the endosomal-autophagic-lysosomal system. *J. Cell Sci.* **2023**, *136*, jcs259875. [CrossRef]
111. Xu, Y.; Du, S.; Marsh, J.A.; Horie, K.; Sato, C.; Ballabio, A.; Karch, C.M.; Holtzman, D.M.; Zheng, H. TFEB regulates lysosomal exocytosis of tau and its loss of function exacerbates tau pathology and spreading. *Mol Psych.* **2021**, *26*, 5925–5939. [CrossRef]
112. Huang, A.S.; Tong, B.C.K.; Wu, A.J.; Chen, X.T.; Sreenivasmurthy, S.G.; Zhu, Z.; Liu, J.; Su, C.F.; Li, M.; Cheune, K.H. Rectifying Attenuated Store-Operated Calcium Entry as a Therapeutic Approach for Alzheimer's Disease. *Curr. Alzheimer Res.* **2020**, *17*, 1072–1087. [CrossRef] [PubMed]
113. Popugaeva, E.; Pchitskaya, E.; Bezprozvanny, I. Dysregulation of Intracellular Calcium Signaling in Alzheimer's Disease. *Antioxid. Redox Signal* **2018**, *29*, 1176–1188. [CrossRef]
114. Zhang, H.C.; Xie, W.Y.; Feng, Y.; Wei, J.L.; Yang, C.B.; Luo, P.; Yang, Y.F.; Zhao, P.; Jiang, X.F.; Liang, W.B.; et al. Stromal Interaction Molecule 1-Mediated Store-Operated Calcium Entry Promotes Autophagy Through AKT/Mammalian Target of Rapamycin Pathway in Hippocampal Neurons After Ischemic Stroke. *Neuroscience* **2023**, *514*, 67–78. [CrossRef] [PubMed]
115. Hu, Y.D.; Tang, C.L.; Jiang, J.Z.; Lv, H.Y.; Wu, Y.B.; Qin, X.D.; Shi, S.; Zhao, B.; Zhu, X.N.; Xia, Z.Y. Neuroprotective Effects of Dexmedetomidine Preconditioning on Oxygen-glucose Deprivation-reoxygenation Injury in PC12 Cells via Regulation of Ca^{2+}-STIM1/Orai1 Signaling. *Curr. Med. Sci.* **2020**, *40*, 699–707. [CrossRef] [PubMed]
116. Zhang, H.; Sun, S.; Wu, L.; Pchitskaya, E.; Zakharova, O.; Fon Tacer, K.; Bezprozvanny, I. Store-Operated Calcium Channel Complex in Postsynaptic Spines: A New Therapeutic Target for Alzheimer's Disease Treatment. *J. Neurosci.* **2016**, *36*, 11837–11850. [CrossRef] [PubMed]
117. Ma, S.; Xu, J.; Zheng, Y.; Li, Y.; Wang, Y.; Li, H.; Fang, Z.; Li, J. Qian Yang Yu Yin granule improves hypertensive renal damage: A potential role for TRPC6-CaMKKbeta-AMPK-mTOR-mediated autophagy. *J. Ethnopharmacol.* **2023**, *302*, 115878. [CrossRef]
118. Zhou, J.; Song, J.; Wu, S. Autophagic degradation of stromal interaction molecule 2 mediates disruption of neuronal dendrites by endoplasmic reticulum stress. *J. Neurochem.* **2019**, *151*, 351–369. [CrossRef]
119. Zhou, J.; Wu, S. Impairment of Store-operated Calcium Entry: Implications in Alzheimer's Neurodegeneration. *Curr. Alzheimer. Res.* **2020**, *17*, 1088–1094. [CrossRef]

life

MDPI

Review

Biophysical Aspects of Neurodegenerative and Neurodevelopmental Disorders Involving Endo-/Lysosomal CLC Cl^-/H^+ Antiporters

Maria Antonietta Coppola [1,2], Abraham Tettey-Matey [1], Paola Imbrici [2], Paola Gavazzo [1], Antonella Liantonio [2] and Michael Pusch [1,3,*]

[1] Istituto di Biofisica, Consiglio Nazionale delle Ricerche, 16149 Genova, Italy; maria.coppola@uniba.it (M.A.C.); masirlow@gmail.com (A.T.-M.); paola.gavazzo@ibf.cnr.it (P.G.)
[2] Department of Pharmacy–Drug Sciences, University of Bari "Aldo Moro", 70125 Bari, Italy; paola.imbrici@uniba.it (P.I.); antonella.liantonio@uniba.it (A.L.)
[3] RAISE Ecosystem, 16149 Genova, Italy
* Correspondence: michael.pusch@ibf.cnr.it

Abstract: Endosomes and lysosomes are intracellular vesicular organelles with important roles in cell functions such as protein homeostasis, clearance of extracellular material, and autophagy. Endolysosomes are characterized by an acidic luminal pH that is critical for proper function. Five members of the gene family of voltage-gated ChLoride Channels (CLC proteins) are localized to endolysosomal membranes, carrying out anion/proton exchange activity and thereby regulating pH and chloride concentration. Mutations in these vesicular CLCs cause global developmental delay, intellectual disability, various psychiatric conditions, lysosomal storage diseases, and neurodegeneration, resulting in severe pathologies or even death. Currently, there is no cure for any of these diseases. Here, we review the various diseases in which these proteins are involved and discuss the peculiar biophysical properties of the WT transporter and how these properties are altered in specific neurodegenerative and neurodevelopmental disorders.

Keywords: CLC proteins; endosome; lysosome; chloride transport; developmental

Citation: Coppola, M.A.; Tettey-Matey, A.; Imbrici, P.; Gavazzo, P.; Liantonio, A.; Pusch, M. Biophysical Aspects of Neurodegenerative and Neurodevelopmental Disorders Involving Endo-/Lysosomal CLC Cl^-/H^+ Antiporters. *Life* **2023**, *13*, 1317. https://doi.org/10.3390/life13061317

Academic Editor: Carlo Musio

Received: 22 May 2023
Revised: 30 May 2023
Accepted: 31 May 2023
Published: 2 June 2023

1. Introduction—The CLC Family

Physiologically, the most abundant anion is chloride. It is an important substrate of many transport proteins, being carried across the membrane as a single anion or coupled with other ions, and is important, for example, for the regulation of the membrane potential, intracellular vesicles acidification and cell volume regulation [1].

In humans, the CLC family is formed by nine members, which had initially been supposed to be all chloride channels, because of their sequence homology with the founding member, the Torpedo electroplax channel ClC-0 [2]. The discovery that the bacterial *Escherichia coli* ecClC-1 homologue is not a passive chloride channel but a stoichiometrically coupled secondary active 2 Cl^-/1 H^+ antiporter has dramatically changed the point of view of the entire CLC group [3]. Based on sequence homology, three branches of human CLCs have been distinguished. The first one includes the plasma membrane-localized chloride channels ClC-1, ClC-2 and the two isoforms ClC-Ka and ClC-Kb. The second branch is formed by ClC-3, ClC-4 and ClC-5, while the third branch contains ClC-6 and ClC-7. ClC-3 to -7 are all Cl^-/H^+ exchangers and are localized to the intracellular membranes of endosomes and/or lysosomes [1].

All CLC family members share the same dimeric architecture that is unique to this protein family. Except for ClC-6 and ClC-7 [4], the other CLC proteins can form homo- or hetero-dimers with members of the same branch [1]. Biochemical studies and single-channel analysis on the first cloned *Torpedo* ClC-0, mutants [2,5,6] and biochemical and low-resolution structural analysis of ecClC-1 [7,8] suggested a homodimeric "double-barreled" architecture, with physically separated anion transport pathways in each protomer. This architecture has been fully confirmed by the determination of ecClC-1 and *Salmonella typhimurium* stClC crystal structures [9,10]. The structures revealed the presence of distinct anion binding sites, formed by residues that are also highly conserved in human CLCs. The sites are denominated Sext, Scen and Sint, with Sext being occupied by the presumably negatively charged side chain of the "gating glutamate" E148 [9,10]. Each monomer presents 18 α-helices (from A to R) of which 17 are partially embedded in the membrane. The two subunits interact in a tight manner and the architecture follows an inverted and parallel orientation [1]. Two C-terminal tandem cystathionine-β-synthase (CBS) domains are present in most eukaryotic CLC proteins [11,12], but are absent in ecClC-1. The two CBS domains may have a role in the so-called common gating process (that will be discussed in more detail below) and confer unique features to the CLC members [1]. Dutzler and colleagues determined the crystal structures of isolated CBS domains of *Torpedo* ClC-0 [13], human ClC-5 [14] and human ClC-Ka [15]. CBS domains are present in many different protein families, where they are often implicated in the sensing of adenine nucleotides [11,16]. Structurally, so far, ATP has been found to be bound in the isolated domains of ClC-5 and in the full-length structure of ClC-7 [17], but not in isolated domains of ClC-0 and ClC-Ka and not in full-length structures of bovine ClC-K or human ClC-1 [18–20].

Single-channel recordings of the *Torpedo* ClC-0 channel displayed two kinds of gating mechanisms that regulate the open probability (Po) of the channel: a *"fast"* or *"protopore"* gate that acts independently on single pores determines the closing or opening state of each pore of the double-barreled structure [1]. The fast gate is mainly determined by the gating glutamate (E166 in ClC-0), in that its neutralization renders CLC-0 channels voltage independent. Protonation of the gating glutamate and its competition with permeant ions underlie the anion and pH dependent protopore gating of most CLC channels [10,21–23]. Conversely, a second mechanism, termed a *slow* or *common* gate, operates on both pores simultaneously and is still not well understood [1].

Some CLC proteins require association with a small ancillary subunit for proper function or membrane expression. In particular, the kidney ClC-Ka and ClC-Kb channels require association with the barttin subunit [24]. In glia cells, ClC-2 associates with GlialCAM, a protein with the typical architecture of a cell adhesion molecule, which is mutated in megalencephalic leukoencephalopathy with subcortical cysts (MLC) [25]. It leads to clustering of ClC-2 at glial cell–cell contacts and alters biophysical functions of the ClC-2 channel [25,26]. The complex of ClC-7 with its subunit Ostm1 is mandatory for mutual stabilization [4,27].

The plasma membrane localized chloride channels belonging to the first branch of the CLC family are expressed in a tissue-dependent manner that is different for each member according to their physiological role. All channel CLCs are involved in various human genetic diseases, as reviewed in detail elsewhere [1,28,29].

ClC-3 through ClC-7, which are the focus of this review, function as Cl^-/H^+ exchangers and are localized to intracellular endosomes and/or lysosomes (Figure 1, Table 1). Initially, when the transporter function of the intracellular CLCs was not yet known, it was proposed that they act as charge-shunting chloride channels to assist the luminal acidification of endosomes and lysosomes intracellular organelles [30–34]. Indeed, the maintenance of an acidic pH of the lumen of endo-/lysosomes is required for their proper physiological function. The proton pumping V-ATPase is electrogenic and thus generates an electrical potential difference that would impede acidification if not neutralized by anionic cotransport and/or cationic counter-transport. Somewhat surprisingly and counterintuitively, model calculations show that a 2 Cl^-/H^+ exchange activity, contributing to a more inside-negative voltage, allows a more acidic steady-state luminal pH compared to a shunting Cl^- channel [35,36].

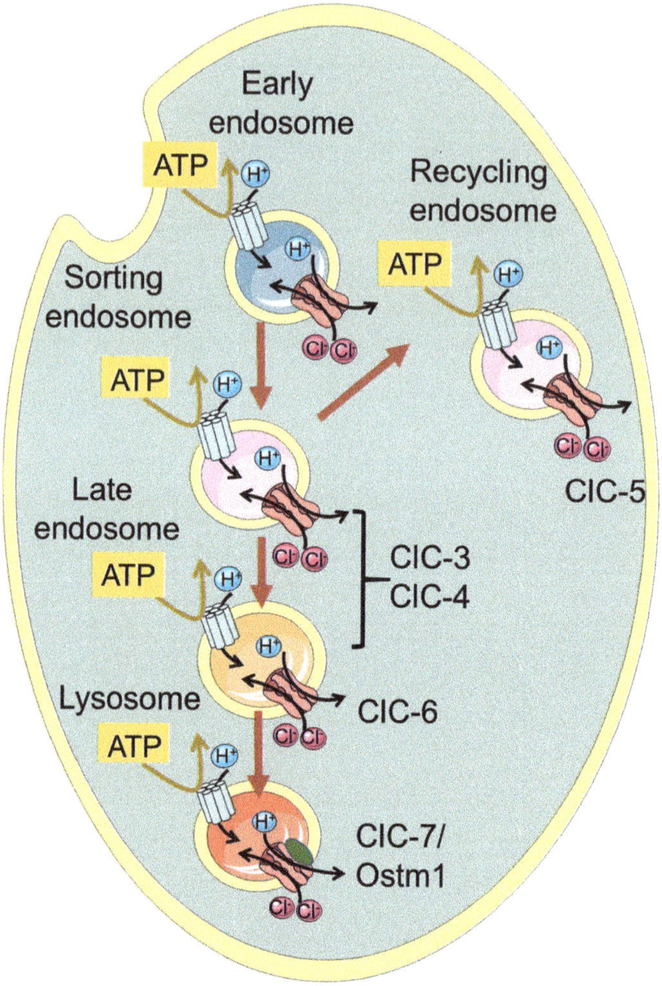

Figure 1. Schematic illustration of localization of vesicular CLCs in the endo-/lysosomal pathway.

Among the endo-/lysosomal CLCs, ClC-5 is rather specifically expressed in the kidney with a predominant presence in epithelial cells of the proximal tubule, where it is involved in endocytic uptake [30,31,33]. Indeed, mutations causing impaired ClC-5 transport activity are associated with Dent's disease, a kidney disorder characterized by the primary symptom of low molecular weight proteinuria, and a series of secondary symptoms including kidney stones and renal failure, caused by defective endocytosis in the proximal tubule [33,37]. ClC-7, together with its subunit Ostm1 [27], is rather ubiquitously expressed in the body and is localized to lysosomes and in the ruffled border of osteoclasts functioning as a $2Cl^-/H^+$ antiporter [1]. Accordingly, impaired bone resorption in osteoclast, caused by a functionally defective ClC-7/Ostm1 complex, causes osteopetrosis, a disease characterized by stiff and fragile bones [38].

Table 1. List of the CLC genes, subcellular localization and CLC related neurological diseases.

Protein	Gene	Cellular Localization	Neurological Disorder	Symptoms	References
ClC-3	*CLCN3*	Sorting and late endosomes	Global developmental delay	Intellectual disability, agenesis of the corpus callosum, epilepsy, visual impairment, hypotonia, anxiety, dysmorphic facial features	[39]
ClC-4	*CLCN4*	Sorting and late endosomes	*CLCN4*-related X linked intellectual disability syndrome	Intellectual disability, epilepsy, autism, growth and feeding difficulties, epilepsy, movement disorders, gastrointestinal conditions, dysmorphic facial features	[40–42]
ClC-6	*CLCN6*	Late endosomes			
			Early Onset Neurodegeneration	Severe neurodegeneration, severe generalized hypotonia and respiratory insufficiency, brain atrophy	[43]
			Kufs' disease	Adult-onset neuronal ceroid Lipofuscinosis, movement and cognitive function impairment	[44,45]
			West syndrome	Severe developmental delay, autism, movement disorder, microcephaly, facial dysmorphism, visual impairment	[46]
ClC-7/ Ostm1	*CLCN7/ OSTM1*	Lysosomes			
			Autosomal Recessive Osteopetrosis	Osteopetrosis, lysosomal storage disease, neurodegeneration, visual impairment	[38,47,48]
			Gain of function *CLCN7* related disease	Delayed myelination and development, organomegaly, and hypopigmentation	[49]

A large phenotypic spectrum of neuronal diseases is associated with mutations in the genes encoding ClC-3/-4/-6 and ClC-7, as will be described in detail in the following paragraphs (see Table 1).

For all vesicular CLCs, an unsolved question pertains to the direction of exchanger transport. Despite being physiologically localized to endo-/lysosomes, ClC-3 to -6 can reach the plasma membrane when heterologously expressed in HEK293 cells, allowing the investigation of their biophysical properties using the patch clamp technique [45,50–53]. For ClC-7, the elimination of N-terminal lysosomal targeting motifs leads to plasma membrane expression [4,54]. ClC-3 to -5 all exhibit extreme outward rectification of currents with very little or nonresolvable activation kinetics [50–53,55]. This current direction corresponds to the transport of luminal Cl^- out of lysosomes with a parallel influx of cytosolic H^+. However, it remains unclear whether the direction of transport is physiologically relevant and whether CLC exchangers work synergistically with V-ATPase, contributing to luminal acidification.

Additionally, ClC-6 and ClC-7 exhibit strongly outwardly rectifying currents, which are, however, characterized by slow activation kinetics and measurable inward "tail" currents [4,45].

2. ClC-3 and ClC-4

The second branch of the CLC family comprises ClC-3, -4 and 5. These three endosomal transporters share high sequence similarity and have similar functional properties [1,29]. Among the human CLCs, they are the most similar to the *Escherichia coli* ecClC-1 homologue. The renal-specific ClC-5 is found mostly in recycling endosomes and its physiological role will not be discussed in detail [1]. ClC-3 and ClC-4 are localized to sorting endosomes, and

ClC-3 is probably localized in late endosomes as well [1,29]. ClC-3 has also been proposed to play a role in synaptic vesicles. This is, however, still controversial and will not be discussed in detail here (see [29] for a discussion).

Functionally, these transporters are characterized by an extremely outwardly rectifying current–voltage relationship with almost instantaneous activation at positive voltages [50–52,55–58] (Figure 2A). The extreme outward rectification precluded the determination of transport stoichiometry by reversal potentials measurements [59]. Using a fluorescent assay, a 2 Cl^-/1 H^+ stoichiometry was determined for ClC-5, which is probably similar for ClC-3 and ClC-4 [55]. The outward rectification at least partially reflects a gating process, as evidenced by a single point mutation (D76H) that conferred detectable inward tail currents to ClC-5 [60]. The mutant also allowed us to measure reversal potentials for ClC-5 for the first time, confirming the 2:1 Cl^-/H^+ transport stoichiometry [60].

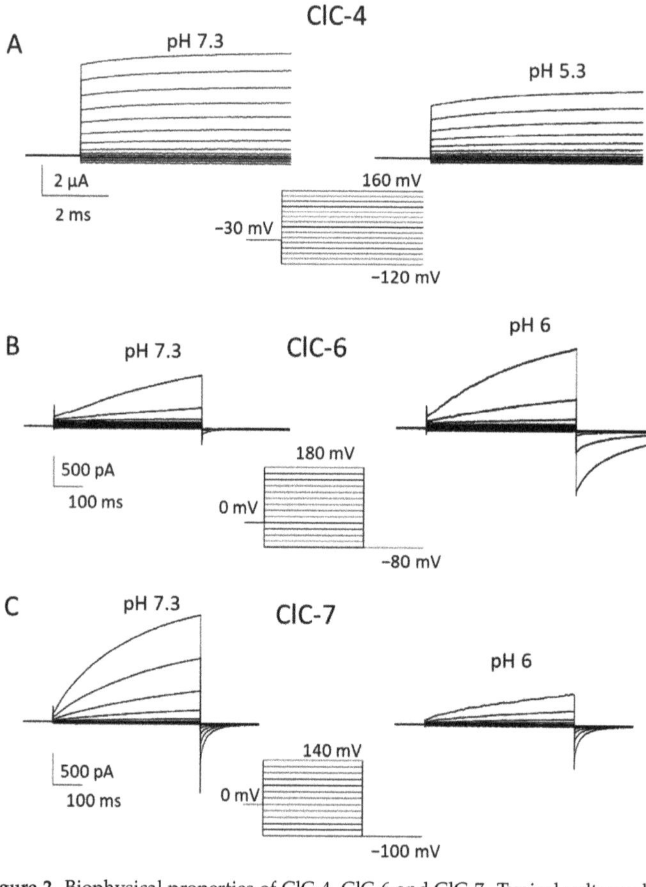

Figure 2. Biophysical properties of ClC-4, ClC-6 and ClC-7. Typical voltage-clamp current traces of ClC-4 ((**A**), expressed in *Xenopus* oocytes), ClC-6 ((**B**), expressed in HEK cells), and ClC-7^{PM} ((**C**), expressed in HEK cells), elicited by the indicated voltage-clamp protocols, measured using the two-electrode voltage clamp method (**A**) or the whole-cell recording mode of the patch clamp technique (**B**,**C**) at neutral (left panels) and acidic (right panels) pH. Extracellular solutions were 100 mM NaCl, 10 mM Hepes or MES, 5 mM $MgSO_4$ (**A**) or 150 mM NaCl, 10 mM Hepes or MES, 4 mM $MgSO_4$ (**B**,**C**). In B and C the pipette solution contained 130 mM NaCl, 10 mM Hepes, 2 mM EGTA, 2 mM $MgSO_4$. ClC-3 is similar to ClC-4.

Most evidence on the physiological roles of ClC-3 and ClC-4 was obtained from mouse models and their involvement in human genetic diseases. ClC-3 KO mice display severe postnatal neurodegeneration with almost total loss of the hippocampus after 3 months [34,61,62]. Yet, ClC-3 KO mice have a normal life span. Neurodegeneration is unlikely to be caused by impaired synaptic function [1,29]. Another set of mouse models showed that ClC-4 protein stability relies on the presence of ClC-3 via heterodimerization [63,64]. While ClC-4 KO mice have no overt phenotype, the double KO of ClC-3 and ClC-4 has a more severe phenotype than CLC-3 KO mice [64]. Notably, differently from humans, *Clcn4* is not X-linked in laboratory mice. For this reason, a rat model might be more useful for the investigation of mechanisms underlying *CLCN4*-related disease, because in rats, as in humans, *Clcn4* is X-linked.

While ClC-4 KO mice have no overt phenotype, in 2013 and 2016, patients (mostly pediatric) with a range of neurodevelopmental and psychiatric complications have been described with X-linked *CLCN4* variants [40,42,65] (see Table 1). In heterologous expression, these and some novel variants [66] showed variable loss of function effects. It is important to note that complete loss of ClC-4 protein leads to non-syndromic intellectual disability in males and no disease in heterozygous females. In contrast, de novo and inherited missense variants can lead to severe syndromic neurological disease in males as well as in females, suggesting a dominant effect. In a recent study, a large number of *CLCN4* families was investigated, describing a large spectrum of clinical phenotypes and studying > 50 missense variants in heterologous expression [41]. Novel biophysical mechanisms were discovered for new and already described variants. These included a toxic gain of function characterized by the presence of negative currents at acidic extracellular (luminal) pH, and a shift in the voltage dependence of gating to more positive voltages [41]. Both effects can be expected to exert dominant negative effects in ClC-3/ClC-4 heterodimers.

Almost simultaneously came the discovery of the first variants in *CLCN3* that cause global developmental delay, intellectual disability and neurodevelopmental disorders [39] (Table 1). Detailed functional analysis revealed a toxic gain of function for two missense variants, similar to the above-described effects in some *CLCN4* variants [39].

3. ClC-6

The third branch of the CLC family comprises ClC-6 and ClC-7, which both function as Cl^-/H^+ exchangers [1,45]. Even though the expression of ClC-6 mRNA appears to be ubiquitous in many tissues [67], biochemical analysis detected native ClC-6 protein predominantly in neurons, where it localizes to late endosomes and partially lysosomes [44]. For a long time, the biophysical profile of ClC-6 remained completely unknown. In the first attempts at heterologous expression, no currents attributable to ClC-6 could be detected [67], possibly caused by the intracellular localization of most of the overexpressed protein [68].

A subtype of lysosomal storage disease, referred to as neuronal ceroid lipofuscinosis (NCL), was observed in ClC-6 *knockout* mice presenting a mild phenotype with features of reduced pain sensitivity, probably due to strong accumulation of materials in axon initial segments, mild cognitive abnormalities and no impact on their span life [44]. This evidence suggested that *CLCN6* variants could be involved in human NCL [44]. Indeed, in a sample of 75 adult-onset variants, including late-onset forms of NLC and Kufs' disease, two individuals were found to be heterozygous for *CLCN6* missense variants (V580M and T628R) [44]. However, no functional analysis had been performed at the time of that study because the transporter had not been successfully functionally characterized.

Preliminary electrophysiological characterization was obtained when the N-terminus of ClC-6 tagged with GFP (GFP-ClC-6) was reported to enhance its cell surface localization [69]. However, the reported currents were small and barely above background levels.

In 2020, the same de novo variant in *CLCN6* leading to the amino acid change Y553C was reported in three pediatric patients with no parental correlation [43]. The patients exhibited a severe syndrome characterized by early-onset neurodegeneration. The mutated tyrosine, located in the extracellular P-Q loop, is highly conserved among CLC antiporters.

Electrophysiological measurements of ClC-6^{Y553C} revealed large currents activated at positive voltages (≥60 mV), representing a clear gain of function effect [43]. HEK cells overexpressing the mutant showed a dramatic vacuolization [43]. The luminal pH of these organelles did not reach low values, suggesting that the mutant impairs endolysosomal ionic homeostasis in patients.

In 2022, Zifarelli et al. were able to obtain robust recordings of WT ClC-6 currents by applying very large positive voltages beyond 140 mV [45,70]. Interestingly, the magnitude and properties of currents were independent of the N-terminal GFP tag. Similar to ClC-7, ClC-6-mediated currents showed slow activation at positive voltages; however, they required voltages of at least 140 mV to measure appreciable amplitudes [45] (Figure 2B). The necessity of the large voltages explained why these currents had escaped detection so far. Currents represent the coupled Cl^-/H^+ antiport, as assayed by reversal potential measurements [45]. Similar to ClC-7 [71], ClC-6 also exhibits so called "transient" or "capacitive" currents, whose possible physiological role is, however, obscure [45]. Interestingly, neutralizing the so-called proton glutamate did not completely abolish transport currents [45]. In contrast to most other CLC proteins, ClC-6-mediated currents were enhanced at acidic pH (6.3) compared to neutral pH [45] (Figure 2B), a finding that is of likely physiological relevance. In light of these novel findings on the functional properties of ClC-6, Zifarelli et al. performed a reexamination of the disease-causing variant ClC-6^{Y553C}, concluding that the mutation causes a gain of function by "shifting" the voltage dependence of ClC-6 gating to less positive voltages [45]. Possibly, Y553, being localized at the subunit interface of the homodimer, is involved in the common gating mechanism that acts on both ion-transporting units of the dimer.

The discovery of suitable recording protocols allowed Zifarelli et al. to study the functional impact of the two above-mentioned variants found in Kufs' disease patients. While T628R was indistinguishable from WT ClC-6, precluding firm conclusions regarding its causative nature for the disease, variant V580M showed a clearly reduced function, suggesting a causal relationship [45]. Since the variant was found in heterozygosity, it might exert a dominant negative effect in WT/mutant heterodimers.

Patients with the completely unrelated West syndrome, characterized by epilepsy, among other symptoms, have been reported to carry the ClC-6 E200A variant [46]. E200 is the critical gating glutamate of the exchanger and it is known that its neutralization in all studied vCLCs, including ClC-6, eliminates H^+ transport and transforms it into an uncoupled ohmic chloride channel [69]. In a heterologous expression system, ClC-6^{E200A} caused an impairment of the autophagosome-mediated degradation system, likely because the fusion with lysosomes was compromised [46].

The three classes of disease related *CLCN6* mutations, i.e., gain of function (Y553C), reduction of function (V580M), and uncoupling (E200A), have different effects on the functional properties of the ClC-6 antiporter, leading to clinical phenotypes with different degrees of severity. In particular, ClC-6^{Y553C} causing a gain of function is associated with drastic neurodegeneration, whereas ClC-6^{E200A} could be defined as a loss of function in the respect of the uncoupling transport generated and related to a mild phenotype.

The recent remarkable progress that has been made regarding the functional analysis of ClC-6 activity will allow us to decode further mechanisms underlying disease caused by defective ClC-6 proteins.

4. ClC-7

Belonging to the third mammalian CLC branch, ClC-7 shares 45% of sequence homology with ClC-6. It was cloned in parallel with ClC-6 in 1995 [67], but could not be functionally analyzed for a long time. Intriguingly, ClC-7 is the only subcellular CLC member to be present almost exclusively in lysosomes [44]. Moreover, it has also been found in the ruffled border of osteoclasts, where it participates in bone resorption [38]. Unlike the other CLC transporters, ClC-7 requires association with a type I transmembrane protein, called Ostm1, for proper function and stability [4,27].

Similarly to ClC-6, no information about electrophysiological ClC-7 characterization has been available for a long time, due to its intracellular localization upon heterologous expression [1]. Ion flux studies with isolated mouse lysosomes showed that ClC-7 is the dominant anion permeation pathway of lysosomal membranes and that it performs 2 Cl^-/1 H^+ antiport activity [72]. A breakthrough was achieved by Stauber and Jentsch, who discovered the sorting motifs that mediate lysosomal targeting [54]. In particular, they found that when four leucine residues localized in the N-terminal portion are changed to alanine, the transporter is at least partially targeted to the plasma membrane [54]. Notably, Ostm1 follows ClC-7 in its expression location. ClC-7^{PM}, the ClC-7 variant in which the two dileucine motifs are mutated to alanine, elicited robust transmembrane, outwardly rectifying voltage-activated currents [4] (Figure 2C). Even though some electrophysiological properties of ClC-7 are similar to that of other vesicular CLCs, including the inhibitory effect of acidic pH, ClC-7 differs substantially from ClC-3 to -5. Most importantly, ClC-7^{PM} exhibits very slow activation kinetics in the seconds time range [4] (Figure 2C). This slow "gating" phenomenon is strictly linked to conformational changes in the proteins, where the interactions between transmembrane as well as cytoplasmic domains play a key role [73]. In addition to the transport currents, Pusch and Zifarelli discovered that the transporter also exhibits rather large "transient" or "capacitive" currents that reflect charge rearrangements within the protein. These are most likely mediated by movements of the gating glutamate and chloride binding/unbinding events [71]. Similar currents have been observed in ClC-5 and ClC-3 [52,74,75]. The transient currents probably have no physiological role, but represent a biophysical feature that can be useful in deciphering molecular mechanisms of gating and transport. Interestingly, while in ClC-5, neutralization of the so-called proton glutamate completely abolished transport currents, leaving only transient currents [74,75], in ClC-7, residual transport currents were observed in the corresponding E312A mutant [71].

The physiological role of ClC-7 remained unclear for a long time. The first insights were obtained with a mouse KO model that was characterized by severe osteopetrosis [38]. The involvement of ClC-7 in bone resorption was confirmed by the presence of *CLCN7* mutations in a human patient with malignant osteopetrosis [38]. Further evidence came from the identification of a spontaneous Ostm1 mutation to be associated with the onset of a severe osteopetrosis in *gray lethal* mice presenting a fur color defect [76]. In Clcn$7^{-/-}$ mice, even though the number of osteoclasts was normal, their ability to reabsorb calcified bone was impaired [38]. Interestingly, however, no impact on lysosomal acidification was observed, suggesting that the osteoclasts' ability to acidify intracellular vesicles was preserved in Clcn$7^{-/-}$ mice [38]. The life span of the KO mice was limited to 6–7 weeks.

Importantly, in addition to osteopetrosis, ClC-$7^{-/-}$ mice also presented severe lysosomal storage associated with central nervous system and retinal degeneration [47]. Using a lacZ fusion protein, the expression profile of ClC-7 was determined in the nervous tissue of WT e KO mice revealing the hippocampus CA3 region, the cortex and the cerebellum as the main regions experiencing neuronal loss in KO mice [47]. Electron microscopy analysis revealed the presence of autofluorescent lipopigment in the regions affected by neurodegeneration and deficient of CLC-7 [47]. This factor, together with the detection of microglial activation and astrogliosis, represents three important hallmarks of neuronal ceroid lipofuscinosis (NCL) [47]. Importantly, no significant difference in pH values in lysosomes of cultured neurons and fibroblasts was found in Clnc$7^{-/-}$ mice, but rather a reduction in lysosomal Cl concentration was observed [47]. Several pieces of evidence suggest that osteopetrosis and neurodegeneration are independent outcomes. First, an osteopetrotic mouse model with a mutation in the a3 subunit of V-type H^+-ATPase (oc/oc mice) does not show retinal or neurodegeneration [47]. Moreover, the osteoporotic phenotype in Clnc$7^{-/-}$ mice could be rescued by transgenically expressing ClC-7 in osteoclasts and macrophages under the control of tartrate-resistant acid phosphatase (TRAP) promoter [47]. This treatment achieved a lifespan increase, but it was not enough to ensure their survival due to the enduring neurological problems [47]. Surprisingly, the same approach failed when TRAP promoter-mediated Ostm1 expression was applied to rescue osteopetrosis in gray lethal (gl) mice which also displayed neuronal loss [77]. The severity of the phenotype and the short life span of the mice were serious problems, preventing a better understanding of the mechanisms underlying the progression of neurodegeneration in lysosomal pathologies [1]. The first information was collected when Wartosch et al. designed a floxed *Clcn7* mouse model allowing tissue-specific ClC-7 depletion [78]. No difference in lifespan between neuron-specific Clcn7 KO and WT mice was observed. Moreover, neuron-specific Clcn7 KO mice had no osteopetrotic phenotype; thus, the quality of life of these mice was improved compared to Clcn7$^{-/-}$ [78]. Importantly, it was observed that neuronal loss occurs in regions, where ClC-7 had been disrupted and neurodegeneration started in the CA3 region of the hippocampus as in constitutive Clcn7$^{-/-}$. Accordingly, in previous studies, astrogliosis and microglia activation were observed in the regions lacking ClC-7. Impaired lysosomal protein degradation was suggested after the detection of increased levels of LC3-II, a marker of autophagy [78].

A somewhat surprising finding was that several, mostly dominantly inherited, *CLCN7* variants causing osteopetrosis (but not neurodegeneration) produce a significant acceleration of gating kinetics [4,79,80]. It is unclear how this biophysical defect is related to ClC-7 malfunction.

More recently, a completely different disease characterized by delayed myelination and development, organomegaly and hypopigmentation was found in two children who both carried the de novo Y715C variant [49]. Surprisingly, none of the patients showed osteopetrosis (see Table 1). The variant, located in the C-terminus, was associated with larger currents when directed to the plasma membrane, representing a clear gain of function effect. Interestingly, the lysosomes of patient fibroblasts were enlarged and had a lower pH (0.2 pH units) than control lysosomes [49]. Even more excitingly, in the ClC-7 structure, Y715 is relatively close to the bound PI(3)P molecule (see below) and Leray et al. recently reported that intracellular phosphatidylinositol-3,5-bisphosphate (PI(3,5)P2) appears to tonically inhibit ClC-7 function, and that the Y715C variant was insensitive to PI(3,5)P2 [81]. The regulation of CLC-transporters by these signaling molecules is clearly an exciting aspect that needs to be understood in more detail.

The structure of the ClC-7/Ostm1 complex has been determined by two independent groups [17,82]. The structures revealed that the heavily N-glycosylated luminal region of Ostm1 forms a sort of cap on the luminal portion of ClC-7, preventing its degradation by lysosomal proteases [17,82]. Importantly, the mutual protein stabilization between ClC-7 and Ostm1 is suggested by the observation that the gray lethal mouse line, which lacks Ostm1, showed very weak ClC-7 staining; similarly, in Clcn7$^{-/-}$ mice, there was only weak Ostm1 staining [27,76]. Both cryo-EM structures revealed strong intramolecular interactions between the cytosolic N-terminal portion and the CBS domains [17,82]. This important feature is also conserved in ClC-6 (Hite, personal communication), but the role of these interactions in other CLCs remains unclear because no information about the cytosolic structure of other CLC members is available. Similarly to ClC-5, the ClC-7 CBS domains bind ATP and additionally, a Mg^{2+} ion was found to be bound [17,82]. However, ATP had no effect on transport activity and its role remains to be understood [4]. Moreover, in the structure of ClC-7 an endolysosomal phosphatidylinositol 3-phosphate (PI3P) lipid was found to be bound at the interface between the CBS and the membrane domains [17].

5. Conclusions

While significant progress has been made in elucidating the functional properties of vesicular CLC transporters and their involvement in various neurological diseases, highlighting their importance in nervous system development and homeostasis, their precise physiological role is still largely unknown. Even for the most studied ClC-7, it is still disputed whether it is primarily necessary for proper luminal acidification or the regulation of the luminal chloride concentration. Most recent evidence favors the idea that ClC-7 is responsible for achieving a high luminal chloride concentration, which is important for phagosomal clearance [83]. Less clear are the roles of ClC-3 and ClC-4 in endosomes and the possible involvement of ClC-3/ClC-4 dimers in human genetic diseases. For all vesicular CLCs, and in particular for ClC-6, the significance of the activation at highly positive voltages remains enigmatic, since similar voltages are not expected to be achieved in endosomes. However, it is clear that ClC-3 and ClC-4 need to be inactive at negative voltages, since even a small amount of activity at these voltages, caused by gate disrupting mutations, leads to severe disease for both transporters [39,41]. In general, it appears that gain of function mutations lead to more severe phenotypes than loss-of function mutations for all vesicular CLCs [39,41,43,49]. In this respect, specific pharmacological inhibitors of vesicular CLCs are highly desirable. Unfortunately, thus far, no useful pharmacological tools are available for any of them.

Author Contributions: M.P. and M.A.C. conceived the idea and prepared the figures. M.A.C., A.T.-M., P.I., P.G., A.L. and M.P. contributed to writing. All authors have read and agreed to the published version of the manuscript.

Funding: This research was partially funded by the European Union—NextGenerationEU (Missione 4 Componente 2, "Dalla ricerca all'impresa", Innovation Ecosystem RAISE "Robotics and AI for Socio-economic Empowerment", ECS00000035). However, the views and opinions expressed are those of the authors alone and do not necessarily reflect those of the European Union or the European Commission. Neither the European Union nor the European Commission can be held responsible for them. In addition, this research was partially funded by the Fondazione AIRC per la Ricerca sul Cancro (grant # IG 21558), PRIN-MIUR 2017 Prot. 20174TB8KW, Fondazione Telethon (grant # GMR22T102) granted to M.P. and Fondazione Telethon/Cariplo (grant # GJC22008) to M.P.

Data Availability Statement: Data of Figure 2 are available upon reasonable request.

Acknowledgments: We would like to thank Francesca Quartino and Alessandro Barbin for their excellent technical assistance. This work was partially carried out within the framework of the project "RAISE—Robotics and AI for Socio-economic Empowerment" and has been supported by the European Union—NextGenerationEU.

Conflicts of Interest: The authors declare no conflict of interest.

References

1. Jentsch, T.J.; Pusch, M. CLC chloride channels and transporters: Structure, function, physiology, and disease. *Physiol. Rev.* **2018**, *98*, 1493–1590. [CrossRef]
2. Jentsch, T.J.; Steinmeyer, K.; Schwarz, G. Primary structure of *Torpedo marmorata* chloride channel isolated by expression cloning in *Xenopus* oocytes. *Nature* **1990**, *348*, 510–514. [CrossRef] [PubMed]
3. Accardi, A.; Miller, C. Secondary active transport mediated by a prokaryotic homologue of ClC Cl^- channels. *Nature* **2004**, *427*, 803–807. [CrossRef] [PubMed]
4. Leisle, L.; Ludwig, C.F.; Wagner, F.A.; Jentsch, T.J.; Stauber, T. ClC-7 is a slowly voltage-gated $2Cl^-/1H^+$-exchanger and requires Ostm1 for transport activity. *EMBO J.* **2011**, *30*, 2140–2152. [CrossRef] [PubMed]
5. Middleton, R.E.; Pheasant, D.J.; Miller, C. Homodimeric architecture of a ClC-type chloride ion channel. *Nature* **1996**, *383*, 337–340. [CrossRef] [PubMed]
6. Ludewig, U.; Pusch, M.; Jentsch, T.J. Two physically distinct pores in the dimeric ClC-0 chloride channel. *Nature* **1996**, *383*, 340–343. [CrossRef]
7. Maduke, M.; Pheasant, D.J.; Miller, C. High-level expression, functional reconstitution, and quaternary structure of a prokaryotic ClC-type chloride channel. *J. Gen. Physiol.* **1999**, *114*, 713–722. [CrossRef] [PubMed]
8. Mindell, J.A.; Maduke, M.; Miller, C.; Grigorieff, N. Projection structure of a ClC-type chloride channel at 6.5 Å resolution. *Nature* **2001**, *409*, 219–223. [CrossRef]
9. Dutzler, R.; Campbell, E.B.; Cadene, M.; Chait, B.T.; MacKinnon, R. X-ray structure of a ClC chloride channel at 3.0 Å reveals the molecular basis of anion selectivity. *Nature* **2002**, *415*, 287–294. [CrossRef]
10. Dutzler, R.; Campbell, E.B.; MacKinnon, R. Gating the selectivity filter in ClC chloride channels. *Science* **2003**, *300*, 108–112. [CrossRef]
11. Ponting, C.P. CBS domains in CIC chloride channels implicated in myotonia and nephrolithiasis (kidney stones). *J. Mol. Med.* **1997**, *75*, 160–163.
12. Estévez, R.; Jentsch, T.J. CLC chloride channels: Correlating structure with function. *Curr. Opin. Struct. Biol.* **2002**, *12*, 531–539. [CrossRef]
13. Meyer, S.; Dutzler, R. Crystal structure of the cytoplasmic domain of the chloride channel ClC-0. *Structure* **2006**, *14*, 299–307. [CrossRef]

14. Meyer, S.; Savaresi, S.; Forster, I.C.; Dutzler, R. Nucleotide recognition by the cytoplasmic domain of the human chloride transporter ClC-5. *Nat. Struct. Mol. Biol.* **2007**, *14*, 60–67. [CrossRef]
15. Markovic, S.; Dutzler, R. The structure of the cytoplasmic domain of the chloride channel ClC-Ka reveals a conserved interaction interface. *Structure* **2007**, *15*, 715–725. [CrossRef]
16. Bateman, A. The structure of a domain common to archaebacteria and the homocystinuria disease protein. *Trends Biochem. Sci.* **1997**, *22*, 12–13. [CrossRef]
17. Schrecker, M.; Korobenko, J.; Hite, R.K. Cryo-EM structure of the lysosomal chloride-proton exchanger CLC-7 in complex with OSTM1. *eLife* **2020**, *9*, e59555. [CrossRef] [PubMed]
18. Park, E.; Campbell, E.B.; MacKinnon, R. Structure of a CLC chloride ion channel by cryo-electron microscopy. *Nature* **2017**, *541*, 500–505. [CrossRef] [PubMed]
19. Park, E.; MacKinnon, R. Structure of the CLC-1 chloride channel from Homo sapiens. *eLife* **2018**, *7*, e36629. [CrossRef] [PubMed]
20. Wang, K.; Preisler, S.S.; Zhang, L.; Cui, Y.; Missel, J.W.; Gronberg, C.; Gotfryd, K.; Lindahl, E.; Andersson, M.; Calloe, K.; et al. Structure of the human ClC-1 chloride channel. *PLoS Biol.* **2019**, *17*, e3000218. [CrossRef]
21. Pusch, M.; Ludewig, U.; Rehfeldt, A.; Jentsch, T.J. Gating of the voltage-dependent chloride channel CIC-0 by the permeant anion. *Nature* **1995**, *373*, 527–531. [CrossRef] [PubMed]
22. Traverso, S.; Elia, L.; Pusch, M. Gating competence of constitutively open CLC-0 mutants revealed by the interaction with a small organic Inhibitor. *J. Gen. Physiol.* **2003**, *122*, 295–306. [CrossRef] [PubMed]
23. Zifarelli, G.; Murgia, A.R.; Soliani, P.; Pusch, M. Intracellular proton regulation of ClC-0. *J. Gen. Physiol.* **2008**, *132*, 185–198. [CrossRef]
24. Estévez, R.; Boettger, T.; Stein, V.; Birkenhäger, R.; Otto, E.; Hildebrandt, F.; Jentsch, T.J. Barttin is a Cl^- channel beta-subunit crucial for renal Cl^- reabsorption and inner ear K^+ secretion. *Nature* **2001**, *414*, 558–561. [CrossRef] [PubMed]
25. Jeworutzki, E.; López-Hernández, T.; Capdevila-Nortes, X.; Sirisi, S.; Bengtsson, L.; Montolio, M.; Zifarelli, G.; Arnedo, T.; Müller, C.S.; Schulte, U.; et al. GlialCAM, a protein defective in a leukodystrophy, serves as a ClC-2 Cl^- channel auxiliary subunit. *Neuron* **2012**, *73*, 951–961. [CrossRef]
26. Jeworutzki, E.; Lagostena, L.; Elorza-Vidal, X.; Lopez-Hernandez, T.; Estevez, R.; Pusch, M. GlialCAM, a CLC-2 Cl^- channel subunit, activates the slow gate of CLC chloride channels. *Biophys. J.* **2014**, *107*, 1105–1116. [CrossRef]
27. Lange, P.F.; Wartosch, L.; Jentsch, T.J.; Fuhrmann, J.C. ClC-7 requires Ostm1 as a beta-subunit to support bone resorption and lysosomal function. *Nature* **2006**, *440*, 220–223. [CrossRef]
28. Stauber, T.; Weinert, S.; Jentsch, T.J. Cell biology and physiology of CLC chloride channels and transporters. *Compr. Physiol.* **2012**, *2*, 1701–1744. [CrossRef]
29. Bose, S.; He, H.; Stauber, T. Neurodegeneration upon dysfunction of endosomal/lysosomal CLC chloride transporters. *Front. Cell Dev. Biol.* **2021**, *9*, 639231. [CrossRef]
30. Günther, W.; Lüchow, A.; Cluzeaud, F.; Vandewalle, A.; Jentsch, T.J. ClC-5, the chloride channel mutated in Dent's disease, colocalizes with the proton pump in endocytotically active kidney cells. *Proc. Natl. Acad. Sci. USA* **1998**, *95*, 8075–8080. [CrossRef]
31. Günther, W.; Piwon, N.; Jentsch, T.J. The ClC-5 chloride channel knock-out mouse—An animal model for Dent's disease. *Pflügers Arch.* **2003**, *445*, 456–462. [CrossRef]
32. Jentsch, T.J.; Günther, W. Chloride channels: An emerging molecular picture. *Bioessays* **1997**, *19*, 117–126. [CrossRef] [PubMed]
33. Piwon, N.; Günther, W.; Schwake, M.; Bösl, M.R.; Jentsch, T.J. ClC-5 Cl^--channel disruption impairs endocytosis in a mouse model for Dent's disease. *Nature* **2000**, *408*, 369–373. [CrossRef]
34. Stobrawa, S.M.; Breiderhoff, T.; Takamori, S.; Engel, D.; Schweizer, M.; Zdebik, A.A.; Bösl, M.R.; Ruether, K.; Jahn, H.; Draguhn, A.; et al. Disruption of ClC-3, a chloride channel expressed on synaptic vesicles, leads to a loss of the hippocampus. *Neuron* **2001**, *29*, 185–196. [CrossRef] [PubMed]
35. Weinert, S.; Jabs, S.; Supanchart, C.; Schweizer, M.; Gimber, N.; Richter, M.; Rademann, J.; Stauber, T.; Kornak, U.; Jentsch, T.J. Lysosomal pathology and osteopetrosis upon loss of H^+-driven lysosomal Cl^- accumulation. *Science* **2010**, *328*, 1401–1403. [CrossRef]
36. Ishida, Y.; Nayak, S.; Mindell, J.A.; Grabe, M. A model of lysosomal pH regulation. *J. Gen. Physiol.* **2013**, *141*, 705–720. [CrossRef] [PubMed]
37. Lloyd, S.E.; Pearce, S.H.; Fisher, S.E.; Steinmeyer, K.; Schwappach, B.; Scheinman, S.J.; Harding, B.; Bolino, A.; Devoto, M.; Goodyer, P.; et al. A common molecular basis for three inherited kidney stone diseases. *Nature* **1996**, *379*, 445–449. [CrossRef]
38. Kornak, U.; Kasper, D.; Bösl, M.R.; Kaiser, E.; Schweizer, M.; Schulz, A.; Friedrich, W.; Delling, G.; Jentsch, T.J. Loss of the ClC-7 chloride channel leads to osteopetrosis in mice and man. *Cell* **2001**, *104*, 205–215. [CrossRef]
39. Duncan, A.R.; Polovitskaya, M.M.; Gaitan-Penas, H.; Bertelli, S.; VanNoy, G.E.; Grant, P.E.; O'Donnell-Luria, A.; Valivullah, Z.; Lovgren, A.K.; England, E.M.; et al. Unique variants in CLCN3, encoding an endosomal anion/proton exchanger, underlie a spectrum of neurodevelopmental disorders. *Am. J. Hum. Genet.* **2021**, *108*, 1450–1465. [CrossRef]
40. Hu, H.; Haas, S.A.; Chelly, J.; Van Esch, H.; Raynaud, M.; de Brouwer, A.P.; Weinert, S.; Froyen, G.; Frints, S.G.; Laumonnier, F.; et al. X-exome sequencing of 405 unresolved families identifies seven novel intellectual disability genes. *Mol. Psychiatry* **2016**, *21*, 133–148. [CrossRef]

41. Palmer, E.E.; Pusch, M.; Picollo, A.; Forwood, C.; Nguyen, M.H.; Suckow, V.; Gibbons, J.; Hoff, A.; Sigfrid, L.; Megarbane, A.; et al. Functional and clinical studies reveal pathophysiological complexity of CLCN4-related neurodevelopmental condition. *Mol. Psychiatry* **2023**, *28*, 668–697. [CrossRef] [PubMed]
42. Palmer, E.E.; Stuhlmann, T.; Weinert, S.; Haan, E.; Van Esch, H.; Holvoet, M.; Boyle, J.; Leffler, M.; Raynaud, M.; Moraine, C.; et al. De novo and inherited mutations in the X-linked gene CLCN4 are associated with syndromic intellectual disability and behavior and seizure disorders in males and females. *Mol. Psychiatry* **2016**, *23*, 222–230. [CrossRef] [PubMed]
43. Polovitskaya, M.M.; Barbini, C.; Martinelli, D.; Harms, F.L.; Cole, F.S.; Calligari, P.; Bocchinfuso, G.; Stella, L.; Ciolfi, A.; Niceta, M.; et al. A Recurrent Gain-of-Function Mutation in CLCN6, Encoding the ClC-6 Cl^-/H^+-Exchanger, Causes Early-Onset Neurodegeneration. *Am. J. Hum. Genet.* **2020**, *107*, 1062–1077. [CrossRef]
44. Poët, M.; Kornak, U.; Schweizer, M.; Zdebik, A.A.; Scheel, O.; Hoelter, S.; Wurst, W.; Schmitt, A.; Fuhrmann, J.C.; Planells-Cases, R.; et al. Lysosomal storage disease upon disruption of the neuronal chloride transport protein ClC-6. *Proc. Natl. Acad. Sci. USA* **2006**, *103*, 13854–13859. [CrossRef]
45. Zifarelli, G.; Pusch, M.; Fong, P. Altered voltage-dependence of slowly activating chloride-proton antiport by late endosomal ClC-6 explains distinct neurological disorders. *J. Physiol.* **2022**, *600*, 2147–2164. [CrossRef]
46. He, H.; Cao, X.; Yin, F.; Wu, T.; Stauber, T.; Peng, J. West syndrome caused by a chloride/proton exchange-uncoupling CLCN6 mutation related to autophagic-lysosomal dysfunction. *Mol. Neurobiol.* **2021**, *58*, 2990–2999. [CrossRef] [PubMed]
47. Kasper, D.; Planells-Cases, R.; Fuhrmann, J.C.; Scheel, O.; Zeitz, O.; Ruether, K.; Schmitt, A.; Poët, M.; Steinfeld, R.; Schweizer, M.; et al. Loss of the chloride channel ClC-7 leads to lysosomal storage disease and neurodegeneration. *EMBO J.* **2005**, *24*, 1079–1091. [CrossRef]
48. Pangrazio, A.; Pusch, M.; Caldana, E.; Frattini, A.; Lanino, E.; Tamhankar, P.M.; Phadke, S.; Lopez, A.G.; Orchard, P.; Mihci, E.; et al. Molecular and clinical heterogeneity in CLCN7-dependent osteopetrosis: Report of 20 novel mutations. *Hum. Mutat.* **2010**, *31*, E1071–E1080. [CrossRef] [PubMed]
49. Nicoli, E.R.; Weston, M.R.; Hackbarth, M.; Becerril, A.; Larson, A.; Zein, W.M.; Baker, P.R., 2nd; Burke, J.D.; Dorward, H.; Davids, M.; et al. Lysosomal storage and albinism due to effects of a de novo CLCN7 variant on lysosomal acidification. *Am. J. Hum. Genet.* **2019**, *104*, 1127–1138. [CrossRef]
50. Steinmeyer, K.; Schwappach, B.; Bens, M.; Vandewalle, A.; Jentsch, T.J. Cloning and functional expression of rat CLC-5, a chloride channel related to kidney disease. *J. Biol. Chem.* **1995**, *270*, 31172–31177. [CrossRef]
51. Friedrich, T.; Breiderhoff, T.; Jentsch, T.J. Mutational analysis demonstrates that ClC-4 and ClC-5 directly mediate plasma membrane currents. *J. Biol. Chem.* **1999**, *274*, 896–902. [CrossRef]
52. Guzman, R.E.; Grieschat, M.; Fahlke, C.; Alekov, A.K. ClC-3 is an intracellular chloride/proton exchanger with large voltage-dependent nonlinear capacitance. *ACS Chem. Neurosci.* **2013**, *4*, 994–1003. [CrossRef] [PubMed]
53. Guzman, R.E.; Miranda-Laferte, E.; Franzen, A.; Fahlke, C. Neuronal ClC-3 splice variants differ in subcellular localizations, but mediate identical transport functions. *J. Biol. Chem.* **2015**, *290*, 25851–25862. [CrossRef] [PubMed]
54. Stauber, T.; Jentsch, T.J. Sorting motifs of the endosomal/lysosomal CLC chloride transporters. *J. Biol. Chem.* **2010**, *285*, 34537–34548. [CrossRef] [PubMed]
55. Zifarelli, G.; Pusch, M. Conversion of the 2 Cl^-/1 H^+ antiporter ClC-5 in a NO_3^-/H^+ antiporter by a single point mutation. *Embo J.* **2009**, *28*, 175–182. [CrossRef]
56. Li, X.; Shimada, K.; Showalter, L.A.; Weinman, S.A. Biophysical properties of ClC-3 differentiate it from swelling-activated chloride channels in Chinese hamster ovary-K1 cells. *J. Biol. Chem.* **2000**, *275*, 35994–35998. [CrossRef]
57. Zifarelli, G.; Pusch, M. Intracellular regulation of human ClC-5 by adenine nucleotides. *EMBO Rep.* **2009**, *10*, 1111–1116. [CrossRef]
58. De Stefano, S.; Pusch, M.; Zifarelli, G. Extracellular determinants of anion discrimination of the Cl^-/H^+ antiporter CLC-5. *J. Biol. Chem.* **2011**, *286*, 44134–44144. [CrossRef]
59. Picollo, A.; Pusch, M. Chloride/proton antiporter activity of mammalian CLC proteins ClC-4 and ClC-5. *Nature* **2005**, *436*, 420–423. [CrossRef]
60. De Stefano, S.; Pusch, M.; Zifarelli, G. A single point mutation reveals gating of the human ClC-5 Cl^-/H^+ antiporter. *J. Physiol.* **2013**, *591*, 5879–5893. [CrossRef]
61. Dickerson, L.W.; Bonthius, D.J.; Schutte, B.C.; Yang, B.; Barna, T.J.; Bailey, M.C.; Nehrke, K.; Williamson, R.A.; Lamb, F.S. Altered GABAergic function accompanies hippocampal degeneration in mice lacking ClC-3 voltage-gated chloride channels. *Brain Res.* **2002**, *958*, 227–250. [CrossRef]
62. Yoshikawa, M.; Uchida, S.; Ezaki, J.; Rai, T.; Hayama, A.; Kobayashi, K.; Kida, Y.; Noda, M.; Koike, M.; Uchiyama, Y.; et al. CLC-3 deficiency leads to phenotypes similar to human neuronal ceroid lipofuscinosis. *Genes Cells* **2002**, *7*, 597–605. [CrossRef] [PubMed]
63. Guzman, R.E.; Bungert-Plumke, S.; Franzen, A.; Fahlke, C. Preferential association with ClC-3 permits sorting of ClC-4 into endosomal compartments. *J. Biol. Chem.* **2017**, *292*, 19055–19065. [CrossRef] [PubMed]
64. Weinert, S.; Gimber, N.; Deuschel, D.; Stuhlmann, T.; Puchkov, D.; Farsi, Z.; Ludwig, C.F.; Novarino, G.; Lopez-Cayuqueo, K.I.; Planells-Cases, R.; et al. Uncoupling endosomal CLC chloride/proton exchange causes severe neurodegeneration. *EMBO J.* **2020**, *39*, e103358. [CrossRef] [PubMed]

65. Veeramah, K.R.; Johnstone, L.; Karafet, T.M.; Wolf, D.; Sprissler, R.; Salogiannis, J.; Barth-Maron, A.; Greenberg, M.E.; Stuhlmann, T.; Weinert, S.; et al. Exome sequencing reveals new causal mutations in children with epileptic encephalopathies. *Epilepsia* **2013**, *54*, 1270–1281. [CrossRef]
66. He, H.; Guzman, R.E.; Cao, D.; Sierra-Marquez, J.; Yin, F.; Fahlke, C.; Peng, J.; Stauber, T. The molecular and phenotypic spectrum of CLCN4-related epilepsy. *Epilepsia* **2021**, *62*, 1401–1415. [CrossRef] [PubMed]
67. Brandt, S.; Jentsch, T.J. ClC-6 and ClC-7 are two novel broadly expressed members of the CLC chloride channel family. *FEBS Lett.* **1995**, *377*, 15–20. [CrossRef]
68. Ignoul, S.; Simaels, J.; Hermans, D.; Annaert, W.; Eggermont, J. Human ClC-6 is a late endosomal glycoprotein that associates with detergent-resistant lipid domains. *PLoS ONE* **2007**, *2*, e474. [CrossRef]
69. Neagoe, I.; Stauber, T.; Fidzinski, P.; Bergsdorf, E.Y.; Jentsch, T.J. The late endosomal ClC-6 mediates proton/chloride countertransport in heterologous plasma membrane expression. *J. Biol. Chem.* **2010**, *285*, 21689–21697. [CrossRef]
70. Kobertz, W.R. Want to hear ClC-6 sing? Push your amp to eleven. *J. Physiol.* **2022**, *600*, 2019–2020. [CrossRef]
71. Pusch, M.; Zifarelli, G. Large transient capacitive currents in wild-type lysosomal Cl^-/H^+ antiporter ClC-7 and residual transport activity in the proton glutamate mutant E312A. *J. Gen. Physiol.* **2021**, *153*, e202012583. [CrossRef] [PubMed]
72. Graves, A.R.; Curran, P.K.; Smith, C.L.; Mindell, J.A. The Cl^-/H^+ antiporter ClC-7 is the primary chloride permeation pathway in lysosomes. *Nature* **2008**, *453*, 788–792. [CrossRef] [PubMed]
73. Ludwig, C.F.; Ullrich, F.; Leisle, L.; Stauber, T.; Jentsch, T.J. Common gating of both CLC transporter subunits underlies voltage-dependent activation of the $2Cl^-/1H^+$ exchanger ClC-7/Ostm1. *J. Biol. Chem.* **2013**, *288*, 28611–28619. [CrossRef] [PubMed]
74. Smith, A.J.; Lippiat, J.D. Voltage-dependent charge movement associated with activation of the CLC-5 $2Cl^-/1H^+$ exchanger. *Faseb J.* **2010**, *24*, 3696–3705. [CrossRef]
75. Zifarelli, G.; De Stefano, S.; Zanardi, I.; Pusch, M. On the mechanism of gating charge movement of ClC-5, a human Cl^-/H^+ antiporter. *Biophys. J.* **2012**, *102*, 2060–2069. [CrossRef]
76. Chalhoub, N.; Benachenhou, N.; Rajapurohitam, V.; Pata, M.; Ferron, M.; Frattini, A.; Villa, A.; Vacher, J. Grey-lethal mutation induces severe malignant autosomal recessive osteopetrosis in mouse and human. *Nat. Med.* **2003**, *9*, 399–406. [CrossRef]
77. Pata, M.; Héraud, C.; Vacher, J. OSTM1 bone defect reveals an intercellular hematopoietic crosstalk. *J. Biol. Chem.* **2008**, *283*, 30522–30530. [CrossRef]
78. Wartosch, L.; Fuhrmann, J.C.; Schweizer, M.; Stauber, T.; Jentsch, T.J. Lysosomal degradation of endocytosed proteins depends on the chloride transport protein ClC-7. *FASEB J.* **2009**, *23*, 4056–4068. [CrossRef]
79. Sartelet, A.; Stauber, T.; Coppieters, W.; Ludwig, C.F.; Fasquelle, C.; Druet, T.; Zhang, Z.; Ahariz, N.; Cambisano, N.; Jentsch, T.J.; et al. A missense mutation accelerating the gating of the lysosomal Cl^-/H^+-exchanger ClC-7/Ostm1 causes osteopetrosis with gingival hamartomas in cattle. *Dis. Model Mech.* **2014**, *7*, 119–128. [CrossRef]
80. Di Zanni, E.; Palagano, E.; Lagostena, L.; Strina, D.; Rehman, A.; Abinun, M.; De Somer, L.; Martire, B.; Brown, J.; Kariminejad, A.; et al. Pathobiologic Mechanisms of Neurodegeneration in Osteopetrosis Derived From Structural and Functional Analysis of 14 ClC-7 Mutants. *J. Bone Miner. Res.* **2021**, *36*, 531–545. [CrossRef]
81. Leray, X.; Hilton, J.K.; Nwangwu, K.; Becerril, A.; Mikusevic, V.; Fitzgerald, G.; Amin, A.; Weston, M.R.; Mindell, J.A. Tonic inhibition of the chloride/proton antiporter ClC-7 by PI(3,5)P2 is crucial for lysosomal pH maintenance. *eLife* **2022**, *11*, e74136. [CrossRef] [PubMed]
82. Zhang, S.; Liu, Y.; Zhang, B.; Zhou, J.; Li, T.; Liu, Z.; Li, Y.; Yang, M. Molecular insights into the human CLC-7/Ostm1 transporter. *Sci. Adv.* **2020**, *6*, eabb4747. [CrossRef] [PubMed]
83. Wu, J.Z.; Zeziulia, M.; Kwon, W.; Jentsch, T.J.; Grinstein, S.; Freeman, S.A. ClC-7 drives intraphagosomal chloride accumulation to support hydrolase activity and phagosome resolution. *J. Cell Biol.* **2023**, *222*, e202208155. [CrossRef] [PubMed]

life

MDPI

Article

$GABA_B$ Receptors Tonically Inhibit Motoneurons and Neurotransmitter Release from Descending and Primary Afferent Fibers

Ximena Delgado-Ramírez [1], Nara S. Alvarado-Cervantes [1], Natalie Jiménez-Barrios [1], Guadalupe Raya-Tafolla [1], Ricardo Felix [2], Vladimir A. Martínez-Rojas [1,*] and Rodolfo Delgado-Lezama [1,*]

1 Department of Physiology, Biophysics and Neuroscience, Centre for Research and Advanced Studies of the National Polytechnic Institute (Cinvestav), Avenida IPN 2508, Col. Zacatenco, Mexico City 07360, Mexico
2 Department of Cell Biology, Cinvestav, Mexico City 07360, Mexico
* Correspondence: vamartinezr@cinvestav.mx (V.A.M.-R.); rdelgado@cinvestav.mx (R.D.-L.)

Abstract: Motoneurons receive thousands of excitatory and inhibitory synapses from descending tracts and primary afferent fibers. The excitability of these neurons must be precisely regulated to respond adequately to the requirements of the environment. In this context, $GABA_A$ and $GABA_B$ receptors regulate motoneuron synaptic strength. $GABA_A$ and $GABA_B$ receptors are expressed on primary afferent fibers and motoneurons, while in the descending afferent fibers, only the $GABA_B$ receptors are expressed. However, it remains to be known where the GABA that activates them comes from since the GABAergic interneurons that make axo-axonic contacts with primary afferents have yet to be identified in the descending afferent terminals. Thus, the main aim of the present report was to investigate how $GABA_B$ receptors functionally modulate synaptic strength between Ia afferent fibers, excitatory and inhibitory descending fibers of the dorsolateral funiculus, and spinal motoneurons. Using intracellular recordings from the spinal cord of the turtle, we provide evidence that the $GABA_B$ receptor antagonist, CGP55845, not only prevents baclofen-induced depression of EPSPs but also increases motoneuron excitability and enhances the synaptic strength between the afferent fibers and motoneurons. The last action of CGP55845 was similar in excitatory and inhibitory descending afferents. Interestingly, the action of baclofen was more intense in the Ia primary afferents than in the descending afferents. Even more, CGP55845 reversed the EPSP depression induced by the increased concentration of ambient GABA produced by interneuron activation and GABA transporter blockade. Immunofluorescence data corroborated the expression of $GABA_B$ receptors in the turtle's spinal cord. These findings suggest that $GABA_B$ receptors are extrasynaptic and tonically activated on descending afferent fibers and motoneurons by GABA released from astrocytes and GABAergic interneurons in the cellular microenvironment. Finally, our results also suggest that the antispastic action of baclofen may be due to reduced synaptic strength between descending fibers and motoneurons.

Keywords: $GABA_B$ receptors; spinal cord; dorsolateral funiculus afferents; motoneurons

Citation: Delgado-Ramírez, X.; Alvarado-Cervantes, N.S.; Jiménez-Barrios, N.; Raya-Tafolla, G.; Felix, R.; Martínez-Rojas, V.A.; Delgado-Lezama, R. $GABA_B$ Receptors Tonically Inhibit Motoneurons and Neurotransmitter Release from Descending and Primary Afferent Fibers. *Life* **2023**, *13*, 1776. https://doi.org/10.3390/life13081776

Academic Editor: Carlo Musio

Received: 29 June 2023
Revised: 8 August 2023
Accepted: 17 August 2023
Published: 20 August 2023

1. Introduction

The remarkable variety of behaviors that result from brain activity becomes evident through movements coordinated by motoneurons. Therefore, exquisite control of these cells' excitability is required to activate the minimum necessary to meet the motor demands. Motoneurons receive many monosynaptic inputs from the ventromedial cord (MF), the reticular formation (RF), the vestibular nucleus (VN), and the dorsolateral funiculus (DLF), which contain rubrospinal and propriospinal fibers that elicit excitatory postsynaptic potentials (EPSPs) in flexor motoneurons, inhibitory postsynaptic potentials (IPSPs), and/or EPSPs in extensor motoneurons and primary afferents [1–7]. Therefore, for the motor activity to be completed accurately, the synaptic strength of motoneurons must be precisely regulated at both the pre- and postsynaptic levels. On the other hand, it is well known that the γ-aminobutyric acid (GABA) is the primary inhibitory neurotransmitter of the nervous system, which acts through $GABA_A$ ionotropic and $GABA_B$ metabotropic receptors [8]. Both receptors are expressed in motoneurons and primary afferent fibers, whereas descending afferent fibers contacting motoneurons only express $GABA_B$ receptors [1,6]. Interestingly, in descending fibers, the function of $GABA_B$ receptors has been elucidated using its agonist, baclofen. In this condition, EPSPs recorded in motoneurons are depressed without affecting the input resistance and membrane time constant [1,3,6]. Therefore, in these synapses, the action of baclofen occurs at the presynaptic level by inhibiting the release of the transmitter.

Likewise, anatomical and functional evidence indicates that Ia fibers make GABAergic axo-axonic contacts that activate synaptic $GABA_A$ receptors that mediate short-term presynaptic inhibition associated with primary afferent depolarization (PAD), which is abolished by the application of an antagonist of the $GABA_A$ receptors [5,9,10]. It has been shown that sustained activation of GABAergic interneurons produces long-term inhibition that is removed by the blockade of $GABA_B$ receptors, activated by a spillover of GABA from axo-axonic synapses [5], suggesting that these receptors are extrasynaptically located in Ia fibers. However, these axo-axonal contacts have not been detected in the descending afferent fibers [1]; therefore, the origin of the GABA that functionally activates the $GABA_B$ receptors is presently unknown.

In addition, it has been reported that environmental GABA can tonically activate $GABA_B$ receptors in primary afferent fibers, depressing EPSPs [7,11,12]. Also, neither baclofen nor CGP affects the input resistance of postsynaptic neurons [7,12], suggesting that presynaptic $GABA_B$ receptors mediate the effects on EPSPs. However, when activated by baclofen, $GABA_B$ receptors reduce the excitability of dorsal horn neurons and motoneurons [13,14] by inhibiting L-type calcium channels, indicating that $GABA_B$ receptors modulate the active properties of these neurons. Thus, the question arises as to how $GABA_B$ receptors are functionally activated in motoneurons and descending fibers of the DLF. One source of GABA may be GABAergic interneurons, as reported in Ia fibers [5]. Another possibility is the astrocytes, as observed in the spinal cord's dorsal horn and cerebellum, where GABA is released via the protein Bestrophin-1 (Best-1) [15,16].

Interestingly, the activity of $GABA_B$ receptors is relevant in patients with spasticity. A widely used pharmacological treatment to reduce this condition is the intrathecal application of baclofen. Synaptic strength studies indicate that baclofen has a greater inhibitory action on Ia fibers [1,6], which has led to the proposal that its antispastic action is due to the inhibition of excitatory transmitter release from the primary afferent fibers. However, in patients with spasticity treated with baclofen, it has been observed that the drug did not affect GABA-mediated presynaptic inhibition, suggesting that baclofen may act on motoneurons but not on Ia afferent fibers [17].

Therefore, the main objective of this report was to investigate how $GABA_B$ receptors are activated in the DLF terminals and motoneurons and whether these receptors are tonically active. Likewise, we investigated whether $GABA_B$ receptors are extrasynaptic and where the GABA that activates these receptors is produced and released.

2. Materials and Methods

2.1. Preparation

Forty adult turtles (*Trachemys scripta* spp., 15–20 cm carapace length) were anesthetized with pentobarbital (100 mg/kg, i.p.). The plastron was opened, and the blood was flushed by intraventricular perfusion with Ringer solution (~10 °C) of the following composition (in mM): 120 NaCl, 5 KCl, 15 $NaHCO_3$, 3 $CaCl_2$, 2 $MgCl_2$, and 20 glucose, saturated with 2% CO_2 and 98% O_2 to attain pH 7.5. The lumbar spinal enlargement was isolated by a laminectomy and cut transversally to obtain 2–3 mm thick slices. The slices were placed in a recording chamber for the experiments and perfused with Ringer solution (20–22 °C). The spinal cord, in continuity with the dorsal and ventral roots, was dissected and cut transversally to obtain two segments to stimulate Ia primary afferent fibers. At the end of the dissection, the animals were euthanized by decapitation. All experimental procedures followed the guidelines set out in the Journal of Physiology for ethical matters [18] and were conducted with the approval of the Cinvestav-IPN Experimental Ethics Committee, following the current Mexican Norm for the Care and Use of Animals for Scientific Purposes. The animals were provided by the National Mexican Turtle Center located in Mazunte, Oaxaca, Mexico, with authorization (DGVS-03821/0907) from the Federal Mexican Government Ministry of Environment and Natural Resources (Secretaría de Medio Ambiente y Recursos Naturales, Semarnat).

2.2. Spinal Cord and DRG Immunostaining

Immediately after the dissection, dorsal root ganglia (DRG) and the lumbar enlargement of the turtle spinal cord were fixed in 4% paraformaldehyde in phosphate-buffered saline (PBS) for 24 h and suspended in 10%, 20%, and 30% sucrose at 4 °C in PBS for 24 h. A total of 20 μm DRG and 30 μm spinal cord slices were obtained using a cryostat (CM1520, Leica Microsystems, Wetzlar, Germany). Tissue sections were permeabilized in 0.3% Triton X-100 in PBS for 10 min, placed in 1% SDS in PBS for 5 min, and blocked in 5% Tween20 and BSA 2% in PBS for 2 h. DRG and spinal cord sections were incubated with mouse anti-$GABA_B$ receptor 1 (1:200; ab55051, Abcam) and rabbit NeuN (1:500; #12943, Cell Signaling Technology, MA, USA) for 24 h at 4 °C. After washing three times in PBS, sections were incubated with the corresponding secondary antibodies, Alexa Fluor® 488 AffiniPure Donkey Anti-Rabbit IgG (1:200; 711-545-152, Jackson InmunoResearch, PA, USA) and Alexa Fluor® 594 AffiniPure Donkey Anti-Mouse IgG (1:200; 715-585-150, Jackson InmunoResearch), for 2 h. Then, slices were incubated for 20 min with Hoechst (1:2000; H1398, Invitrogen, MA, USA) and mounted with VECTASHIELD® Antifade Mounting Medium (H-1000, Vector Laboratories, CA, USA). Samples were examined using an epifluorescence microscope (BA410E, Motic, Kowloon, Hong Kong) and Image-Pro Premier software. Images were acquired with the 10× and 40× objectives.

2.3. Intracellular Recordings

The input resistance and the postsynaptic potentials (EPSPs and IPSPs) were monitored by applying a negative rectangular current pulse (500 ms duration) every 5 s before the paired-pulse stimulation of the DLF or dorsal root (Figure 1A). The EPSP and IPSP were evoked in motoneurons by electrical stimulation (1.3xTh) of the DLF and the dorsal root every 5 s in the presence of strychnine and 6-Cyano-7-nitroquinoxaline-2, 3-dione (CNQX). Here, it is worth mentioning that IPSPs could be fully discriminated by the AMPA receptor blockade with CNQX (at 20 μM) because the expression of NMDA receptors has not been reported in motoneurons, and therefore, AMPA receptors are considered to be the only ones responsible for mediating EPSPs in these cells [19]. In the first case, a tungsten bipolar electrode (World Precision Instruments, Sarasota, FL, USA) was used. In the second case, the glass pipette suctioned the dorsal root. The tungsten electrode and the glass pipette were connected to a source of constant current (Neuro Data Instruments Corp., TX, USA). The threshold was defined as the minimum stimulus current intensity that elicits a measurable postsynaptic potential (PSP). The PSPs were recorded with an Axoclamp 2B amplifier (Molecular Devices, Union City, CA, USA) with a bandwidth of 0.1 Hz to 10 KHz, digitized at 10 KHz, and stored in a hard disk for off-line analysis.

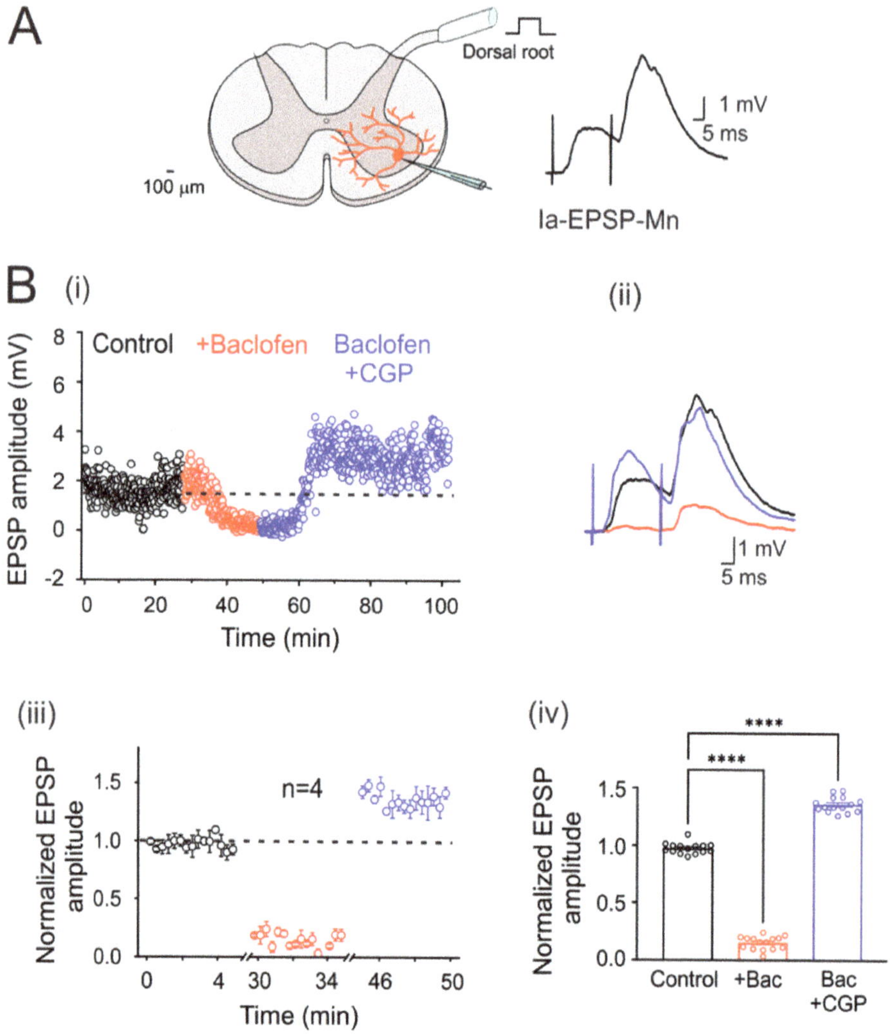

Figure 1. $GABA_B$ receptors are tonically active in the Ia primary afferent fibers. (**A**) Schematic showing the cross-section of the lumbar spinal cord with a motoneuron recorded intracellularly (red). The EPSPs were elicited by electrical stimulation (1.3xTh) of the dorsal root. (**B**) (**i**) Time course of the EPSP amplitude recorded in a motoneuron in the control bath solution (2 µM strychnine; black circles), baclofen (red circles), and baclofen plus CGP55845 (blue circles). (**ii**) Representative traces of the EPSPs recorded in the three experimental conditions. (**iii**) Segmented time course of normalized steady-state EPSP amplitude, relative to control values, recorded during the last 5 min in each condition of steady-state response in each condition with respect to the mean control EPSP amplitude. (**iv**) Bar graph summarizing the mean values of the normalized EPSPs shown in (**iii**) for four motoneurons under all experimental conditions. Asterisks denote statistically significant differences in the values of the EPSPs ($p < 0.0001$ by one-way ANOVA).

2.4. Drugs

$GABA_B$ receptors were activated with baclofen (2 μM) and blocked with CGP55845 (5 μM). Glycinergic receptors were blocked by strychnine (2 μM) and ionotropic glutamatergic AMPA receptors with CNQX (20 μM). Nipecotic acid (100 μM) and CaCCihn-A01 (20 μM) were used to block the activity of GABA transporters and Bestrophin-1 channels, respectively. All drugs used in this study were purchased from Sigma Chemical Company (St. Louis, MO, USA). Baclofen was dissolved in DMSO according to the manufacturer's instructions and stored at 4 °C. From this initial dilution, aliquots (of 10 mM) were prepared and kept at 4 °C. From these aliquots, fresh working solutions were prepared just before each experiment.

2.5. Statistical Analysis

The time course response of the PSP of each motoneuron to the application of the drugs was not the same; therefore, to evaluate the effects for each group of neurons, 4 consecutive EPSPs were recorded every 5 s over 5 min at the steady state for each condition (drug and train) and averaged. These values were normalized to the average control value of each neuron and plotted as a function of time and presented in a bar graph. The action potentials were elicited by 5 s depolarizing current injections for the motoneuron excitability test. Waveform kinetics was determined by differentiating the spike voltage over time (dV/dt), and phase graphs were constructed by plotting the membrane voltage versus the first derivative (mV/ms). The excitability was evaluated by plotting the number of action potentials as the function of supra-threshold intracellular depolarizing current pulses starting from the rheobase current. One-way ANOVA and Student's *t*-test determined the statistical differences between the means. Means were considered statistically different when $p < 0.05$. Values were presented as the mean ± S.E.M.

3. Results

3.1. Characterization of Motoneurons

The recorded cells of the spinal ventral horn were classified as motoneurons (n = 41) when, after being stimulated with a pulse of intracellular depolarizing current greater than the rheobase (Figure 1A), they presented adaptation of the action potential firing pattern. The average input resistance of these cells was 18 ± 3.2 MΩ (n = 41), and the membrane time constant was 19 ± 0.5 ms (n = 41), consistent with previously reported values for motoneurons [4,20].

3.2. $GABA_B$ Receptors Are Tonically Active in Ia Primary Afferent Fibers

In order to corroborate that turtle Ia primary afferents were presynaptically inhibited by tonic activation of $GABA_B$ receptors, as in rats, monosynaptic Ia-EPSPs were evoked in motoneurons by dorsal root stimulation at 1.3xTh every 5 s (Figure 1A). Figure 1(Bi) shows the time course of an evoked Ia-EPSP amplitude in a representative neuron in the presence of baclofen alone or baclofen plus CGP55845. Taking the dashed line as the mean value of the control condition, it can be seen that baclofen produced a depression of the EPSP amplitude, as shown in the representative traces on the right (Figure 2(Bii)). Application of CGP55845 not only reversed EPSP depression but also augmented its amplitude. As can be seen in the graph of normalized values (Figure 1(Biii)) and the corresponding bar graph on the right (Figure 1(Biv)), baclofen depressed the amplitude of the EPSPs by 84.6 ± 0.01% (n = 4, ANOVA test; $F_{(2,42)} = 1706$; $p < 0.001$), while CGP55845 facilitated it by 45 ± 0.02% (n = 4, ANOVA test; $F_{(2,42)} = 1706$; $p < 0.001$), concerning the control value. The input resistance (control 16.5 ± 5.4; baclofen 16.4 ± 4.2 MΩ; n = 4, $p > 0.5$) and the time constant (control 19.7 ± 0.8; baclofen 20.2 ± 0.7 ms; n = 4, $p > 0.5$) were not affected by the drug treatment. Furthermore, the amplitude ratio increased by 93.8% in the presence of baclofen. Overall, these findings agree with previous studies performed in rats, suggesting a conserved mechanism of presynaptic inhibition mediated by $GABA_B$ receptors.

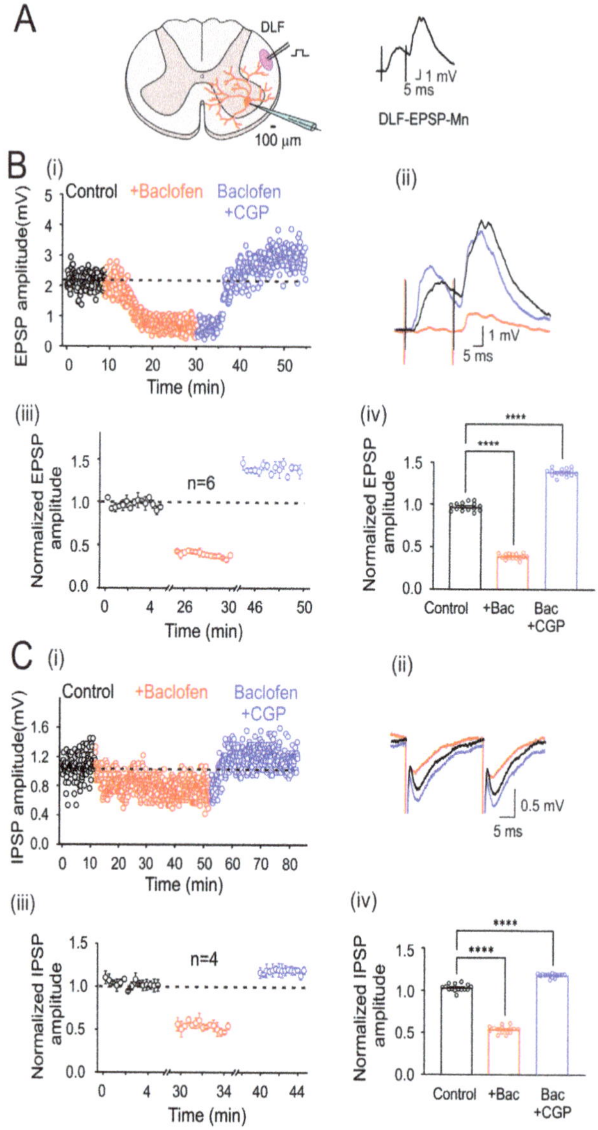

Figure 2. $GABA_B$ receptors are tonically active in excitatory and inhibitory DLF fibers. (**A**) Schematic showing the transverse section of the lumbar spinal cord with a motoneuron recorded intracellularly (red). Here and throughout the subsequent outcomes, the EPSP was elicited by the electrical stimulation (1.3xTh) of the DLF. (**B**) Time course of the EPSP recorded in control bath solution (strychnine 2 μM; black circles), baclofen (red circles), and baclofen plus CGP55845 (blue circles) (**i**). (**C**) IPSP was recorded as in (**B**), but CNQX 20 μM was added instead of strychnine in the control bath solution. Representative EPSP (**B**) and IPSP (**C**) traces recorded under their respective conditions are shown (**ii**). Segmented time course of normalized EPSP (**Biii**) and IPSP (**Ciii**) amplitude recorded during 5 min of steady-state response in each condition with respect to the mean control EPSP amplitude. Bar graph showing the mean values of the EPSPs (**Biv**) and IPSPs (**Civ**), recorded over 5 min. Asterisks denote statistically significant differences in the values of the EPSPs and IPSPs ($p < 0.0001$ by one-way ANOVA).

3.3. $GABA_B$ *Receptors Are Tonically Active in the DLF Terminals Synapsing Motoneurons*

EPSPs in motoneurons were evoked by electrical stimulation of the DLF at 1.3xTh (0.3 ms) every 5 s (Figure 2A). However, since the DLF stimulation evokes EPSPs and IPSPs, the glycine receptor antagonist strychnine (2 μM) was added to the bathing solution (control) to isolate EPSPs. Figure 2(Bi and Bii) show the time course and averaged traces of evoked DLF-EPSPs in a representative motoneuron, in the control condition and the presence of baclofen alone and baclofen plus CGP55845. As observed with the Ia-EPSPs, baclofen depressed the DLF-EPSPs, while CGP55845 not only reversed depression but also facilitated the EPSPs. As shown in the graph of normalized values (Figure 2(Biii)) and the corresponding bar graph on the right (Figure 2(Biv)), baclofen depressed EPSP amplitude by 59 ± 0.01% (n = 6, ANOVA test; $F_{(2,42)} = 2204$; $p < 0.001$), and CGP55845 facilitated it by 43 ± 0.01% (n = 6, ANOVA test; $F_{(2,42)} = 2204$; $p < 0.001$), concerning the control, in six recorded neurons. On average, the amplitude of the DLF-EPSPs was reduced by baclofen to a lesser extent than the Ia-EPSPs.

Next, to record IPSPs in motoneurons elicited by DLF stimulation, an AMPA receptor antagonist (CNQX, 20 μM) was added to the bath solution to block the EPSPs. Figure 2(Ci) shows the time course of evoked DLF-IPSPs in a representative motoneuron under the control condition, in the presence of baclofen alone, or baclofen plus CGP55845. Figure 2(Cii) shows representative traces of the IPSPs recorded under the three conditions. The addition of CGP55845 had two apparent effects. First, it reversed the action of baclofen, and second, it facilitated the amplitude of DLF-IPSPs. As shown in the normalized value plot (Figure 2(Ciii)) and the corresponding bar graph on the right (Figure 2(Civ)), baclofen depressed the IPSP amplitude by 46 ± 0.01% (n = 4, ANOVA test; $F_{(2,42)} = 1314$; $p < 0.001$), while CGP55845 facilitated it by 18 ± 0.01% (n = 4, ANOVA test; $F_{(2,42)} = 1314$; $p < 0.001$), concerning the control. As observed with the DLF-EPSPs, the DLF-IPSPs were less affected than the Ia-EPSPs.

It should be noted that in all motoneurons recorded, the input resistance (control 14.8 ± 1.7 MΩ; baclofen 15.4 ± 2.4 MΩ; n = 10; p = 0.83) and the time constant of the membrane (control 18.9 ± 0.2; baclofen 19.2 ± 0.2 ms) were not affected by baclofen, suggesting that the depression of the EPSPs and IPSPs was caused by $GABA_B$ receptor-mediated presynaptic inhibition of DLF afferent fiber terminals.

3.4. *On the Origin of GABA*

It is well known that motoneurons are innervated by GABAergic and glycinergic interneurons. Therefore, we next decided to investigate whether electrical stimulation of premotor interneurons with a train of 100 pulses at 100 Hz could increase ambient GABA concentration, eventually activating $GABA_B$ receptors. Figure 3i shows the time course of EPSP amplitude in a representative motoneuron in the control condition after one train of activating pulses (AT) and after a second train in the presence of CGP55845. Interestingly, the depression of the EPSPs produced by the first train was reversed and facilitated with CGP55845, which prevented the depression after the second train (Figure 3i,ii). As shown in the normalized and averaged bar graphs, stimulation resulted in the depression of the DLF-EPSPs by 18 ± 0.01% (n = 4, ANOVA test; $F_{(3,56)} = 598.5$; $p < 0.001$) for more than 5 min. However, when CGP55845 was applied, DLF-EPSPs were facilitated by 24 ± 0.01% (n = 4, ANOVA test; $F_{(3,56)} = 598.5$; $p < 0.001$), and the second-train decrease was also prevented (Figure 3iii,iv). Considering that DLF terminals do not express $GABA_A$ receptors and only express $GABA_B$ receptors, these results indicate that the pulse train produced GABA release from GABAergic interneurons, increasing the number of active $GABA_B$ receptors expressed on DLF terminals. This conclusion is supported by the observation that in the presence of CGP55845, the amplitude of the basal EPSPs was restored and facilitated, preventing a decrease in amplitude after a second train of pulses.

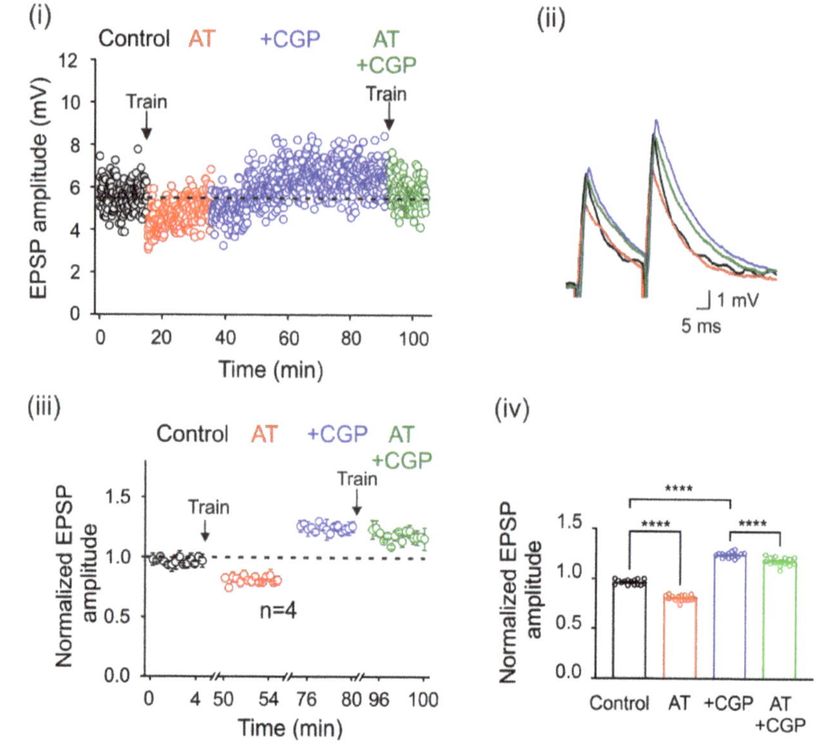

Figure 3. Spillover of GABA tonically activates GABA$_B$ receptors. GABA spillover tonically activates GABA$_B$ receptors. (**i**) Time course of the EPSPs elicited by DLF stimulation at 1.3xTh in the control bath solution (3 μM strychnine), after a train (AT) of 100 electrical pulses (100 Hz) applied to the spinal cord premotor area, in the presence of CGP55845 and again after a second train of pulses. (**ii**) Superimposed traces from representative EPSPs recorded under the four experimental conditions are shown. The color of the individual traces corresponds to the conditions described in (**ii**). Normalized EPSP amplitude was recorded over 5 min at steady state in each condition relative to the mean EPSP amplitude in the control. The number of independent motoneurons evaluated under each condition is shown below the symbols in (**iii**). (**iv**) Bar graph showing the mean value of the EPSPs recorded for 5 min for four motoneurons. Asterisks indicate statistically significant differences ($p < 0.0001$ by one-way ANOVA).

3.5. Regulation of Ambient GABA Concentration with GABA Transporters

It is also known that neurons and astrocytes express transporters that can uptake synaptically released GABA. Therefore, we next decided to investigate whether the blockade of these transporters may affect environmental GABA that tonically activates GABA$_B$ receptors. Figure 4Ai shows that the application of nipecotic acid (100 μM) decreases the amplitude of EPSPs compared to the control condition in one representative motoneuron. This effect was reversed and facilitated by adding CGP55845 to the bath solution (Figure 4(Ai,ii)). A similar effect was observed in five neurons that presented a mean EPSP depression of 21 ± 0.01%, concerning the control ($n = 5$, $p < 0.05$; Student's *t*-test). As previously observed, CGP55845 reversed its decrease and facilitated EPSPs by 19 ± 0.02%, concerning the control values ($n = 5$, ANOVA test; $F_{(2,42)} = 226.7$; $p < 0.001$) (Figure 4(Aiii,iv)). These results show that GABA uptake by transporters may regulate the environmental neurotransmitter concentration and, therefore, the number of tonically activated GABA receptors.

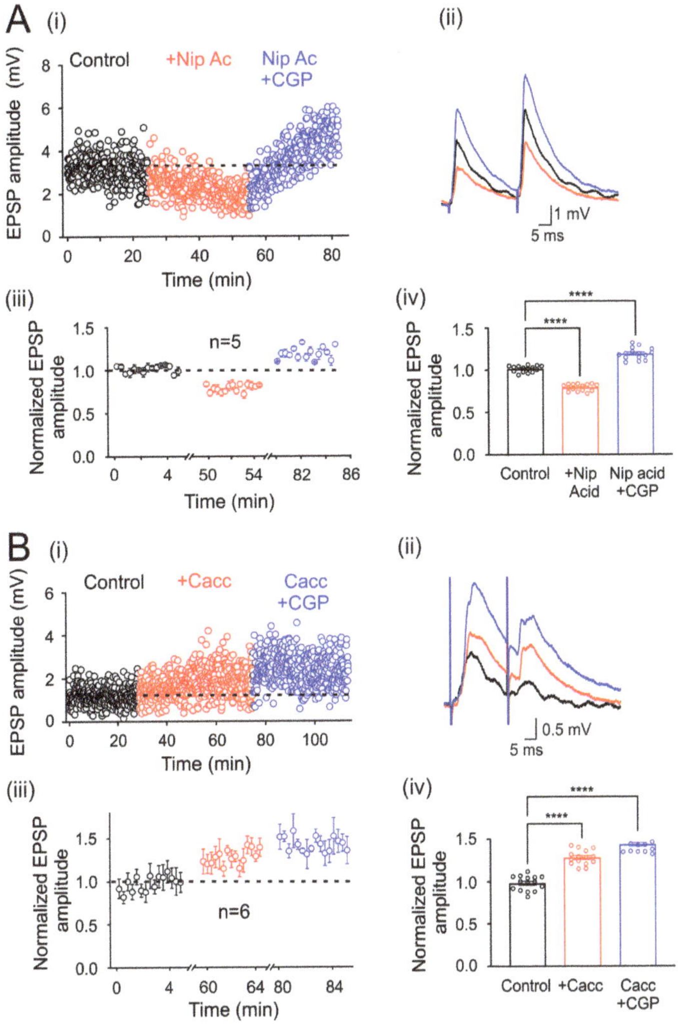

Figure 4. Modulation of environmental GABA concentration. (**Ai**) Time course of the EPSPs evoked every 5 s by DLF stimulation at 1.3xTh in the control bath solution (3 μM strychnine), in the presence of nipecotic acid (NA) and NA plus CGP55845 (**i**). (**Bi**) Time course of the EPSPs evoked every 5 s by DLF stimulation at 1.3xTh in the control bath solution (3 μM strychnine), in the presence of CaCCinh (Cacc) and Cacc plus CGP55845 (**i**). (**A**) and (**B**) (**ii**) show the traces of representative EPSPs recorded in the different conditions. (**iii**) Amplitude of the normalized EPSPs recorded for 5 min at steady state in each experimental condition to the mean control amplitude. (**iv**) Bar graph showing mean values of EPSPs recorded over 5 min for six motoneurons in each condition. Asterisks indicate statistically significant differences ($p < 0.0001$ by one-way ANOVA).

3.6. GABA May Be Released through Best-1 Channels

It has been reported that extrasynaptic $GABA_A$ receptors are tonically active by ambient GABA released by astrocytes through the Best-1 protein channel [16]. Therefore, in order to know whether the ambient GABA that activates the extrasynaptic $GABA_B$ receptors (expressed in the DLF-motoneuron synapse) is released through Best-1 channels, we decided to block this channel with CaCCinh, an antagonist of the Best-1 protein. As is shown in Figure 4(Bi–iv), the EPSP amplitude was facilitated by CaCCinh (20 μM) in 27 ± 0.02%, concerning control ($n = 6$, $p < 0.001$, Student's *t*-test) while, CGP55845 produced an additional facilitation of 15 ± 0.021% ($n = 6$, ANOVA test; $F_{(2,42)} = 107.2$; $p < 0.001$). Therefore, the facilitation of the EPSPs could be due to the Best-1-mediated decrease in the ambient GABA concentration, resulting in a lower amount of tonically active $GABA_B$ receptors.

3.7. $GABA_B$ *Receptors Tonically Inhibit Motoneuron Excitability*

Experimental evidence indicates that baclofen application does not affect the passive properties of motoneurons. However, Svirskis and Hounsgaard (1998) showed that activation of $GABA_B$ receptors with baclofen inhibits an inward calcium current through L-type channels [14]. Therefore, we next decided to investigate whether these receptors are modulating the firing pattern of these cells. To this end, excitability was determined by applying supra-threshold depolarizing current pulses under control conditions and in the presence of CGP55845. The results of these series of experiments show that the blockade of $GABA_B$ receptors with CGP55845 did not modify the kinetics of the single action potential (Figure 5A), but significantly increased the firing frequency (Figure 5B,C) ($n = 12$, Student's *t*-test; $t_{(7)} = 17.4$; $p < 0.001$). These results suggest that $GABA_B$ receptors expressed in motoneurons are tonically decreasing the excitability of motoneurons.

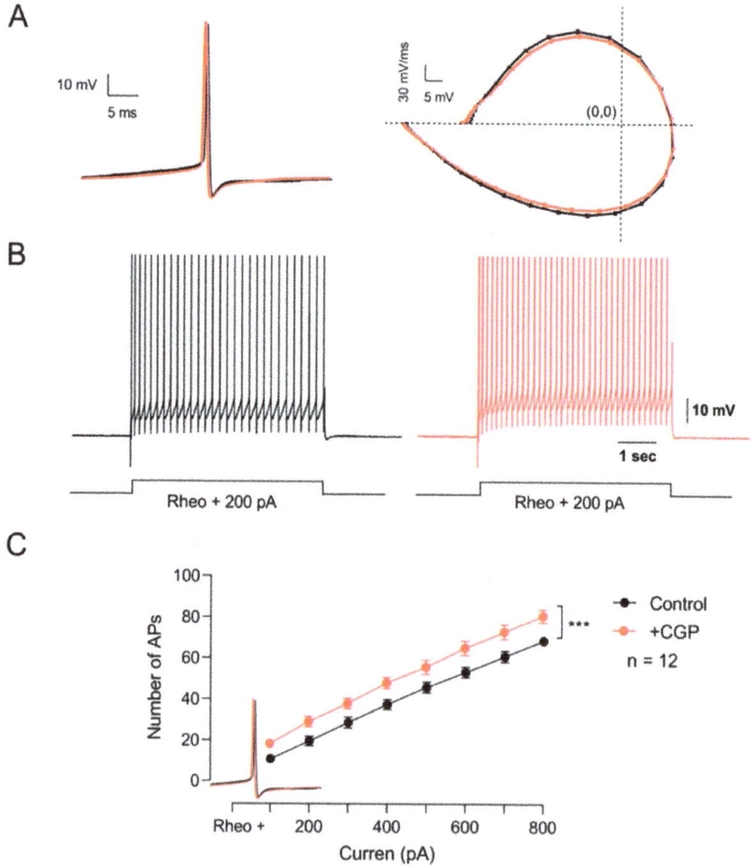

Figure 5. $GABA_B$ receptors tonically depress the excitability of motor neurons. (**A**) On the left, representative action potentials recorded in the control bath solution (black) and the presence of CGP55845 (red). The phase

plot of membrane voltage versus the first derivative (mV/ms) of the action potential is shown on the right. (**B**) Firing pattern of the action potential elicited with the same depolarizing current in the control condition (black) and the presence of CGP55845 (red). (**C**) Graph of the number of action potentials as a function of the injection of depolarizing intracellular current recorded in the control bath solution (black) and the presence of CGP55845 (red). Asterisks indicate statistically significant differences ($p < 0.001$ by Student's *t*-test).

3.8. GABA$_B$ Receptor Expression in the Spinal Cord and Dorsal Root Ganglia of the Turtle

Finally, immunohistochemical staining was performed on cross-sections of the spinal cord and dorsal root ganglia (DRG) to determine the expression of GABA$_B$ receptors in these regions. The results of this analysis show that GABA$_B$ immunostaining is more prominent in the dorsal horn than in the ventral horn of the spinal cord (upper panel of Figure 6). Interestingly, in the ventral horn (lower panels of Figure 6), the signal was evident in the largest cells that, due to their size, may correspond to motoneurons. In Figure 6, immunostaining for GABA$_B$ in the DRG (upper panel) is present in almost all neurons regardless of size, which is best appreciated in the magnification of the boxed region shown in the bottom panels of Figure 6.

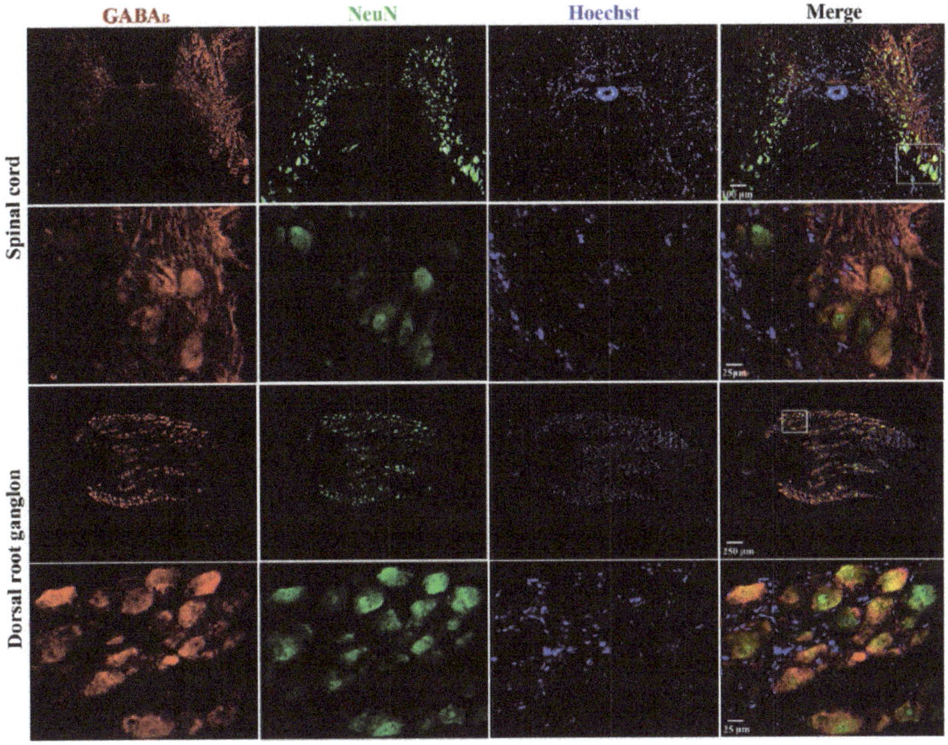

Figure 6. Expression of GABA$_B$ receptors in the spinal cord and dorsal root ganglia of the turtle. Images shown correspond to cross-sections of the spinal cord and DRG immunostained with antibodies against GABA$_B$ receptors in red, neuronal nuclei (NeuN) in green, and nuclei (Hoechst) in blue. In both cases, the images in the upper row were obtained with the 10× objective, and those in the lower row were obtained with the 40× objective. A magnified image of a selected region is also shown, as indicated. The combined image suggests colocalization (yellow/orange) of GABA$_B$ receptors in neuronal cell bodies. It should be noted that Hoechst spots might not overlap (colocalize) with the

fluorescence of anti-NeuN antibodies because the light intensity to activate Hoechst fluorescence in large cells is not strong enough to see it in its entirety, as if it occurs in neurons of small size. For this same reason, the nuclei of satellite (small) glial cells surrounding the soma of neurons in the DRG stain adequately. In addition, negative controls are presented as a Supplemental Figure S1.

4. Discussion

In the present report, we show that the $GABA_B$ receptors are tonically activated by ambient GABA, inhibiting the excitability of motoneurons and the transmitter release from excitatory and inhibitory DLF terminals synapsing motoneurons. Likewise, we corroborated that these receptors tonically inhibit Ia afferent fibers. We also show that the number of tonically activated $GABA_B$ receptors depends on the ambient GABA concentration, which is determined by its release from interneurons and astrocytes. Therefore, extrasynaptic $GABA_B$ receptors may have a relevant role in motor control.

4.1. $GABA_B$ Receptors Tonically Inhibit Transmitter Release from DLF Terminals and Ia Afferents

The role of $GABA_B$ receptors in modulating the synaptic strength of excitatory DLF terminals and motoneurons was investigated in the presence of strychnine to block the inhibitory potentials evoked by the DLF terminals of propriospinal origin [2]. In these experimental conditions, baclofen was found to decrease the EPSPs and IPSPs elicited by stimulation of the DLF and Ia fibers, respectively. In both cases, CGP55845 prevented the decrease and facilitated the EPSPs and IPSPs. Notably, the observed effects on the EPSPs and IPSPs did not alter the motoneurons' input resistance or time constant. This finding agrees with previous reports in Ia, Aβ, Aδ, and C afferent fibers, where a blockade of $GABA_B$ receptors facilitated the EPSPs evoked in motoneurons and dorsal horn neurons [7,11,12].

Likewise, the blockade of $GABA_B$ receptors in primary afferent fibers and DLF terminals did not affect miniature EPSCs (mEPSCs), miniature IPSCs (mIPSCs), or membrane conductance [6,11]. Before our work, the function of these receptors in the afferent fibers of descending tracts, such as the ventromedial, dorsal, and dorsolateral funiculus, had been evidenced only by their activation with baclofen, producing a significant inhibition of transmitter release [1,3,4,6].

Interestingly, in excitatory DLF afferents, $GABA_B$ receptors may act by inhibiting Ca^{2+} channels of the N- and P/Q-type via the activation of Gi/o proteins [21]. On the other hand, the expression of $GABA_A$ receptors in the DLF terminals has been underestimated because, unlike Ia fibers, the vestibulospinal and reticulospinal terminals that synapse with motoneurons appear not to be under the presynaptic control of muscle-type afferents I and II [1]. Moreover, there is no evidence of axo-axonic GABAergic synapses in these afferents [22]. Therefore, DLF terminals may only express $GABA_B$ receptors located at extrasynaptic sites. Another relevant result of our work is that the blockade of the $GABA_B$ receptors produced a facilitation of EPSPs and IPSPs, which indicates that these receptors, as in primary afferent fibers, are tonically active by the ambient GABA.

4.2. The Cellular Origin of GABA

Though the descending afferent fibers express $GABA_B$ receptors, GABAergic axo-axonal contacts, such as those in primary afferents that release GABA, have not yet been identified. Here, we show that the simultaneous stimulation of interneurons in the premotor region produces a depression of EPSPs that is prevented by applying CGP55845, suggesting the involvement of $GABA_B$ receptors. This further suggests that synaptically released GABA must have diffused from the synaptic gap to reach extrasynaptic receptors. This result is consistent with what has been described for CA3 pyramidal cells of the hippocampus, where extrasynaptic $GABA_B$ receptors are activated by GABA released by the simultaneous activation of neighboring interneurons [23]. Likewise, in the spinal cord, activation of these receptors expressed in the primary afferents was produced by tetanic stimulation of the hindlimb flexor muscles, leading to long-lasting presynaptic inhibition of

Ia fibers, which is prevented with the application of CPG [5]. Thus, in the spinal cord, diffusion of synaptically released GABA can regulate synaptic strength between primary and descending afferent fibers with motoneurons, which is relevant to motor control. Activation of $GABA_B$ receptors by GABA diffusion indicates that these receptors are extrasynaptic. Consequently, the functioning of neural networks in the spinal cord, just as it occurs in the hippocampus, requires the action of GABA released from the synaptic and diffusion systems. The diffusion of neurotransmitters is known as spillover and, together with its related receptors, is called the neurotransmitter diffusion system [24].

The depression of the EPSPs produced by the blockade of the GABA transporters and their reversal and facilitation by CGP55845 suggest that in the process, more $GABA_B$ receptors were activated than those already tonically active, which, when blocked, facilitated the EPSPs in response to the increase in GABA concentration. This enhanced action of GABA produced by blocking GABA uptake has been reported in neurons of the spinal cord, cerebellum, cortex, and hippocampus [25,26]. Particularly in the spinal cord, the blockade of GABA transporters with nipecotic acid prolongs the neuronal response to the iontophoretic application of GABA [25]. Therefore, GABA transporters expressed in neurons and glial cells may play an important role in the uptake of neurotransmitters in the synaptic cleft and in controlling its extrasynaptic concentration [27].

Another potential source of GABA is astrocytes. EPSP facilitation in the DLF in the presence of CaCCinh, a Best-1 channel blocker, indicates a reduction in active $GABA_B$ receptors due to decreased ambient GABA concentration. This is consistent with the reduction in the tonic currents recorded in cerebellar granule cells in the presence of the Best-1 antagonist [16]. Interestingly, GABA and Best-1 channels have previously been shown to be expressed in Bermann glia and lamellar astrocytes adjacent to granule cell bodies [16]. In addition, glutamate has been shown to induce GABA release from astrocytes in the spinal cord, as recorded using HEK293 cells heterologously expressing GABA receptors as GABA sensors [15]. Likewise, GABA transporters expressed in neurons and glial cells regulate the extracellular concentration of GABA by capturing and releasing it, as occurs in the Bermann glia [27,28].

4.3. The Motoneuron Excitability Is Controlled Tonically by $GABA_B$ Receptors

In all motoneurons recorded, baclofen depressed the EPSPs and IPSPs evoked by DLF and Ia afferent fibers without affecting their passive properties. The same action of baclofen was reported in the recordings of EPSPs evoked in motoneurons by activation of descending afferent fibers and Ia [1,5,6] and those recorded in dorsal horn neurons of the spinal cord evoked by activation of the primary afferents Aβ, Ad, and C [7,11,12]. Furthermore, it has been shown that the size of the quantum of miniature potentials did not change, indicating that the action of baclofen was at the presynaptic level [6]. However, in these reports, the active properties of postsynaptic neurons were not evaluated [6,7,11,12]. Our findings show that in motoneurons, $GABA_B$ receptors are tonically reducing excitability, since blockade with CGP55845 increases the number of action potentials. This result suggests that, as in afferent fibers, $GABA_B$ receptors are tonically active in motoneurons, reducing their excitability due to the inhibition of the L-type calcium current. This action of $GABA_B$ receptors was reported by Svirskis and Hounsgaard, who showed that in motoneurons, the activation of the inward calcium current through L-type channels was inhibited in the presence of baclofen [14].

On the other hand, immunofluorescence assays have shown that $GABA_B$ receptors are located in both the dorsal and ventral horns of the turtle spinal cord, with greater expression dorsally and 60–90% lower density in laminae IX and X than in the superficial laminae of the rat spinal cord [29]. Probably the high density of $GABA_B$ receptors in the dorsal horn represents its expression not only in the soma and neuropil of interneurons but also in primary afferents of the Aδ-type and C-type [30]. The positively stained cells in the ventral horn may correspond to motoneurons due to their size and localization [31]. In the DRG, we found a high expression level of $GABA_B$ receptors in all cells regardless of size. This result is similar to that reported for the DRG in rats [29]. Therefore, the images indicate that $GABA_B$ receptors are expressed in the postsynaptic neurons throughout the spinal cord, where they control its excitability.

4.4. $GABA_B$ Receptors and Spasticity

Spasticity is a condition present in patients with injury of the brain and spinal cord, defined as a motor disorder with a velocity-dependent increase in tonic stretch reflexes with exacerbated tendon jerks caused by hyperexcitability of the stretch reflex, as a component of the upper motoneuron syndrome [32,33]. For a long time, baclofen has been used to reduce spasticity, although it is presently unknown which neuronal element is acting.

Based on the evidence showing that the main action of baclofen is to decrease transmitter release from Ia afferent and descending afferent fibers that synapse with motoneurons [1], it has been proposed that the antispastic action of baclofen is due to the presynaptic inhibition of Ia fibers. However, it has also been shown that, in patients with spasticity, neither of the two tests of presynaptic inhibition, postactivation depression, or reciprocal disynaptic inhibition of the soleus H reflex were affected by baclofen administration. In addition, primary afferent action potentials did not change after baclofen administration. However, in both cases, baclofen depressed the H reflex. Based on these results, the antispastic action of baclofen could be attributed to a decrease in motoneuron excitability [17].

Interestingly, in all studies performed on animal preparations, the reported effect of baclofen is the inhibition of transmitter release from primary Ia afferents and excitatory descending afferents of the ventromedial and dorsolateral funicles [1,3,4,7,21] without affecting the passive properties of motoneurons. In these reports, only the passive properties were explored, either by recording the input resistance or by determining the time constant of the EPSP decay phase, without studying the action of baclofen on the active properties of motoneurons. Our results indicate that $GABA_B$ receptors tonically inhibit motoneuron excitability. This could be explained by the inhibition of L-type calcium currents by baclofen [14]. This postsynaptic action of baclofen was also observed in the dorsal horn neurons of the spinal cord, where activation of $GABA_B$ receptors inhibits the wind-up and potential plateau that underlies a sustained burst of action potentials [13]. Our results support the idea that the antispastic action of baclofen may be produced by the reduction in motoneuron excitability due to the inactivation of the presynaptic N- and P/Q-type calcium channels in the descending excitatory fibers [21] and the postsynaptic L-type calcium channels expressed in motoneurons.

In conclusion, this paper shows evidence that $GABA_B$ receptors are tonically activated by environmental GABA, thereby reducing synaptic strength between DLF afferent fibers and motoneurons by inhibiting transmitter release and postsynaptic excitability. Furthermore, we found that the release from interneurons and astrocytes determines the environmental GABA concentration.

Supplementary Materials: The following supporting information can be downloaded at: https://www.mdpi.com/article/10.3390/life13081776/s1, Figure S1: Negative control of immunofluorescence experiments of the expression of $GABA_B$ receptors in the turtle's spinal cord and dorsal root ganglia. (A) Images shown correspond to cross-sections of the spinal cord (A) and DRG (B) immunostained without primary antibodies against $GABA_B$ receptors and neuronal nuclei (NeuN), but in the presence of the fluorescent DNA dye Hoechst (blue signal). The images in the upper and lower rows were obtained with the 4× and 10× objectives, respectively.

Author Contributions: Conceptualization, X.D.-R., R.D.-L. and V.A.M.-R.; methodology, X.D.-R., N.S.A.-C., G.R.-T., N.J.-B., V.A.M.-R. and R.D.-L.; formal analysis, X.D.-R., N.S.A.-C., N.J.-B., V.A.M.-R., R.F., R.D.-L., R.D.-L. and V.A.M.-R. Wrote the original draft of the work. All authors have read and agreed to the published version of the manuscript.

Funding: This work was partially supported by Conahcyt grants PN5098 (to RDL). XCD (420935), NJB (935751), and NSAC (936253) are Conahcyt PhD fellows. A postdoctoral fellowship from Conahcyt to VMR is also acknowledged (Estancias Posdoctorales por México 2022, 396733).

Institutional Review Board Statement: Animal care and experimental procedures were conducted with the approval of the Cinvestav-IPN Experimental Ethics Committee and in accordance with the current Mexican Norm for the Care and Use of Animals for Scientific Purposes. The animals were provided by the National Mexican Turtle Center located in Mazunte, Oaxaca, Mexico, with authorization (DGVS-03821/0907) from the Federal Mexican Government Ministry of Environment and Natural Resources (Secretaría de Medio Ambiente y Recursos Naturales, Semarnat).

Informed Consent Statement: Not applicable.

Data Availability Statement: Derived data supporting the findings of this study are available from the corresponding authors upon request.

Acknowledgments: The authors thank the two anonymous reviewers for their valuable comments and suggestions, which substantially improved the manuscript.

Conflicts of Interest: The authors declare no conflict of interest.

References

1. Jiménez, I.; Rudomin, P.; Enriquez, M. Differential Effects of (-)-Baclofen on Ia and Descending Monosynaptic EPSPs. *Exp. Brain Res.* **1991**, *85*, 103–113. [CrossRef]
2. Kostyuk, P.G.; Vasilenko, D.A.; Lang, E. Propriospinal Pathways in the Dorsolateral Funiculus and Their Effects on Lumbosacral Motoneuronal Pools. *Brain Res.* **1971**, *28*, 233–249. [CrossRef] [PubMed]
3. Curtis, D.R.; Malik, R. The Differential Effects of Baclofen on Segmental and Descending Excitation of Spinal Interneurones in the Cat. *Exp. Brain Res.* **1985**, *58*, 333–337. [CrossRef]
4. Delgado-Lezama, R.; Aguilar, J.; Cueva-Rolón, R. Synaptic Strength between Motoneurons and Terminals of the Dorsolateral Funiculus Is Regulated by GABA Receptors in the Turtle Spinal Cord. *J. Neurophysiol.* **2004**, *91*, 40–47. [CrossRef] [PubMed]
5. Curtis, D.R.; Lacey, G. Prolonged GABA(B) Receptor-Mediated Synaptic Inhibition in the Cat Spinal Cord: An in Vivo Study. *Exp. Brain Res.* **1998**, *121*, 319–333. [CrossRef]
6. Edwards, F.R.; Harrison, P.J.; Jack, J.J.; Kullmann, D.M. Reduction by Baclofen of Monosynaptic EPSPs in Lumbosacral Motoneurones of the Anaesthetized Cat. *J. Physiol.* **1989**, *416*, 539–556. [CrossRef]
7. Peshori, K.R.; Collins, W.F.; Mendell, L.M. EPSP Amplitude Modulation at the Rat Ia-Alpha Motoneuron Synapse: Effects of GABAB Receptor Agonists and Antagonists. *J. Neurophysiol.* **1998**, *79*, 181–189. [CrossRef] [PubMed]
8. Sivilotti, L.; Nistri, A. GABA Receptor Mechanisms in the Central Nervous System. *Prog. Neurobiol.* **1991**, *36*, 35–92. [CrossRef] [PubMed]
9. Eccles, J.C.; Schmidt, R.; Willis, W.D. Pharmacological Studies on Presynaptic Inhibition. *J. Physiol.* **1963**, *168*, 500–530. [CrossRef] [PubMed]
10. Bautista, W.; Aguilar, J.; Loeza-Alcocer, J.E.; Delgado-Lezama, R. Pre- and Postsynaptic Modulation of Monosynaptic Reflex by GABAA Receptors on Turtle Spinal Cord. *J. Physiol.* **2010**, *588 Pt 14*, 2621–2631. [CrossRef]
11. Yang, K.; Ma, H. Blockade of GABA(B) Receptors Facilitates Evoked Neurotransmitter Release at Spinal Dorsal Horn Synapse. *Neuroscience* **2011**, *193*, 411–420. [CrossRef]
12. Salio, C.; Merighi, A.; Bardoni, R. GABAB Receptors-Mediated Tonic Inhibition of Glutamate Release from Aβ Fibers in Rat Laminae III/IV of the Spinal Cord Dorsal Horn. *Mol. Pain* **2017**, *13*. [CrossRef]
13. Russo, R.E.; Nagy, F.; Hounsgaard, J. Inhibitory Control of Plateau Properties in Dorsal Horn Neurones in the Turtle Spinal Cord in Vitro. *J. Physiol.* **1998**, *506 Pt 3*, 795–808. [CrossRef] [PubMed]
14. Svirskis, G.; Hounsgaard, J. Transmitter Regulation of Plateau Properties in Turtle Motoneurons. *J. Neurophysiol.* **1998**, *79*, 45–50. [CrossRef] [PubMed]
15. Christensen, R.K.; Delgado-Lezama, R.; Russo, R.E.; Lind, B.L.; Alcocer, E.L.; Rath, M.F.; Fabbiani, G.; Schmitt, N.; Lauritzen, M.; Petersen, A.V.; et al. Spinal Dorsal Horn Astrocytes Release GABA in Response to Synaptic Activation. *J. Physiol.* **2018**, *596*, 4983–4994. [CrossRef] [PubMed]
16. Lee, S.; Yoon, B.E.; Berglund, K.; Oh, S.J.; Park, H.; Shin, H.S.; Augustine, G.J.; Lee, C.J. Channel-Mediated Tonic GABA Release from Glia. *Science* **2010**, *330*, 790–796. [CrossRef] [PubMed]
17. Ørsnes, G.B.; Sørensen, P.S.; Larsen, T.K.; Ravnborg, M. Effect of Baclofen on Gait in Spastic MS Patients. *Acta Neurol. Scand.* **2000**, *101*, 244–248. [CrossRef] [PubMed]
18. Drummond, G.B. Reporting Ethical Matters in The Journal of Physiology: Standards and Advice. *J. Physiol.* **2009**, *587*, 713–719. [CrossRef]
19. Moghaddasi, M.; Velumian, A.A.; Zhang, L.; Fehlings, M.G. An ex vivo preparation of mature mice spinal cord to study synaptic transmission on motoneurons. *J. Neurosci. Methods* **2007**, *159*, 1–7. [CrossRef]
20. Hounsgaard, J.; Mintz, I. Calcium Conductance and Firing Properties of Spinal Motoneurones in the Turtle. *J. Physiol.* **1988**, *398*, 591–603. [CrossRef]

21. Castro, A.; Aguilar, J.; Elias, D.; Felix, R.; Delgado-Lezama, R. G-Protein-Coupled GABAB Receptors Inhibit Ca^{2+} Channels and Modulate Transmitter Release in Descending Turtle Spinal Cord Terminal Synapsing Motoneurons. *J. Comp. Neurol.* **2007**, *503*, 642–654. [CrossRef] [PubMed]
22. Glusman, S.; Vázquez, G.; Rudomín, P. Ultrastructural Observations in the Frog Spinal Cord in Relation to the Generation of Primary Afferent Depolarization. *Neurosci. Lett.* **1976**, *2*, 137–145. [CrossRef] [PubMed]
23. Scanziani, M. GABA Spillover Activates Postsynaptic GABAB Receptors to Control Rhythmic Hippocampal Activity. *Neuron* **2000**, *25*, 673–681. [CrossRef]
24. Kullmann, D.M. Spillover and Synaptic Cross Talk Mediated by Glutamate and GABA in the Mammalian Brain. *Prog. Brain Res.* **2000**, *125*, 339–351. [CrossRef] [PubMed]
25. Curtis, D.R.; Game, C.J.A.; Lodge, D. The in Vivo Inactivation of GABA and Other Inhibitory Amino Acids in the Cat Nervous System. *Exp. Brain Res.* **1976**, *25*, 413–428. [CrossRef]
26. Roepstorff, A.; Lambert, J.D.C. Factors Contributing to the Decay of the Stimulus-Evoked IPSC in Rat Hippocampal CA1 Neurons. *J. Neurophysiol.* **1994**, *72*, 2911–2926. [CrossRef]
27. Semyanov, A.V. Diffusional Extrasynaptic Neurotransmission via Glutamate and GABA. *Neurosci. Behav. Physiol.* **2005**, *35*, 253–266. [CrossRef] [PubMed]
28. Barakat, L.; Wang, D.; Bordey, A. Carrier-Mediated Uptake and Release of Taurine from Bergmann Glia in Rat Cerebellar Slices. *J. Physiol.* **2002**, *541 Pt 3*, 753. [CrossRef]
29. Towers, S.; Princivalle, A.; Billinton, A.; Edmunds, M.; Bettler, B.; Urban, L.; Castro-Lopes, J.; Bowery, N.G. GABAB Receptor Protein and MRNA Distribution in Rat Spinal Cord and Dorsal Root Ganglia. *Eur. J. Neurosci.* **2000**, *12*, 3201–3210. [CrossRef]
30. Désarmenien, M.; Feltz, P.; Occhipinti, G.; Santangelo, F.; Schlichter, R. Coexistence of GABAA and GABAB Receptors on A Delta and C Primary Afferents. *Br. J. Pharmacol.* **1984**, *81*, 327. [CrossRef]
31. Andres, C.; Aguilar, J.; González-Ramírez, R.; Elias-Viñas, D.; Felix, R.; Delgado-Lezama, R. Extrasynaptic A6 Subunit-Containing GABAA Receptors Modulate Excitability in Turtle Spinal Motoneurons. *PLoS ONE* **2014**, *9*, e115378. [CrossRef]
32. Lance, J.W. Spasticity: Disordered Motor Control. Symposium synopsis. In *Symposia Specialists, Miami*; Feldman, R.G., Young, R.R., Koella, W.W., Eds.; Medical Publishers: Chicago, IL, USA, 1980; pp. 485–500.
33. Nielsen, J.B.; Crone, C.; Hultborn, H. The Spinal Pathophysiology of Spasticity—From a Basic Science Point of View. *Acta Physiol.* **2007**, *189*, 171–180. [CrossRef] [PubMed]

life

MDPI

Review

Bidirectional Regulation of $GABA_A$ Reversal Potential in the Adult Brain: Physiological and Pathological Implications

Haram R. Kim [1] and Marco Martina [1,2,*]

1 Department of Neuroscience, Feinberg School of Medicine, Northwestern University, 300 E. Superior, Chicago, IL 60611, USA; haram.kim@northwestern.edu
2 Department of Psychiatry, Feinberg School of Medicine, Northwestern University, 300 E. Superior, Chicago, IL 60611, USA
* Correspondence: m-martina@northwestern.edu

Abstract: In physiological conditions, the intracellular chloride concentration is much lower than the extracellular. As $GABA_A$ channels are permeable to anions, the reversal potential of $GABA_A$ is very close to that of Cl^-, which is the most abundant free anion in the intra- and extracellular spaces. Intracellular chloride is regulated by the activity ratio of NKCC1 and KCC2, two chloride-cation cotransporters that import and export Cl^-, respectively. Due to the closeness between $GABA_A$ reversal potential and the value of the resting membrane potential in most neurons, small changes in intracellular chloride have a major functional impact, which makes $GABA_A$ a uniquely flexible signaling system. In most neurons of the adult brain, the $GABA_A$ reversal potential is slightly more negative than the resting membrane potential, which makes $GABA_A$ hyperpolarizing. Alterations in $GABA_A$ reversal potential are a common feature in numerous conditions as they are the consequence of an imbalance in the NKCC1-KCC2 activity ratio. In most conditions (including Alzheimer's disease, schizophrenia, and Down's syndrome), $GABA_A$ becomes depolarizing, which causes network desynchronization and behavioral impairment. In other conditions (neonatal inflammation and neuropathic pain), however, $GABA_A$ reversal potential becomes hypernegative, which affects behavior through a potent circuit deactivation.

Keywords: transporter; disease; brain oscillations; excitation; inhibition

Citation: Kim, H.R.; Martina, M. Bidirectional Regulation of $GABA_A$ Reversal Potential in the Adult Brain: Physiological and Pathological Implications. *Life* **2024**, *14*, 143. https://doi.org/10.3390/life14010143

Academic Editor: Carlo Musio

Received: 8 December 2023
Revised: 16 January 2024
Accepted: 17 January 2024
Published: 19 January 2024

1. Introduction

The balance between excitation and inhibition (E/I balance) is critical for maintaining optimal function of the brain [1–4]. At the cellular level, a range of inhibitory mechanisms are designed to precisely control the excitation so that the neurons convey accurate information through the network [5–8]. In the brain, GABA (γ-aminobutyric acid) is the primary inhibitory neurotransmitter for both long-range and local inhibitory transmission. E/I imbalance can be caused by alterations of excitation, inhibition, or both. Altered E/I balance has been suggested as a causal factor for many symptoms associated with neuropsychiatric diseases, including fragile X syndrome [9], Alzheimer's disease [10], and schizophrenia [11,12], and in most cases, it is attributable to altered GABAergic signaling [13,14].

In the brain, GABA exercises its effects by acting on two major types of receptors, the ionotropic ($GABA_AR$) receptor and the metabotropic ($GABA_BR$) receptor. In contrast to the $GABA_B$ receptors, which set the basal tone of inhibition through indirectly opening potassium channels via G-protein activation, $GABA_A$ receptors immediately alter the ion permeability of the membrane. When the $GABA_AR$ is activated, monovalent anions (mostly chloride but also nitrate and bicarbonate ions [15]) pass through the transmembrane channel structure of receptors. This movement of anions generates the postsynaptic current (PSC). The PSC amplitude depends on the number and availability of receptors and the intracellular concentration of anions, as well as the amount of GABA released from presynaptic terminals. Although multiple mechanisms regulate signaling through $GABA_A$ receptors, including the modulation of the subunit composition of $GABA_A$ receptors and the alteration in GABA metabolism, an increasing amount of work suggests that in most pathological conditions, the main alterations in $GABA_A$ function are caused by changes in the relative activity of NKCC1 and KCC2, two chloride/potassium cotransporters that import and export chloride, respectively, and, thus, set the reversal potential of the current.

Over the past decades, it has become clear that despite its primary role as an inhibitory signal, the $GABA_A$ current can also be excitatory. Apart from the well-known role of depolarizing $GABA_A$ in developmental stages, which has been extensively investigated, numerous pathological states are associated with the reappearance of a depolarizing effect of GABA in the adult brain. Additionally, we recently showed that alterations leading to hypernegative $GABA_A$ reversal potential also have an important functional impact on the adult cortex [16]. Here, we review the bidirectional regulation of $GABA_A$ reversal potential in pathological conditions of the adult brain and discuss how the altered $GABA_A$ function may impact network function.

2. $GABA_A$ Reversal Potential

2.1. $GABA_A$ Reversal Potential and Intracellular Chloride

In physiological conditions, the concentration of intracellular chloride ($[Cl^-]_i$ = 5~30 mM) is considerably lower compared to extracellular chloride ($[Cl^-]_o$ = ~140 mM) [17]. As $GABA_A$ channels are permeable to anions, the reversal potential of $GABA_A$ is very close to that of Cl^-, which is the most abundant free anion in physiological solutions. A few studies have measured $[Cl^-]_i$ directly in physiological conditions using the visualization of chloride ions [18,19], but their reliability in assessing absolute chloride concentrations remains questionable. The most reliable method to estimate $[Cl^-]_i$ remains the measurement of $GABA_A$ current reversal potential. The reversal potential of $GABA_A$ can be obtained by measuring the $GABA_A$ currents at different membrane voltages. Using the perforated patch clamp recording technique, which preserves the native concentration of anions, $[Cl^-]_i$ is calculated using the Nernst equation from the measured reversal potential.

At reversal potential, the net Cl^- flux across the membrane is zero, and the effect of $GABA_A$ activation is entirely due to shunting inhibition (the conductance increase that diminishes cellular excitability due to Ohm's law). In most neurons of the adult brain, the $GABA_A$ reversal potential is slightly more negative than the resting membrane potential, which makes $GABA_A$ hyperpolarizing. It is important to stress that because, in most cases, the $GABA_A$ reversal potential is close to the resting potential, even small changes in $GABA_A$ reversal can shift the effect of $GABA_A$ activation from hyperpolarizing to depolarizing.

Thus, small increases in $[Cl^-]_i$ can cause the $GABA_A$ current reversal potential to become depolarized relative to the resting potential. However, depolarizing reversal potential is not synonymous with an excitatory effect. It is only when this depolarizing effect exceeds the inhibitory shunting effect that $GABA_A$ becomes excitatory. Due to the vicinity, in most neurons, between the value of $GABA_A$ reversal potential and resting membrane potential, the opposite case is also true, and decreased $[Cl^-]_i$ boosts the influx of negatively charged ions and, therefore, strengthens the inhibitory effect of the current.

Interestingly, the internal chloride concentration in neurons is differentially regulated across developmental stages and cell types, even in the normal brain. This sets the $GABA_A$ current clearly apart from the currents mediated by ionotropic glutamate receptors (AMPAR and NMDAR), which have reversal potentials (~0 mV) that are much less sensitive to changes in sodium (Na^+) or potassium (K^+) concentration and never change from excitatory to inhibitory. This "flexibility" of the $GABA_A$ current, which is regulated by the level of $[Cl^-]_I$, sets $GABA_A$ apart from other signaling systems.

2.2. Chloride/Potassium Cotransporters

Neuronal $[Cl^-]_i$ is primarily set by the activity of NKCC1 and KCC2, two chloride/potassium cotransporters that act in opposite directions. While NKCC1 increases

$[Cl^-]_i$, KCC2 promotes Cl^- extrusion [20]. The actions of these two transporters are differently balanced in different locations and conditions and depend on transporter density and distribution as well as modulation (such as phosphorylation).

NKCC1 is widely expressed in cells within and outside the brain. NKCC1 transcripts and proteins are also present in non-neuronal cells like oligodendrocytes, microglia, and astrocytes [21]. KCC2 expression, on the contrary, is neuron-selective across numerous CNS structures, including the spinal cord, retina, neocortex, and most subcortical structures [22–24].

The structures of both transporters were well described [25–29]. NKCC1 cotransports Na^+ and K^+ inwardly together with two Cl^-, thus increasing $[Cl^-]_i$. In contrast, KCC2 exports K^+ and Cl^- and thus decreases $[Cl^-]_i$. As in mature neurons of the adult brain the activity of KCC2 exceeds that of NKCC1, $[Cl^-]_i$ is maintained at low levels, which results in the $GABA_A$ current being inhibitory and hyperpolarizing.

Increased NKCC1 activity, on the other hand, results in a depolarizing shift in the $GABA_A$ current reversal potential by elevating $[Cl^-]_i$. In this case, the reversal potential of the $GABA_A$ current becomes depolarized compared with the resting membrane potential (see Section 2.1). In contrast, increased KCC2 activity lowers $[Cl^-]_i$ and hyperpolarizes $GABA_A$ reversal potential to values considerably lower than the resting membrane potential, producing a stronger inhibitory effect.

3. Regulation of $GABA_A$ Reversal Potential in Physiological Conditions

Depolarizing $GABA_A$ has been discovered and intensively studied in immature neurons for over 30 years, and it is evolutionally conserved across vertebrate species and in *Caenorhabditis elegans* [30]. Ben-Ari (2002) suggested that there are several potential biological advantages to having GABA mediate excitation in immature neurons because, compared with the excitatory glutamatergic current, the depolarizing GABA current could be less excitotoxic for postsynaptic neurons and more energetically efficient as it does not require pumping ions against large transmembrane gradients. Strong evidence also supports the idea that the reversal potential of the $GABA_A$ current is heavily regulated, even in a normally functioning adult brain, which suggests the idea that the flexibility of the $GABA_A$ current effect is exploited as a general mechanism to that allows precise regulation of synaptic transmission. As there are excellent published reviews on the $GABA_A$ shift in developmental stages [17,31,32], in this section, we will only briefly summarize it and focus on the $GABA_A$ reversal potential regulation in the adult brain.

3.1. Developmental Regulation

The presence of depolarizing $GABA_A$ current is the norm in the brain during early development when $GABA_A$ is the main fast excitatory neurotransmitter and key to establishing normal circuitry [33,34]. Neonatal neurons express a substantial amount of NKCC1, while KCC2 expression is very low. At this stage, $GABA_A$ is depolarizing, as $[Cl^-]_i$ is 20–30 mM, which implies a $GABA_A$ reversal potential of −40 to −50 mV [17], while the resting potential is similar to that of mature neurons (~−70 mV) [35]. This depolarizing $GABA_A$ has a critical role in synaptic formation and in the establishment of the neuronal network. $GABA_A$ currents become inhibitory around postnatal Day 12 in rodents [36] and around 2 years of age in humans, largely due to increased KCC2 expression leading to lower intracellular Cl^- [37–40].

3.2. Cell Type and Spatial Heterogeneity of Chloride Transporters

In the adult brain, NKCC1 is strongly expressed in all types of non-neuronal cells—oligodendrocytes, microglia, and astrocytes [21,41,42]. It is suggested that its splice variants (NKCC1a/1b) are involved distinctively in the localization of NKCC1. NKCC1a is predominant in glial cells, and NKCC1b is found in neurons [21].

The presence of NKCC1 in glial cells increases the complexity of $GABA_A$ signaling. Astrocytes affect the activity of transporters by changing the extracellular potassium and chloride concentration around the neurons [43,44]. Additionally, astrocytes directly contribute to synaptic transmission by releasing glutamate and other modulators in an activity-regulated calcium-dependent or -independent fashion [45–47]. Beyond the effect on membrane potential, astrocytic NKCC1 has been shown to regulate mitochondrial calcium levels in astrocytes [48]. So, NKCC1 regulation in non-neuronal cells has a major impact on brain function. A discussion of these effects is beyond the scope of this paper; interested readers can find detailed reviews on this topic [49,50].

Even when restricting the analysis to neuronal expression, changes in NKCC1/KCC2 activity are typically confined to select brain areas. For example, in a pharmacological model of cognitive impairment associated with schizophrenia, NKCC1 is increased in the ventral but not in the dorsal subregion of the medial prefrontal cortex (mPFC) [51], and in a rodent model of neuropathic pain, KCC2 expression is increased in the mPFC but not in the neighboring anterior cingulate cortex [16]. NKCC1 and KCC2 can also be differentially regulated in different layers within the same brain areas, as shown in mouse granule cells and outer and middle molecular layers following entorhinal denervation [52].

Interestingly, in the adult mouse hippocampus, the reversal potential of the $GABA_A$ current in parvalbumin-positive interneurons is sensitive to KCC2 and NKCC1 blockades in a similar way to in pyramidal neurons, yet it is ~7 mV more depolarized than in pyramidal neurons [53], suggesting the differential expression of transporters between these cell types. Together with the observation that non-neuronal cells abundantly express NKCC1, these data point to the necessity of single-cell resolution studies when evaluating NKCC1/KCC2 and show that using tissue-level approaches, such as Western blot analysis of homogenized brain tissue, may easily miss cell-type selective alterations.

3.3. Compartmental Distribution of Chloride Transporters in Neurons

3.3.1. Somatodendritic Compartment

Both NKCC1 and KCC2 are distributed differently across cellular compartments. NKCC1 mRNA is detected and abundant in most cells of the adult brain [21], but protein expression in mature neurons is low, which has hampered studying the subcellular distribution of this transporter.

Recently, a super-resolution microscopy study in dissociated neurons took advantage of hemagglutinin-tagged transporters in hippocampal neurons and showed that NKCC1 isoforms are detected in the somatodendritic compartment [54]. This study also found that the NKCC1-KCC2 ratio is higher in axon initial segments of hippocampal neurons (see Section 3.3.2). Another study found that in the adult brain, two KCC2 isoforms (KCC2a and KCC2b) are expressed, but their expression differs between different cellular compartments (E.G. soma and proximal dendrites versus distal neuronal dendrite [22]). Both dentate gyrus granule cells and CA1 pyramidal cells of the adult hippocampus show different patterns of KCC2 expression from distal/apical dendrites to the soma [55]. This observation also suggests that the NKCC1-KCC2 ratio is differentially regulated along dendrites, possibly providing local control over the efficacy of synaptic activation. Indeed, spatially nonuniform $[Cl^-]_i$ shifts have been described in dendrites [56]. In keeping with the idea of spatial segregation in $[Cl^-]_i$ gradients, KCC2 is enriched in the vicinity of synapses [57–59] by a gephyrin-dependent mechanism, suggesting that $GABA_A$ is hyperpolarizing in these structures. In particular, KCC2 is clustered at the spine periphery rather than at the postsynaptic density (PSD) region [60]. NKCC1 is expressed in dendrites both at perisynaptic and extrasynaptic sites, where, in most cases, it colocalizes with KCC2 [54]. Intriguingly, these authors also showed that, in cultured neurons, NKCC1 and, to a lesser extent, KCC2 proteins are mobile, suggesting that their distribution can be rapidly modified in response to neuronal activity and the activation of individual synapses. This observation suggests that activity-dependent alterations in NKCC1/KCC2 activity contribute to synaptic plasticity. In agreement with this suggestion, a study showed that in hippocampal neurons (both cultured and in acute slices), KCC2 activity is regulated on a timescale of minutes by coincident postsynaptic spiking and synaptic activity by a calcium-dependent mechanism, which leads to persistent changes in synaptic strength [61].

3.3.2. Axonal Structures

In mature neurons, NKCC1 appears to be more abundantly expressed in the axon initial segment (AIS) compared with the somatodendritic compartment, which creates an axo-somatodendritic gradient in intracellular chloride [62]. Additionally, KCC2 expression in the AIS is low [55,63] or absent [54]. This high NKCC1-to-KCC2 ratio maintains high $[Cl^-]_i$ and, therefore, a depolarizing effect of $GABA_A$ in these structures. Accordingly, elegant work showed that the GABAergic input from axo-axonic cells to cortical pyramidal neuron axons is excitatory and elicits action potential firing in postsynaptic axons [63], although it may become hyperpolarizing in older animals [64].

3.4. *Circadian Regulation of $GABA_A$ Reversal Potential*

A very recent paper reported another intriguing aspect of $[Cl^-]_i$ homeostasis in the adult brain, circadian regulation [65]. These authors showed that in the pyramidal neurons of the adult mouse neocortex baseline, $[Cl^-]_i$ undergoes a two-fold increase from day to night and that such an increase is due to changes in the surface expression and phosphorylation of NKCC1 and KCC2. These findings add another layer of complexity to the understanding of GABAergic signaling in the brain and provide further proof of the remarkable flexibility of this system.

4. Regulation of Neuronal $GABA_A$ Reversal Potential in Disease

4.1. *Depolarizing $GABA_A$ Current*

An altered NKCC1-KCC2 activity ratio has been linked with various neuropathological disorders and discussed in previous reviews [66,67]. Depolarizing $GABA_A$ currents in the adult brain have been reported in many pathological conditions such as epilepsy [37,68], Down syndrome [69], schizophrenia [51], and severe stress [70]. The mechanism by which depolarizing GABA currents interferes with brain activity is likely caused by a loss of synchronization in network oscillations (Figure 1). The cellular mechanisms leading to the altered reversal potential include alterations in transporter expression and in their regulation. For example, increased activity of kinase, such as OXSR1 and WNK3, which regulate NKCC1 and KCC2 was reported in the prefrontal cortex of schizophrenia patients [71]. On the other hand, in subchronic phencyclidine (scPCP)-treated mice, an animal model that mimics the cognitive symptoms associated with schizophrenia [72,73], depolarizing $GABA_A$ appears in the prefrontal cortex as a consequence of increased NKCC1 mRNA expression [51]. In line with these data, a gain-of-function NKCC1 mutation has been identified in human schizophrenia patients [74]. Thus, similar to the scenario in Down syndrome [69] and in the spinal cord of neuropathic pain models [75], NKCC1 is the transporter whose activity is mostly altered in schizophrenia. Therefore, the cellular mechanism leading to altered $GABA_A$ reversal potential in the diseased adult brain appears to be different than in the neonatal brain, where the depolarizing $GABA_A$ current is due to low KCC2 expression [39,40]. Importantly, the pharmacological inhibition of NKCC1 rescues cognitive impairment in both scPCP and Down syndrome mice. Additionally, depolarizing GABA has also been suggested as a mechanism leading to a loss of the synapse specificity of LTP in aging animals [76]. Finally, depolarizing GABA may also play a role in Alzheimer's disease, as it was recently shown that NKCC1 expression is increased in the hippocampus of mice injected with amyloid β-peptide [77]. These authors further showed that treatment with beta-amyloid (Aβ1–42) also increases NKCC1 expression in hippocampal cultures. Another group [78] obtained complementary results showing KCC2 downregulation in 5xFAD mice, a widely used animal model of Alzheimer's disease. Importantly, these authors also showed that treating the mice with the KCC2 activator CLP290 rescued cognitive performance.

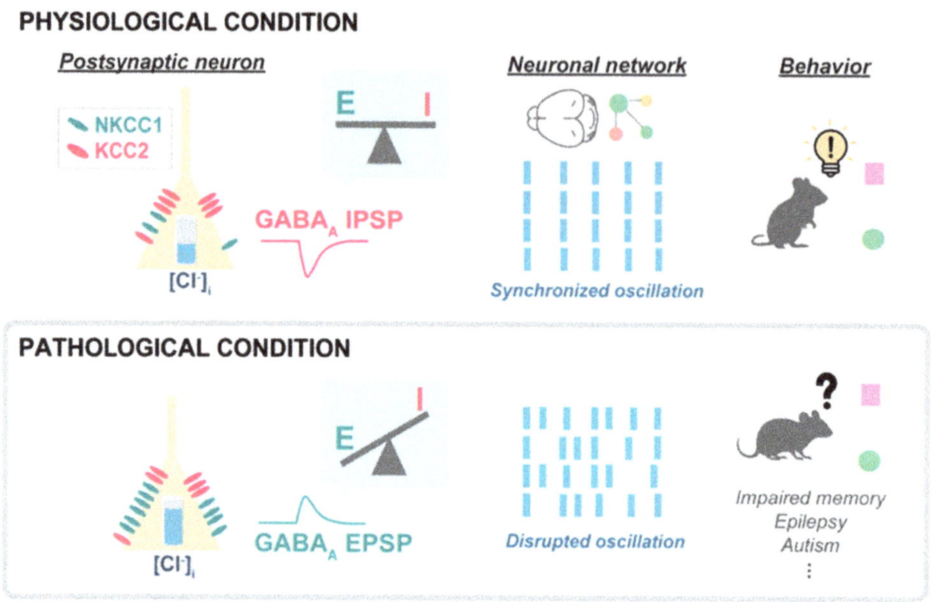

Figure 1. Schematic of the effects of depolarizing GABA$_A$ in disease. In postsynaptic neurons, two types of chloride/potassium cotransporters (NKCC1 and KCC2) regulate intracellular chloride concentration ([Cl$^-$]$_i$). The polarity of the GABA$_A$ current, either depolarizing or hyperpolarizing, is largely determined by the transmembrane chloride concentration gradient. When activation of GABA$_A$ receptors pathologically evokes excitatory (EPSP) instead of inhibitory (IPSP) synaptic potentials due to increased NKCC1-KCC2 activity ratio, the excitatory/inhibitory (E/I) balance is altered, which leads to a disruption of network oscillations and behavioral impairment.

4.2. *Hypernegative GABA$_A$ Current*

So far, the study of GABA$_A$ current reversal potential alterations in pathological conditions has been largely focused on the (re)appearance of excitatory GABA$_A$ in adult animals. However, recent investigations have shown that the opposite regulation can also be detrimental and that excessive GABA$_A$-mediated inhibition due to hypernegative current reversal potential represents another potential mechanism of pathology. For example, adult mice that were exposed to neonatal inflammation (using an LPS injection) show cognitive impairments associated with hypernegative GABA$_A$ reversal potential and increased KCC2 activity in the hippocampus [79]. Interestingly, these authors showed a complex TGF-β1-dependent regulation, suggesting the alteration is not simply due to an overdrive of the normal developmental KCC2 upregulation. Accordingly, we recently showed that a hypernegative GABA$_A$ current can appear in the adult brain independently of any developmental dysregulation. We found that in adult mice 7 days after peripheral nerve injury (the Spared Nerve Injury model of neuropathic pain), the reversal potential of the GABA$_A$ current in the pyramidal neurons of the medial prefrontal cortex is ~10 mV more negative than in sham-operated control mice. This change is due to a large overexpression of KCC2 transcripts and, to a minor extent, to decreased NKCC1 expression [16]. Computational modeling of a cortical network demonstrated that the ~10 mV shift in GABA$_A$ reversal potential is by itself sufficient to dramatically reduce the overall activity of the network, resulting in over 60%-decreased activity of excitatory neurons in the range of natural external inputs (Figure 2). Thus, hypernegative GABA has a major impact on neuronal activity and should be considered whenever increased inhibition is present and no changes in the activity and/or number of interneurons are detected.

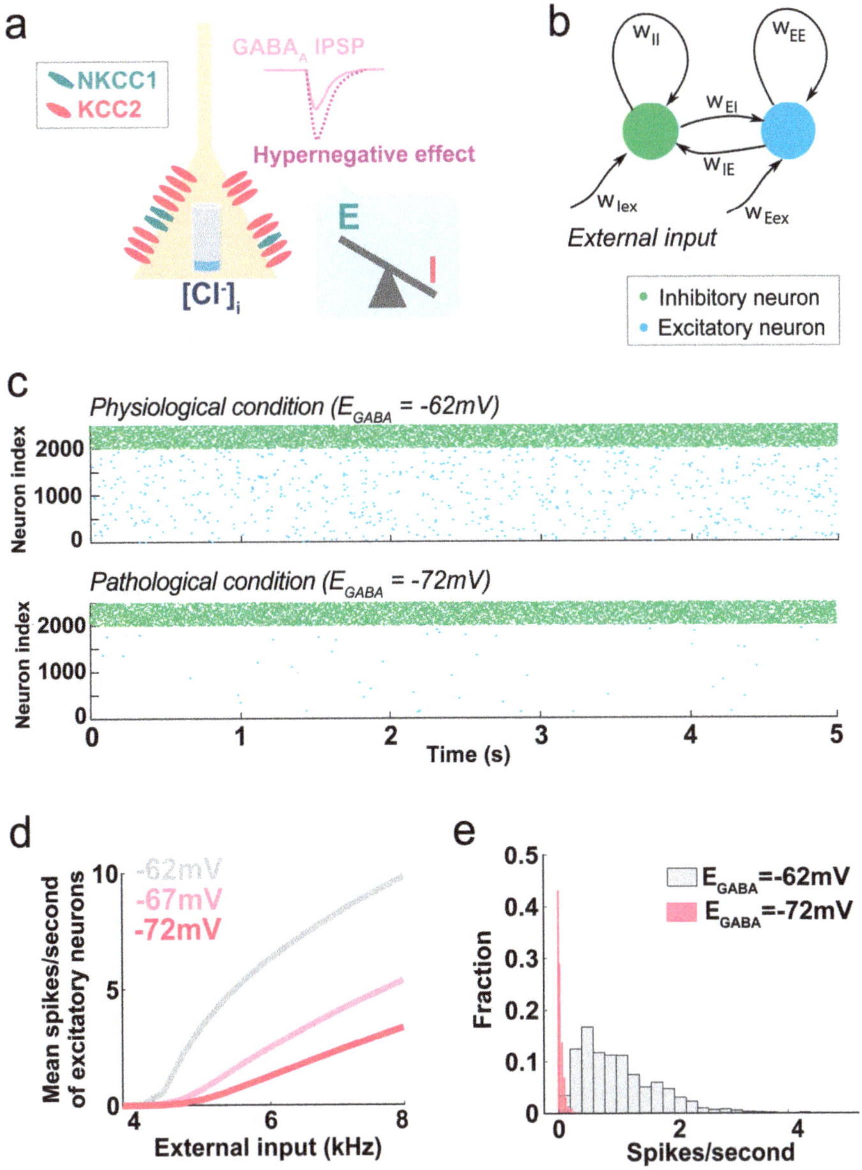

Figure 2. Effect of hypernegative $GABA_A$ current on network function. (**a**) When the neuronal KCC2-NKCC1 activity ratio is increased, intracellular chloride concentration ($[Cl^-]_i$) is lower than the physiological level, and the inhibitory effect of $GABA_A$ current is strengthened, disrupting the E/I balance. (**b–e**) Simulated effects of hypernegative $GABA_A$ current on cortical network activity. In pathological conditions ($GABA_A$ reversal potentials = −72 mV), the spike frequency of the postsynaptic excitatory neurons is severely decreased, even when the frequency of the inhibitory inputs is unaffected. (**b**) Computational model of a cortical network based on local microcircuit of

excitatory–inhibitory recurrent architecture. (**c**) Simulated spontaneous spiking activity of excitatory (blue) and inhibitory (green) cortical neurons with physiological (−62 mV, top panel) or pathologically hyperpolarized (−72 mV, bottom) $GABA_A$ reversal potentials. (**d**) Average firing rate of excitatory neurons with three different $GABA_A$ reversal potentials in response to variable external input frequencies. (**e**) Histogram of firing rates of excitatory neurons when the external input was set at 4.5 kHz. (Adapted with permission from [16]).

4.3. $GABA_A$ Flexibility and Pathology

Why are changes in $GABA_A$ current reversal potential detected in so many different conditions? The proximity of the value of the $GABA_A$ current reversal potential and the resting membrane potential of most neurons is probably the critical factor. Such proximity means that even small changes in the intracellular chloride concentration can switch the current effect from hyperpolarizing to depolarizing; this is a unique property of the $GABA_A$ channel and sets it apart from other synaptic receptors. Another reason might be the intrinsic complexity and plasticity of the regulatory mechanisms. Intracellular chloride concentration changes significantly during development because the regulating machinery is heavily modulated by numerous mechanisms (including epigenetic [80]); this makes this process particularly sensitive to even small functional abnormalities. Therefore, we hypothesize that alterations in $GABA_A$ reversal potential (either in the depolarizing or hyperpolarizing direction) in the cerebral cortex represent a common maladaptive mechanism leading to cognitive impairment in multiple pathological conditions with different etiologies.

5. Network Effects and Clinical Implications of Flexible $GABA_A$ Reversal Potential

Numerous studies show that neuronal oscillations in the prefrontal cortex and hippocampus are closely associated with cognitive tasks. Theta rhythms (4–8 Hz) in the cortex are enhanced during memory tasks in humans and animals [81,82]. Gamma oscillations (25–100 Hz) are also critical for cognitive function, and aberrant gamma oscillations have been observed in multiple cognitive disorders. Additionally, changes in brain oscillations can predict cognitive symptoms in individuals with autism spectrum disorders, and increasing evidence shows that long-range gamma synchronization is reduced in patients with schizophrenia [83] and other conditions characterized by cognitive disability [84]. GABAergic inhibition is essential to the generation of oscillatory networks [85,86]. Accordingly, E/I balance in the PFC and hippocampus is critical for cognitive and social functions [87,88]. However, despite the centrality of E/I imbalance in numerous brain disorders, the cellular mechanisms of E/I imbalance are mostly unknown. As oscillations emerge from reciprocal interactions between populations of excitatory (glutamatergic) and inhibitory (GABAergic) neurons [89], E/I imbalance may be due to multiple types of abnormalities in GABAergic and/or glutamatergic signaling. Consequently, it is possible that differences in the mechanisms underlying E/I imbalance lay behind the large variety and diverse severity of cognitive impairments in different neurodevelopmental disorders. While abundant evidence has been collected over the past several years supporting the role of depolarizing GABA as an important pathogenic mechanism in various diseases, it has only become clear very recently that the hypernegative reversal potential of the $GABA_A$ current also has a major functional impact, although the network mechanisms are likely different. We suggest that the main functional impact of depolarizing GABA is a loss of network synchronization, which results in the disruption of network oscillations (Figure 1). On the other hand, a hypernegative $GABA_A$ current seems to cause a general inhibition of the circuit, roughly equivalent to an increased number of inhibitory synaptic inputs (Figure 2).

Additionally, while both these conditions represent dysfunctions in GABAergic signaling, their potential treatment requires completely different approaches, each with potential drawbacks. The case of hypernegative $GABA_A$ is theoretically straightforward because GABA antagonists are likely beneficial to reduce the effects of the hypernegative $GABA_A$.

The use of such drugs, however, is limited by their toxicity and epileptogenicity. The situation is more complex in the case of depolarizing GABA, where inhibition is insufficient but cannot be improved by the common approach of using GABA agonists such as benzodiazepines or GABA analogs such as valproic acid or gabapentin. The most direct treatment for this condition would obviously be the restoration of the inhibitory effect of $GABA_A$, which can be achieved either pharmacologically or genetically [51,69]. Both these approaches, however, also have major limitations. As depolarizing GABA in disease conditions is, by and large, the consequence of NKCC1 upregulation, NKCC1 antagonists, such as bumetanide, may provide a treatment opportunity. However, these drugs have very poor BBB permeability [90], which limits their effective concentration in the brain parenchyma. Additionally, these drugs do not have any cellular selectivity and not only affect excitatory and inhibitory neurons but also non-neuronal cells, which often express NKCC1 at a higher level than neurons. Thus, the network effects are necessarily broad and hard to predict. The genetic approach may help as NKCC1 expression could be selectively repressed in select cellular populations, but this approach is still far from viable for clinical treatment. An alternative approach may take advantage of KCC2 agonists, which would be neuron-selective, but effective agonists are still missing. However, KCC2 activity is regulated by numerous signaling pathways such as WNK, PKC, and BDNF [20,71,91–93]; therefore, these might also represent interesting pharmacological targets. Even in this case, however, cell-type selectivity would help considerably.

6. Conclusions

The $GABA_A$ current represents a particularly vulnerable target in many neurological conditions. This is mainly driven by two facts: 1—The net effect of $GABA_AR$ activation may be inhibitory or excitatory, depending on the electrochemical gradient for the permeant anions across the membrane of the postsynaptic neurons. Additionally, even when inhibitory, the $GABA_A$ current may dramatically change its efficacy depending on its reversal potential. Consequently, this highly plastic and regulated system is sensitive to even small changes in transmembrane ionic gradients. 2—This functional balance is regulated by mechanisms that are highly plastic because they are designed to change during physiological development. Taken together, these factors explain both the particular vulnerability of this system to even small cellular alterations and the oversized functional effects that can derive from these alterations. In keeping with such strong vulnerability, alterations in $GABA_A$ reversal potential have been recently implicated in numerous psychiatric and neurological conditions, suggesting that regulation of chloride-cation cotransporters represents a viable therapeutic target.

Author Contributions: H.R.K.: literature search, conceptualization of the paper and writing. M.M.: literature search, conceptualization of the paper and writing. All authors have read and agreed to the published version of the manuscript.

Funding: This work was supported by NIH grants NS112292 and DA044121 to M.M.

Conflicts of Interest: The authors declare no conflicts of interest.

References

1. Atallah, B.V.; Scanziani, M. Instantaneous modulation of gamma oscillation frequency by balancing excitation with inhibition. *Neuron* **2009**, *62*, 566–577. [CrossRef] [PubMed]
2. Yizhar, O.; Fenno, L.E.; Prigge, M.; Schneider, F.; Davidson, T.J.; O'Shea, D.J.; Sohal, V.S.; Goshen, I.; Finkelstein, J.; Paz, J.T.; et al. Neocortical excitation/inhibition balance in information processing and social dysfunction. *Nature* **2011**, *477*, 171–178. [CrossRef] [PubMed]
3. Dehghani, N.; Peyrache, A.; Telenczuk, B.; Le Van Quyen, M.; Halgren, E.; Cash, S.S.; Hatsopoulos, N.G.; Destexhe, A. Dynamic Balance of Excitation and Inhibition in Human and Monkey Neocortex. *Sci. Rep.* **2016**, *6*, 23176. [CrossRef]
4. Eichler, S.A.; Meier, J.C. E-I balance and human diseases—From molecules to networking. *Front. Mol. Neurosci.* **2008**, *1*, 2. [CrossRef] [PubMed]
5. D'Amour, J.A.; Froemke, R.C. Inhibitory and excitatory spike-timing-dependent plasticity in the auditory cortex. *Neuron* **2015**, *86*, 514–528. [CrossRef] [PubMed]

6. Isaacson, J.S.; Scanziani, M. How inhibition shapes cortical activity. *Neuron* **2011**, *72*, 231–243. [CrossRef] [PubMed]
7. Liu, B.H.; Li, Y.T.; Ma, W.P.; Pan, C.J.; Zhang, L.I.; Tao, H.W. Broad inhibition sharpens orientation selectivity by expanding input dynamic range in mouse simple cells. *Neuron* **2011**, *71*, 542–554. [CrossRef]
8. Haider, B.; Hausser, M.; Carandini, M. Inhibition dominates sensory responses in the awake cortex. *Nature* **2013**, *493*, 97–100. [CrossRef]
9. Bassell, G.J.; Warren, S.T. Fragile X syndrome: Loss of local mRNA regulation alters synaptic development and function. *Neuron* **2008**, *60*, 201–214. [CrossRef]
10. Snowden, S.G.; Ebshiana, A.A.; Hye, A.; Pletnikova, O.; O'Brien, R.; Yang, A.; Troncoso, J.; Legido-Quigley, C.; Thambisetty, M. Neurotransmitter Imbalance in the Brain and Alzheimer's Disease Pathology. *J. Alzheimers Dis.* **2019**, *72*, 35–43. [CrossRef]
11. Yang, G.J.; Murray, J.D.; Wang, X.J.; Glahn, D.C.; Pearlson, G.D.; Repovs, G.; Krystal, J.H.; Anticevic, A. Functional hierarchy underlies preferential connectivity disturbances in schizophrenia. *Proc. Natl. Acad. Sci. USA* **2016**, *113*, E219–E228. [CrossRef] [PubMed]
12. Kehrer, C.; Maziashvili, N.; Dugladze, T.; Gloveli, T. Altered Excitatory-Inhibitory Balance in the NMDA-Hypofunction Model of Schizophrenia. *Front. Mol. Neurosci.* **2008**, *1*, 6. [CrossRef] [PubMed]
13. Lozano, R.; Hare, E.B.; Hagerman, R.J. Modulation of the GABAergic pathway for the treatment of fragile X syndrome. *Neuropsychiatr. Dis. Treat.* **2014**, *10*, 1769–1779. [CrossRef] [PubMed]
14. Gonzalez-Burgos, G.; Cho, R.Y.; Lewis, D.A. Alterations in cortical network oscillations and parvalbumin neurons in schizophrenia. *Biol. Psychiatry* **2015**, *77*, 1031–1040. [CrossRef] [PubMed]
15. Bormann, J.; Hamill, O.P.; Sakmann, B. Mechanism of anion permeation through channels gated by glycine and gamma-aminobutyric acid in mouse cultured spinal neurones. *J. Physiol.* **1987**, *385*, 243–286. [CrossRef]
16. Kim, H.R.; Long, M.; Sekerkova, G.; Maes, A.; Kennedy, A.; Martina, M. Hypernegative GABA(A) Reversal Potential in Pyramidal Cells Contributes to Medial Prefrontal Cortex Deactivation in a Mouse Model of Neuropathic Pain. *J. Pain*, 2023, *ahead of print*. [CrossRef]
17. Ben-Ari, Y. Excitatory actions of gaba during development: The nature of the nurture. *Nat. Rev. Neurosci.* **2002**, *3*, 728–739. [CrossRef]
18. Fukuda, A.; Tanaka, M.; Yamada, Y.; Muramatsu, K.; Shimano, Y.; Nishino, H. Simultaneous optical imaging of intracellular Cl- in neurons in different layers of rat neocortical slices: Advantages and limitations. *Neurosci. Res.* **1998**, *32*, 363–371. [CrossRef]
19. Kuner, T.; Augustine, G.J. A genetically encoded ratiometric indicator for chloride: Capturing chloride transients in cultured hippocampal neurons. *Neuron* **2000**, *27*, 447–459. [CrossRef]
20. Kahle, K.T.; Delpire, E. Kinase-KCC2 coupling: Cl- rheostasis, disease susceptibility, therapeutic target. *J. Neurophysiol.* **2016**, *115*, 8–18. [CrossRef]
21. Kurki, S.N.; Uvarov, P.; Pospelov, A.S.; Trontti, K.; Hubner, A.K.; Srinivasan, R.; Watanabe, M.; Hovatta, I.; Hubner, C.A.; Kaila, K.; et al. Expression patterns of NKCC1 in neurons and non-neuronal cells during cortico-hippocampal development. *Cereb. Cortex* **2023**, *33*, 5906–5923. [CrossRef]
22. Markkanen, M.; Karhunen, T.; Llano, O.; Ludwig, A.; Rivera, C.; Uvarov, P.; Airaksinen, M.S. Distribution of neuronal KCC2a and KCC2b isoforms in mouse CNS. *J. Comp. Neurol.* **2014**, *522*, 1897–1914. [CrossRef] [PubMed]
23. Vu, T.Q.; Payne, J.A.; Copenhagen, D.R. Localization and developmental expression patterns of the neuronal K-Cl cotransporter (KCC2) in the rat retina. *J. Neurosci.* **2000**, *20*, 1414–1423. [CrossRef] [PubMed]
24. Hubner, C.A.; Stein, V.; Hermans-Borgmeyer, I.; Meyer, T.; Ballanyi, K.; Jentsch, T.J. Disruption of KCC2 reveals an essential role of K-Cl cotransport already in early synaptic inhibition. *Neuron* **2001**, *30*, 515–524. [CrossRef] [PubMed]
25. Chew, T.A.; Orlando, B.J.; Zhang, J.; Latorraca, N.R.; Wang, A.; Hollingsworth, S.A.; Chen, D.H.; Dror, R.O.; Liao, M.; Feng, L. Structure and mechanism of the cation-chloride cotransporter NKCC1. *Nature* **2019**, *572*, 488–492. [CrossRef]
26. Neumann, C.; Rosenbaek, L.L.; Flygaard, R.K.; Habeck, M.; Karlsen, J.L.; Wang, Y.; Lindorff-Larsen, K.; Gad, H.H.; Hartmann, R.; Lyons, J.A.; et al. Cryo-EM structure of the human NKCC1 transporter reveals mechanisms of ion coupling and specificity. *EMBO J.* **2022**, *41*, e110169. [CrossRef] [PubMed]
27. Agez, M.; Schultz, P.; Medina, I.; Baker, D.J.; Burnham, M.P.; Cardarelli, R.A.; Conway, L.C.; Garnier, K.; Geschwindner, S.; Gunnarsson, A.; et al. Molecular architecture of potassium chloride co-transporter KCC2. *Sci. Rep.* **2017**, *7*, 16452. [CrossRef] [PubMed]
28. Xie, Y.; Chang, S.; Zhao, C.; Wang, F.; Liu, S.; Wang, J.; Delpire, E.; Ye, S.; Guo, J. Structures and an activation mechanism of human potassium-chloride cotransporters. *Sci. Adv.* **2020**, *6*, eabc5883. [CrossRef]
29. Hartmann, A.M.; Nothwang, H.G. NKCC1 and KCC2: Structural insights into phospho-regulation. *Front. Mol. Neurosci.* **2022**, *15*, 964488. [CrossRef]
30. Han, B.; Bellemer, A.; Koelle, M.R. An evolutionarily conserved switch in response to GABA affects development and behavior of the locomotor circuit of Caenorhabditis elegans. *Genetics* **2015**, *199*, 1159–1172. [CrossRef]
31. Peerboom, C.; Wierenga, C.J. The postnatal GABA shift: A developmental perspective. *Neurosci. Biobehav. Rev.* **2021**, *124*, 179–192. [CrossRef]
32. Cherubini, E.; Gaiarsa, J.L.; Ben-Ari, Y. GABA: An excitatory transmitter in early postnatal life. *Trends Neurosci.* **1991**, *14*, 515–519. [CrossRef] [PubMed]

33. Leinekugel, X.; Khalilov, I.; McLean, H.; Caillard, O.; Gaiarsa, J.L.; Ben-Ari, Y.; Khazipov, R. GABA is the principal fast-acting excitatory transmitter in the neonatal brain. *Adv. Neurol.* **1999**, *79*, 189–201. [PubMed]
34. Ben-Ari, Y.; Cherubini, E.; Corradetti, R.; Gaiarsa, J.L. Giant synaptic potentials in immature rat CA3 hippocampal neurones. *J. Physiol.* **1989**, *416*, 303–325. [CrossRef] [PubMed]
35. Verheugen, J.A.; Fricker, D.; Miles, R. Noninvasive measurements of the membrane potential and GABAergic action in hippocampal interneurons. *J. Neurosci.* **1999**, *19*, 2546–2555. [CrossRef] [PubMed]
36. Dehorter, N.; Vinay, L.; Hammond, C.; Ben-Ari, Y. Timing of developmental sequences in different brain structures: Physiological and pathological implications. *Eur. J. Neurosci.* **2012**, *35*, 1846–1856. [CrossRef]
37. Dzhala, V.I.; Talos, D.M.; Sdrulla, D.A.; Brumback, A.C.; Mathews, G.C.; Benke, T.A.; Delpire, E.; Jensen, F.E.; Staley, K.J. NKCC1 transporter facilitates seizures in the developing brain. *Nat. Med.* **2005**, *11*, 1205–1213. [CrossRef] [PubMed]
38. Hyde, T.M.; Lipska, B.K.; Ali, T.; Mathew, S.V.; Law, A.J.; Metitiri, O.E.; Straub, R.E.; Ye, T.; Colantuoni, C.; Herman, M.M.; et al. Expression of GABA signaling molecules KCC2, NKCC1, and GAD1 in cortical development and schizophrenia. *J. Neurosci.* **2011**, *31*, 11088–11095. [CrossRef]
39. Rivera, C.; Voipio, J.; Kaila, K. Two developmental switches in GABAergic signalling: The K^+-Cl^- cotransporter KCC2 and carbonic anhydrase CAVII. *J. Physiol.* **2005**, *562 Pt 1*, 27–36. [CrossRef]
40. Yamada, J.; Okabe, A.; Toyoda, H.; Kilb, W.; Luhmann, H.J.; Fukuda, A. Cl^- uptake promoting depolarizing GABA actions in immature rat neocortical neurones is mediated by NKCC1. *J. Physiol.* **2004**, *557 Pt 3*, 829–841. [CrossRef]
41. Su, G.; Kintner, D.B.; Flagella, M.; Shull, G.E.; Sun, D. Astrocytes from Na(+)-K(+)-Cl(−) cotransporter-null mice exhibit absence of swelling and decrease in EAA release. *Am. J. Physiol. Cell Physiol.* **2002**, *282*, C1147–C1160. [CrossRef]
42. Kanaka, C.; Ohno, K.; Okabe, A.; Kuriyama, K.; Itoh, T.; Fukuda, A.; Sato, K. The differential expression patterns of messenger RNAs encoding K-Cl cotransporters (KCC1,2) and Na-K-2Cl cotransporter (NKCC1) in the rat nervous system. *Neuroscience* **2001**, *104*, 933–946. [CrossRef] [PubMed]
43. Kitayama, T. The Role of Astrocytes in the Modulation of K(+)-Cl(−)-Cotransporter-2 Function. *Int. J. Mol. Sci.* **2020**, *21*, 9539. [CrossRef]
44. Nguyen, T.D.; Ishibashi, M.; Sinha, A.S.; Watanabe, M.; Kato, D.; Horiuchi, H.; Wake, H.; Fukuda, A. Astrocytic NKCC1 inhibits seizures by buffering Cl(−) and antagonizing neuronal NKCC1 at GABAergic synapses. *Epilepsia* **2023**, *64*, 3389–3403. [CrossRef]
45. Fellin, T.; Pascual, O.; Gobbo, S.; Pozzan, T.; Haydon, P.G.; Carmignoto, G. Neuronal synchrony mediated by astrocytic glutamate through activation of extrasynaptic NMDA receptors. *Neuron* **2004**, *43*, 729–743. [CrossRef] [PubMed]
46. Covelo, A.; Araque, A. Neuronal activity determines distinct gliotransmitter release from a single astrocyte. *eLife* **2018**, *7*, e32237. [CrossRef] [PubMed]
47. Huda, R.; Chang, Z.; Do, J.; McCrimmon, D.R.; Martina, M. Activation of astrocytic PAR1 receptors in the rat nucleus of the solitary tract regulates breathing through modulation of presynaptic TRPV1. *J. Physiol.* **2018**, *596*, 497–513. [CrossRef] [PubMed]
48. Kintner, D.B.; Luo, J.; Gerdts, J.; Ballard, A.J.; Shull, G.E.; Sun, D. Role of Na^+-K^+-Cl^- cotransport and Na^+/Ca^{2+} exchange in mitochondrial dysfunction in astrocytes following in vitro ischemia. *Am. J. Physiol. Cell Physiol.* **2007**, *292*, C1113–C1122. [CrossRef] [PubMed]
49. Rose, C.R.; Verkhratsky, A. Principles of sodium homeostasis and sodium signalling in astroglia. *Glia* **2016**, *64*, 1611–1627. [CrossRef]
50. Kahle, K.T.; Simard, J.M.; Staley, K.J.; Nahed, B.V.; Jones, P.S.; Sun, D. Molecular mechanisms of ischemic cerebral edema: Role of electroneutral ion transport. *Physiology* **2009**, *24*, 257–265. [CrossRef]
51. Kim, H.R.; Rajagopal, L.; Meltzer, H.Y.; Martina, M. Depolarizing GABA(A) current in the prefrontal cortex is linked with cognitive impairment in a mouse model relevant for schizophrenia. *Sci. Adv.* **2021**, *7*, eaba5032. [CrossRef]
52. Del Turco, D.; Paul, M.H.; Schlaudraff, J.; Muellerleile, J.; Bozic, F.; Vuksic, M.; Jedlicka, P.; Deller, T. Layer-specific changes of KCC2 and NKCC1 in the mouse dentate gyrus after entorhinal denervation. *Front. Mol. Neurosci.* **2023**, *16*, 1118746. [CrossRef] [PubMed]
53. Otsu, Y.; Donneger, F.; Schwartz, E.J.; Poncer, J.C. Cation-chloride cotransporters and the polarity of GABA signalling in mouse hippocampal parvalbumin interneurons. *J. Physiol.* **2020**, *598*, 1865–1880. [CrossRef] [PubMed]
54. Pol, E.; Come, E.; Merlaud, Z.; Gouhier, J.; Russeau, M.; Scotto-Lomassese, S.; Moutkine, I.; Marques, X.; Levi, S. NKCC1 and KCC2 Chloride Transporters Have Different Membrane Dynamics on the Surface of Hippocampal Neurons. *Cells* **2023**, *12*, 2363. [CrossRef] [PubMed]
55. Baldi, R.; Varga, C.; Tamas, G. Differential distribution of KCC2 along the axo-somato-dendritic axis of hippocampal principal cells. *Eur. J. Neurosci.* **2010**, *32*, 1319–1325. [CrossRef] [PubMed]
56. Weilinger, N.L.; Wicki-Stordeur, L.E.; Groten, C.J.; LeDue, J.M.; Kahle, K.T.; MacVicar, B.A. KCC2 drives chloride microdomain formation in dendritic blebbing. *Cell Rep.* **2022**, *41*, 111556. [CrossRef] [PubMed]
57. Al Awabdh, S.; Donneger, F.; Goutierre, M.; Seveno, M.; Vigy, O.; Weinzettl, P.; Russeau, M.; Moutkine, I.; Levi, S.; Marin, P.; et al. Gephyrin Interacts with the K-Cl Cotransporter KCC2 to Regulate Its Surface Expression and Function in Cortical Neurons. *J. Neurosci.* **2022**, *42*, 166–182. [CrossRef] [PubMed]
58. Gulyas, A.I.; Sik, A.; Payne, J.A.; Kaila, K.; Freund, T.F. The KCl cotransporter, KCC2, is highly expressed in the vicinity of excitatory synapses in the rat hippocampus. *Eur. J. Neurosci.* **2001**, *13*, 2205–2217. [CrossRef] [PubMed]

59. Chamma, I.; Heubl, M.; Chevy, Q.; Renner, M.; Moutkine, I.; Eugene, E.; Poncer, J.C.; Levi, S. Activity-dependent regulation of the K/Cl transporter KCC2 membrane diffusion, clustering, and function in hippocampal neurons. *J. Neurosci.* **2013**, *33*, 15488–15503. [CrossRef]
60. Chamma, I.; Chevy, Q.; Poncer, J.C.; Levi, S. Role of the neuronal K-Cl co-transporter KCC2 in inhibitory and excitatory neurotransmission. *Front. Cell Neurosci.* **2012**, *6*, 5. [CrossRef]
61. Woodin, M.A.; Ganguly, K.; Poo, M.M. Coincident pre- and postsynaptic activity modifies GABAergic synapses by postsynaptic changes in Cl- transporter activity. *Neuron* **2003**, *39*, 807–820. [CrossRef]
62. Khirug, S.; Yamada, J.; Afzalov, R.; Voipio, J.; Khiroug, L.; Kaila, K. GABAergic depolarization of the axon initial segment in cortical principal neurons is caused by the Na-K-2Cl cotransporter NKCC1. *J. Neurosci.* **2008**, *28*, 4635–4639. [CrossRef] [PubMed]
63. Szabadics, J.; Varga, C.; Molnar, G.; Olah, S.; Barzo, P.; Tamas, G. Excitatory effect of GABAergic axo-axonic cells in cortical microcircuits. *Science* **2006**, *311*, 233–235. [CrossRef] [PubMed]
64. Rinetti-Vargas, G.; Phamluong, K.; Ron, D.; Bender, K.J. Periadolescent Maturation of GABAergic Hyperpolarization at the Axon Initial Segment. *Cell Rep.* **2017**, *20*, 21–29. [CrossRef] [PubMed]
65. Pracucci, E.; Graham, R.T.; Alberio, L.; Nardi, G.; Cozzolino, O.; Pillai, V.; Pasquini, G.; Saieva, L.; Walsh, D.; Landi, S.; et al. Daily rhythm in cortical chloride homeostasis underpins functional changes in visual cortex excitability. *Nat. Commun.* **2023**, *14*, 7108. [CrossRef] [PubMed]
66. Cherubini, E.; Di Cristo, G.; Avoli, M. Dysregulation of GABAergic Signaling in Neurodevelomental Disorders: Targeting Cation-Chloride Co-transporters to Re-establish a Proper E/I Balance. *Front. Cell Neurosci.* **2021**, *15*, 813441. [CrossRef] [PubMed]
67. Savardi, A.; Borgogno, M.; De Vivo, M.; Cancedda, L. Pharmacological tools to target NKCC1 in brain disorders. *Trends Pharmacol. Sci.* **2021**, *42*, 1009–1034. [CrossRef] [PubMed]
68. Cohen, I.; Navarro, V.; Clemenceau, S.; Baulac, M.; Miles, R. On the origin of interictal activity in human temporal lobe epilepsy in vitro. *Science* **2002**, *298*, 1418–1421. [CrossRef] [PubMed]
69. Deidda, G.; Parrini, M.; Naskar, S.; Bozarth, I.F.; Contestabile, A.; Cancedda, L. Reversing excitatory GABAAR signaling restores synaptic plasticity and memory in a mouse model of Down syndrome. *Nat. Med.* **2015**, *21*, 318–326. [CrossRef]
70. MacKenzie, G.; Maguire, J. Chronic stress shifts the GABA reversal potential in the hippocampus and increases seizure susceptibility. *Epilepsy Res.* **2015**, *109*, 13–27. [CrossRef]
71. Arion, D.; Lewis, D.A. Altered expression of regulators of the cortical chloride transporters NKCC1 and KCC2 in schizophrenia. *Arch. Gen. Psychiatry* **2011**, *68*, 21–31. [CrossRef]
72. Jentsch, J.D.; Roth, R.H. The neuropsychopharmacology of phencyclidine: From NMDA receptor hypofunction to the dopamine hypothesis of schizophrenia. *Neuropsychopharmacology* **1999**, *20*, 201–225. [CrossRef] [PubMed]
73. Steeds, H.; Carhart-Harris, R.L.; Stone, J.M. Drug models of schizophrenia. *Ther. Adv. Psychopharmacol.* **2015**, *5*, 43–58. [CrossRef] [PubMed]
74. Merner, N.D.; Mercado, A.; Khanna, A.R.; Hodgkinson, A.; Bruat, V.; Awadalla, P.; Gamba, G.; Rouleau, G.A.; Kahle, K.T. Gain-of-function missense variant in SLC12A2, encoding the bumetanide-sensitive NKCC1 cotransporter, identified in human schizophrenia. *J. Psychiatr. Res.* **2016**, *77*, 22–26. [CrossRef]
75. Chen, S.R.; Zhu, L.; Chen, H.; Wen, L.; Laumet, G.; Pan, H.L. Increased spinal cord Na(+)-K(+)-2Cl(−) cotransporter-1 (NKCC1) activity contributes to impairment of synaptic inhibition in paclitaxel-induced neuropathic pain. *J. Biol. Chem.* **2014**, *289*, 31111–31120. [CrossRef] [PubMed]
76. Ferando, I.; Faas, G.C.; Mody, I. Diminished KCC2 confounds synapse specificity of LTP during senescence. *Nat. Neurosci.* **2016**, *19*, 1197–1200. [CrossRef]
77. Lam, P.; Vinnakota, C.; Guzmán, B.C.F.; Newland, J.; Peppercorn, K.; Tate, W.P.; Waldvogel, H.J.; Faull, R.L.M.; Kwakowsky, A. Beta-Amyloid (Aβ) Increases the Expression of NKCC1 in the Mouse Hippocampus. *Molecules* **2022**, *27*, 2440. [CrossRef]
78. Keramidis, I.; McAllister, B.B.; Bourbonnais, J.; Wang, F.; Isabel, D.; Rezaei, E.; Sansonetti, R.; Degagne, P.; Hamel, J.P.; Nazari, M.; et al. Restoring neuronal chloride extrusion reverses cognitive decline linked to Alzheimer's disease mutations. *Brain* **2023**, *146*, 4903–4915. [CrossRef]
79. Rong, J.; Yang, Y.; Liang, M.; Zhong, H.; Li, Y.; Zhu, Y.; Sha, S.; Chen, L.; Zhou, R. Neonatal inflammation increases hippocampal KCC2 expression through methylation-mediated TGF-beta1 downregulation leading to impaired hippocampal cognitive function and synaptic plasticity in adult mice. *J. Neuroinflamm.* **2023**, *20*, 15. [CrossRef]
80. Orlov, S.N. NKCC1 as an epigenetically regulated transporter involved in blood pressure elevation with age. *Am. J. Hypertens.* **2011**, *24*, 1264. [CrossRef]
81. Raghavachari, S.; Kahana, M.J.; Rizzuto, D.S.; Caplan, J.B.; Kirschen, M.P.; Bourgeois, B.; Madsen, J.R.; Lisman, J.E. Gating of human theta oscillations by a working memory task. *J. Neurosci.* **2001**, *21*, 3175–3183. [CrossRef]
82. Lee, H.; Simpson, G.V.; Logothetis, N.K.; Rainer, G. Phase locking of single neuron activity to theta oscillations during working memory in monkey extrastriate visual cortex. *Neuron* **2005**, *45*, 147–156. [CrossRef] [PubMed]
83. Mulert, C.; Kirsch, V.; Pascual-Marqui, R.; McCarley, R.W.; Spencer, K.M. Long-range synchrony of gamma oscillations and auditory hallucination symptoms in schizophrenia. *Int. J. Psychophysiol.* **2011**, *79*, 55–63. [CrossRef] [PubMed]
84. Alemany-Gonzalez, M.; Gener, T.; Nebot, P.; Vilademunt, M.; Dierssen, M.; Puig, M.V. Prefrontal-hippocampal functional connectivity encodes recognition memory and is impaired in intellectual disability. *Proc. Natl. Acad. Sci. USA* **2020**, *117*, 11788–11798. [CrossRef] [PubMed]

85. Fricker, D.; Miles, R. EPSP amplification and the precision of spike timing in hippocampal neurons. *Neuron* **2000**, *28*, 559–569. [CrossRef] [PubMed]
86. Cardin, J.A. Inhibitory Interneurons Regulate Temporal Precision and Correlations in Cortical Circuits. *Trends Neurosci.* **2018**, *41*, 689–700. [CrossRef] [PubMed]
87. Ferguson, B.R.; Gao, W.J. PV Interneurons: Critical Regulators of E/I Balance for Prefrontal Cortex-Dependent Behavior and Psychiatric Disorders. *Front. Neural Circuits* **2018**, *12*, 37. [CrossRef] [PubMed]
88. Sohal, V.S.; Rubenstein, J.L.R. Excitation-inhibition balance as a framework for investigating mechanisms in neuropsychiatric disorders. *Mol. Psychiatry* **2019**, *24*, 1248–1257. [CrossRef] [PubMed]
89. Wilson, H.R.; Cowan, J.D. Excitatory and inhibitory interactions in localized populations of model neurons. *Biophys. J.* **1972**, *12*, 1–24. [CrossRef]
90. Loscher, W.; Kaila, K. CNS pharmacology of NKCC1 inhibitors. *Neuropharmacology* **2022**, *205*, 108910. [CrossRef]
91. Kahle, K.T.; Deeb, T.Z.; Puskarjov, M.; Silayeva, L.; Liang, B.; Kaila, K.; Moss, S.J. Modulation of neuronal activity by phosphorylation of the K-Cl cotransporter KCC2. *Trends Neurosci.* **2013**, *36*, 726–737. [CrossRef]
92. Porcher, C.; Medina, I.; Gaiarsa, J.L. Mechanism of BDNF Modulation in GABAergic Synaptic Transmission in Healthy and Disease Brains. *Front. Cell Neurosci.* **2018**, *12*, 273. [CrossRef] [PubMed]
93. Rivera, C.; Li, H.; Thomas-Crusells, J.; Lahtinen, H.; Viitanen, T.; Nanobashvili, A.; Kokaia, Z.; Airaksinen, M.S.; Voipio, J.; Kaila, K.; et al. BDNF-induced TrkB activation down-regulates the K^+-Cl^- cotransporter KCC2 and impairs neuronal Cl^- extrusion. *J. Cell Biol.* **2002**, *159*, 747–752. [CrossRef] [PubMed]

life

MDPI

Review

Ion Channels and Ionotropic Receptors in Astrocytes: Physiological Functions and Alterations in Alzheimer's Disease and Glioblastoma

Annamaria Lia [1], Alessandro Di Spiezio [1,2], Lorenzo Vitalini [3], Manuela Tore [4,5], Giulia Puja [3] and Gabriele Losi [4,5,*]

1 Department Biomedical Science, University of Padova, 35131 Padova, Italy; annamaria.lia@unipd.it (A.L.); alessando.dispiezio@unipd.it (A.D.S.)
2 Neuroscience Institute (CNR-IN), Padova Section, 35131 Padova, Italy
3 Department Life Science, University of Modena and Reggio Emilia, 41125 Modena, Italy; 267888@studenti.unimore.it (L.V.); giulia.puja@unimore.it (G.P.)
4 Institute of Nanoscience (CNR-NANO), Modena Section, 41125 Modena, Italy; manuela.tore@unimore.it
5 Department Biomedical Science, Metabolic and Neuroscience, University of Modena and Reggio Emilia, 41125 Modena, Italy
* Correspondence: gabriele.losi@nano.cnr.it

Abstract: The human brain is composed of nearly one hundred billion neurons and an equal number of glial cells, including macroglia, i.e., astrocytes and oligodendrocytes, and microglia, the resident immune cells of the brain. In the last few decades, compelling evidence has revealed that glial cells are far more active and complex than previously thought. In particular, astrocytes, the most abundant glial cell population, not only take part in brain development, metabolism, and defense against pathogens and insults, but they also affect sensory, motor, and cognitive functions by constantly modulating synaptic activity. Not surprisingly, astrocytes are actively involved in neurodegenerative diseases (NDs) and other neurological disorders like brain tumors, in which they rapidly become reactive and mediate neuroinflammation. Reactive astrocytes acquire or lose specific functions that differently modulate disease progression and symptoms, including cognitive impairments. Astrocytes express several types of ion channels, including K^+, Na^+, and Ca^{2+} channels, transient receptor potential channels (TRP), aquaporins, mechanoreceptors, and anion channels, whose properties and functions are only partially understood, particularly in small processes that contact synapses. In addition, astrocytes express ionotropic receptors for several neurotransmitters. Here, we provide an extensive and up-to-date review of the roles of ion channels and ionotropic receptors in astrocyte physiology and pathology. As examples of two different brain pathologies, we focus on Alzheimer's disease (AD), one of the most diffuse neurodegenerative disorders, and glioblastoma (GBM), the most common brain tumor. Understanding how ion channels and ionotropic receptors in astrocytes participate in NDs and tumors is necessary for developing new therapeutic tools for these increasingly common neurological conditions.

Keywords: astrocytes; ion channels; glia; Alzheimer's disease; glioblastoma

Citation: Lia, A.; Di Spiezio, A.; Vitalini, L.; Tore, M.; Puja, G.; Losi, G. Ion Channels and Ionotropic Receptors in Astrocytes: Physiological Functions and Alterations in Alzheimer's Disease and Glioblastoma. *Life* **2023**, *13*, 2038. https://doi.org/10.3390/life13102038

Academic Editor: Carlo Musio

Received: 4 September 2023
Revised: 3 October 2023
Accepted: 7 October 2023
Published: 11 October 2023

1. Introduction

Astrocytes constitute the most abundant glial cell population in the brain. Astrocytes are heterogeneous and highly complex cells (Verkhratsky and Nedergaard (2018) [1]; Khakh and Deneen (2019) [2]) and they develop in strict and dynamic interaction with neurons. Indeed, astrocytes govern synaptogenesis and synapse maturation and elimination through different active molecules and phagocytic functions (Khakh and Sofroniew (2015) [3]; Allen and Eroglu (2017) [4]). During development and throughout the whole life span, astrocytes play a plethora of crucial functions that include control of tissue homeostasis, metabolism, and neurovascular coupling (Allaman et al. (2011) [5]; Allen and Lyons (2018) [6]; Lia et al. (2023a) [7]).

Neuronal, glial, vascular, and extracellular matrix elements contribute to forming the functional unit responsible for information processing and brain functions that was recently named the "brain active milieu" (Semyanov and Verkhratsky (2021) [8]), which is largely affected in brain diseases. Indeed, in neurodegenerative diseases (NDs), gliosis is a common hallmark, with both astrocytes and microglia playing a major role.

Astrocytes and microglia rapidly transform into a reactive state characterized by morphological, transcriptional, and functional changes that frequently lead to neuroinflammation (Pekny et al. (2016) [9]; Liddelow and Barres (2017) [10]; Sofroniew (2020) [11]; Escartin et al. (2021) [12]). The molecular signature of these dynamic changes is only partially understood. Mixed and even opposing effects may be produced by the same activated glia depending on the specific context and factors that are the focus of several ongoing studies (Escartin et al. (2021) [12]). Reactive astrocytes may release protective cytokines and neurotrophic factors that may reduce cellular damage and, eventually, restore tissue physiological functions, especially after acute events such as stroke or brain trauma.

In chronic NDs, instead, activated astrocytes and microglia persist in a reactive state that may exert detrimental effects by producing excessive reactive oxygen or nitrogen species (ROS or RNS, respectively), as well as proinflammatory cytokines and neurotoxic factors (Guttenplan et al. (2021) [13]; Brandebura et al. (2023) [14]; Patani et al. (2023) [15]).

In addition to NDs, astrocytes are involved in glioblastoma (GBM) genesis and progression and in glioblastoma-related symptoms due to the alterations of local glial and neuronal functions (Brandao et al. (2019) [16]; Yang et al. (2022) [17]).

Astrocytic ion channels, in contrast to neuronal ones, were considered important only for the control of ion tissue homeostasis, particularly for potassium (Kuffler et al. (1966) [18]; Orkand et al. (1966) [19]; Kuffler (1967) [20]). Indeed, astrocytes are electrically non-excitable cells since they do not generate action potentials. For this reason, astrocytes were, for a long time, described as passive cells in contrast with neurons, which were considered the only players in synaptic transmission and information processing. This view has changed considerably in the last few decades. In particular, calcium imaging studies revealed that astrocytes are highly active cells capable of releasing gliotransmitters, such as glutamate, ATP, and D-serine. Gliotransmitters dynamically interact with neurons, modulating synaptic transmission, plasticity, and ultimately, behavior (Araque et al. (2014) [21]; Bazargani and Attwell (2016) [22]; Santello et al. (2019) [23]; Lyon and Allen (2021) [24]; Kofuji and Araque (2021) [25]).

Intracellular calcium transients in astrocytes were initially studied in the soma, where intracellular stores are the main source of these events. Currently, perisynaptic astrocytic processes (PAPs), the ultra-thin processes that intimately contact synapses (Ventura and Harris (1999) [26]; Bushong et al. (2002) [27]), are considered the most active compartment, in which calcium events are far more frequent and complex than in the soma, occurring in localized and heterogeneous events. This calcium activity in small processes relies not only on calcium released from internal stores but also on calcium influx from membrane ion channels (Lia et al. (2021) [28]; Ahmadpour et al. (2021) [29]).

Finally, the membrane potential was very recently shown to be highly dynamic in PAPs (Armbruster et al. (2022) [30]). These voltage transients may affect, for instance, glutamate uptake and, indirectly, neuronal excitability. Accordingly, although not excitable, astrocytes are now considered electrically active cells (McNeill et al. (2021) [31]).

A summary of ion channels and ionotropic receptors expressed in astrocytes is depicted in Figure 1. Notably, the ion channels and ionotropic receptors expressed in astrocytes and neurons are frequently the same, making the study of their specific roles very difficult. The use of genetically encoded molecules or transgenic animals provides an invaluable solution to this problem.

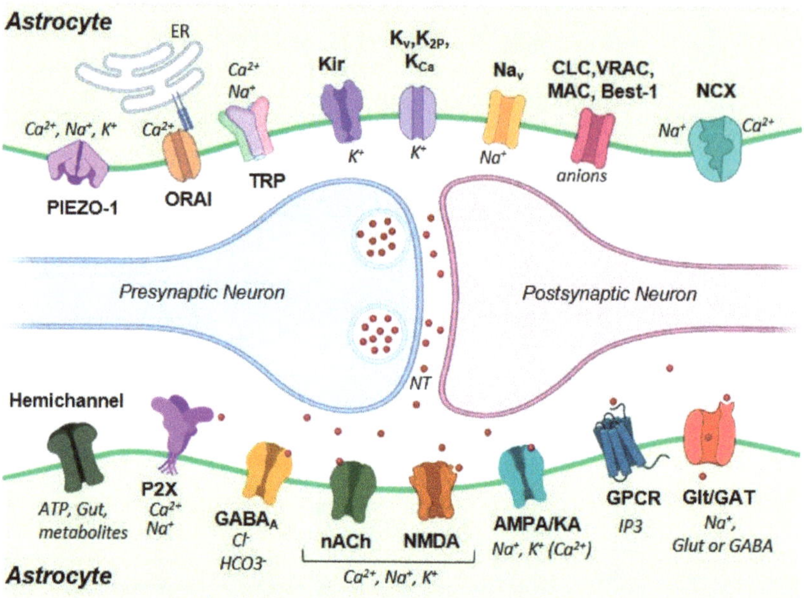

Figure 1. Ion permeable channels and receptors in astrocytes. Schematic view of the main ion channels and ionotropic receptors in astrocyte processes in a tripartite synapse. Ion channels (top), from left to right: tension gated PIEZO-1 channel; endoplasmic reticulum (ER)-STIM1 activated ORAI channels (ORAI-1,-3); TRP channels (TRPV4 is depicted, and others include TRPA1, TRPV1, TRPC1, and TRPC4-6); potassium permeable channels Kir (Kir 4.1, Kir 5.1, and K_{ATP}, which include Kir 2.1-3 and Kir 6.1-2), K_V (K_V1.1, K_V1.6, K_V3.4, and K_V4.3), K_{2P} (TREK1-2 and TWIK1), and K_{Ca} (K_{Ca}1.1, K_{Ca}3.1, and K_{Ca}2.1-3, also known as BK, IK, and SK, respectively); sodium channels (Na_V1.2-3 and Na_V1.5-6); and anion permeable channels (CLC1-3, VRAC, MAC, and Best-1). The Na^+Ca^{2+} exchanger (NCX) is also reported. Ionotropic receptors and others (bottom), from left to right: hemichannels composed of Cx 43, 30, and 26; purinegic P2X receptors ($P2X_7$); $GABA_A$ receptors; cholinergic nicotinic receptors (nACh); and glutamatergic ionotropic receptors (NMDA, AMPA, and KA). Metabotropic receptors coupled to G protein (GPCRs) and electrogenic glutamate and GABA transporters (Glt and GAT1-3, respectively) are also reported. NT, neurotransmitter; Glut, glutamate. Obtained from Biorender.

Ion and water imbalances may affect calcium activity and membrane potential, modulating different intracellular pathways like NF-kB, JAK-STAT3, calcineurin, and NFAT, which are important for astrocytic activation in response to pathological conditions. Furthermore, in NDs, astrocyte modulation of synaptic plasticity is impaired, suggesting a direct involvement of astrocytes in cognitive dysfunctions related to these pathologies (Osborn et al. (2016) [32]). Understanding the functional roles of ion channels and ionotropic receptors in astrocytes, particularly in small processes, may unveil unknown mechanisms of brain function and new therapeutic targets for brain diseases. Indeed, being so deeply affected in NDs, neurons were the focus of the drug research field for many years. Currently, the poor results obtained by neurocentric therapeutic strategies for NDs prompt us to consider glial cells as crucial players and possible new targets. Indeed, astrocytes are largely involved in virtually all brain diseases, including AD, Parkinson's disease, Huntington's disease, amyotrophic lateral sclerosis, epilepsy, tumors, ischemia, and injury (Losi et al. (2012) [33]; Pekny et al. (2016) [9]; Brandao et al. (2019) [16]; Khakh and Goldman (2023) [34]; Patani et al. (2023) [15]).

Here, we will review the most recent advances in this very active field, focusing on astrocytic ion channels in AD (see also Table 1), one of the most diffuse NDs, and in GBM (see also Table 2), a brain tumor that directly involves astrocytes. These two different neurological conditions may serve as examples of how glial cells alterations may contribute to brain pathology. We will first present the role of astrocytic channels in physiological conditions. Then, we will present how their expression and function are modified in AD and GBM. We aim to give an overview of the complex mechanisms underlying two of the most important neurological disorders without forgetting the broader scenario in which many other pathologies show alterations of ion and ionotropic channels in astrocytes.

Table 1. Summary of the main findings reported on ion channels and ionotropic receptors in astrocytes from AD brains.

	Alzheimer's Disease	
Channel or Receptor	**Main Findings**	**References**
TRPA1	Upregulated in the late AD phase.	Lee et al. (2016) [35]
	Pharmacological inhibition or genetic KO ameliorates AD outcomes in mice.	Lee et al. (2016) [35], Paumier et al. (2022) [36]
Piezo1	Pharmacological activation reduces Aβ accumulation and improves plasticity and memory.	Hu et al. (2023) [37]
	Selective KO in microglia exacerbates AD pathology.	
AQP4	AQP4-KO in AD mice increases AB deposition and cognitive deficits.	Xu et al. (2015) [38]
	AQP4 mislocalization has been found in AD patients.	Reeves et al. (2020) [39]
Kir4.1	Kir4.1 KO mice show neuronal hyperexcitability associated with AD.	Nwaobi et al. (2016) [40]
	Kir4.1 is reduced in postmortem AD brains.	Wilcock et al. (2009) [41]
	Dentate gyrus astrocytes around Aβ plaque show higher Kir4.1	Huffels et al. (2022) [42]
K_{ATP}	Kir6.2 subunit is upregulated in AD mice and postmortem AD brains.	Griffith et al. (2016) [43]
	Pharmacological activation of K_{ATP} reduces AD hallmarks and cognitive deficits in AD mice.	Liu et al. (2010) [44]
	Pharmacological inhibition of K_{ATP} increases Aβ deposition in mice.	Macauley et al. (2015) [45]
$K_v3.4$	$K_v3.4$ is upregulated in AD human and mouse brains.	Angulo et al. (2004) [46], Pannaccione et al. (2007) [47], Boscia et al. (2017) [48]
	$K_v3.4$ silencing reduces GFAP expression and Aβ loading in mice.	Boscia et al. (2017) [48]
$K_{Ca}3.1$	$K_{Ca}3.1$ is upregulated in AD patients.	Yi et al. (2016) [49]
	$K_{Ca}3.1$ blockade attenuates neuroinflammation and ameliorates cognitive deficits in AD mice.	Wei et al. (2016) [50], Yi et al. (2016) [49], Yu et al. (2018) [51]
Best1	Mediates abnormal GABA release in AD mice hippocampus and affects synaptic plasticity.	Jo et al. (2014) [52]
	Altered localization in astrocytes from AD mice.	
α7nAChR	Higher expression in astrocytes from AD patients.	Teaktong et al. (2003) [53], Yu et al. (2005) [54]
	Activated by Aβ at physiological or pathological concentrations, affecting synaptic plasticity.	Wang et al. (2002) [55], Pirttimaki et al. (2013) [56], Gulisano et al. (2019) [57]
	α7nAChRsKO mice develop an AD-like pathology	Tropea et al. (2021) [58]

Table 1. *Cont.*

	Alzheimer's Disease	
Channel or Receptor	**Main Findings**	**References**
$P2X_7R$	Upregulated in microglia from both AD mice and postmortem AD brains.	McLarnon et al. (2006) [59]; Martínez-Frailes et al. (2019) [60]
	Upregulated in astrocytes from AD mice.	Jin et al. (2018) [61], Martin et al. (2019) [62]
	$P2X_7R$ KO in AD mice reduces cognitive deficits and Aβ plaques without affecting microglia.	Martin et al. (2019) [62]

Table 2. Summary of the main findings reported on ion channels and ionotropic receptors in GBM.

	Glioblastoma	
Channel or Receptor	**Main Findings**	**References**
VRAC	Highly expressed in GBM cells.	Caramia et al. (2019) [63]
	Promotes migration and resistance to apoptosis but is not necessary for tumor development.	Caramia et al. (2019) [63]; Liu and Stauber (2019) [64]
	May transport anticancer drugs like cisplatin and carboplatin.	Planells-Cases et al. (2015) [65]
gBK	Overexpressed in GBM and contributes to aggressive tumor growth and migration.	Molenaar (2011) [66]
	Inhibition reduces tumor migration only when GBM cells are in their active state.	Brandalise et al. (2020) [67]
Kir4.1	Downregulated in GBM cells.	Tan et al. (2008) [68]; Brandalise et al. (2020) [67]
	The remaining portion of functional Kir4.1 channels may cooperate with gBK channels to promote tumor invasion.	Brandalise et al. (2020) [67]
	Kir4.1 loss depolarizes GBM cells and increases tumor proliferation, while Kir4.1 reintroduction reduces proliferation.	Madadi et al. (2021) [69]
$K_{Ca}3.1$	Elevated levels in GBM cells correlate with poor survival.	Brown et al. (2018) [70]; Hausmann et al. (2023) [71]
	Contribute to GBM cell migration by influencing Ca^{2+} signaling.	Catacuzzeno and Franciolini (2018) [72]
	Silencing and inhibition reduce tumor infiltration and improve survival in mouse models.	Brown et al. (2018) [70]
$K_v1.3$, $K_v1.5$	Inverse correlation between $K_v1.5$ expression and glioma malignancy (no such association was observed for $K_v1.3$).	Preußat et al. (2003) [73]; Arvind et al. (2012) [74]
	$K_v1.3$ inhibition in glial populations regulates astrocyte and microglia reactivity, reducing tumor growth and invasiveness.	Grimaldi et al. (2018) [75]
	$K_v1.3$ inhibitors can induce cell death in glioma cells.	Venturini et al. (2017) [76]
TRPV4	Overexpressed in malignant gliomas like GBM. Negatively correlated with the prognosis.	Yang et al. (2020) [77]
	CBD treatment causes TRPV4-mediated calcium influx triggered mitophagy and consequent glioma cell death.	Huang et al. (2021) [78]

Table 2. *Cont.*

Glioblastoma		
Channel or Receptor	**Main Findings**	**References**
PIEZO1	Overexpressed in malignant gliomas like GBM. Negatively correlates with the prognosis.	Chen et al. (2018) [79]
	ECM stiffness activates PIEZO1 and increases its expression, promoting GBM invasiveness.	
AQPs	AQPs facilitate tumor mobility, survival, and growth through diverse mechanisms.	Varricchio et al. (2021) [80]
	AQP4 overexpression and redistribution in GBM are linked to increased cell migration.	Vandebroek and Yasui (2020) [81]
	T3 hormone may decrease AQP4 in GBM cells, leading to reduced tumor growth and migration.	Costa et al. (2019) [82]
$GABA_AR$	The majority of $GABA_AR$ subunits are downregulated in GBM, except for the δ subunit, which is upregulated.	Smits et al. (2012) [83]; Tantillo et al. (2023) [84]
	GABA signaling decreases GBM proliferation and growth, but it may lead to epilepsy development.	Blanchart et al. (2017) [85]; Radin and Tsirka (2020) [86]; Tantillo et al. (2023) [84]
	$GABA_AR$ activity supports tumor cell quiescence. After tumor resection, its reduced activity may be responsible for GBM recurrence.	Smits et al. (2012) [83]; Blanchart et al. (2017) [85]
	Neurosteroids could modulate GBM cell line biology.	Zamora-Sánchez et al. (2017, 2022) [87,88]; Feng YH. et al. (2022) [89]
nAchR	Few subtypes of nAchR are expressed in GBM cells.	Thompson et al. (2019) [90]; Pucci et al. (2022) [91]
	nAchR controls the survival, proliferation, and infiltrative capacity of GBM.	Thompson et al. (2019) [90]; Pucci et al. (2021) [92]
	Activation of nAchR produces an increase in $[Ca^{2+}]_{int}$ that leads to intracellular pathways activation (Akt, ERK).	Pucci et al. (2022) [91]
	nAchR antagonists decrease the viability, proliferation, and migration of GBM cells.	Spina et al. (2016) [93]; Pucci et al. (2022) [91]; Bavo et al. (2023) [94]
iGluRs	Stimulation of iGluRs leads to GBM progression and contributes to seizures in the peritumoral area.	Venkataramani et al. (2019) [95]; Jung et al. (2020) [96]
	NMDAR, AMPAR, and KAR expressed in GBM can be activated by glutamate released by neurons and by GBM itself.	Lyons et al. (2007) [97]; Nandakumar et al. (2019) [98]; Venkataramani et al. (2019) [95]; Venkatesh et al. (2019) [99]
	Ca^{2+} permeable AMPARs and NMDARs promote tumor progression in the glutamate-rich environment of GBM.	Nandakumar et al. (2019) [98]
	iGluR antagonists may slow GBM progression while limiting neurodegeneration and seizure onset.	Grossman et al. (2009) [100]; Cacciatore et al. (2017) [101]; Yamada et al. (2020) [102]; Albayrak et al. (2021) [103]; Blyufer et al. (2021) [104]; Venkataramani et al. (2021) [105]

Table 2. *Cont.*

Glioblastoma		
Channel or Receptor	**Main Findings**	**References**
Cx43	Cx43 expression is reduced in the tumor core and increased at the tumor–astrocyte interface. Due to its heterogeneous distribution, Cx43 acts both as a tumor suppressor and a promoter of cell migration.	Caltabiano et al. (2010) [106]; Sin et al. (2016) [107]; Uzu et al. (2018) [108]; McCutcheon and Spray (2022) [109]
	Cx43 regulates apoptosis, impacts cell homeostasis, and promotes epileptic activity in the peritumoral zone.	Sin et al. (2016) [107]; Dong et al. (2017) [110]; Xing et al. (2019) [111]
	Cx43 expression in GBM cells is associated with the development of resistance to TMZ treatment.	Gielen et al. (2013) [112]

2. Ion Channels and Ionotropic Receptors in Astrocyte Physiology

2.1. Potassium Channels

The relevance of K^+ channels in astrocyte physiology was discovered nearly sixty years ago in seminal studies that revealed a high permeability to this cation at very negative values of resting membrane potential, close to the Nernst's equilibrium for K^+ (Kuffler et al. (1966) [18]; Orkand et al. (1966) [19]; Kuffler (1967) [20]; Ransom and Goldring (1973) [113]). Astrocytes express a large set of K^+-permeable channels, including voltage-dependent channels (Kir), voltage-gated channels (K_v), voltage-independent two-pore domain channels (K_{2P}), and Ca^{2+}-dependent channels (K_{Ca}) (Seifert et al. (2018) [114]; Verkhratsky and Nedergaard (2018) [1]; McNeill et al. (2021) [31]). Maintenance of K^+ homeostasis in brain tissue is one of the major known astrocytic functions, and it is mediated by different K^+ channels together with active Na^+/K^+ cotransporters.

Astrocyte K^+ channels and transporters mediate the spatial buffering of extracellular K^+ that is internalized in regions of high neuronal activity and redistributed by gradients in nearby, less active regions through gap junctions that connect astrocytes (Kofuji and Newman (2004) [115]; Verkhratsky and Nedergaard (2018) [1]; McNeill et al. (2021) [31]). Of note, an extracellular K^+ increase is not due solely to repolarizing axons but also to the activation of postsynaptic N-methyl-D-aspartate (NMDA) receptors, which are K^+ permeable (Shih et al. (2013) [116]; Tyurikova et al. (2022) [117]), and the activation of γ-aminobutyric acid (GABA) receptors through the involvement of cation chloride cotransporters (Kaila et al. (1997) [118]; Voipio and Kaila (2000) [119]; Viitanen et al. (2010) [120]). Also, for this reason, extracellular K^+ concentrations are particularly dynamic in active brain regions and need to be properly controlled.

Kir channels are inwardly rectifying channels composed of a family of 16 subtypes, classified into 7 sub-families, expressed by distinct KCNJ genes. The main ion channels expressed by astrocytes are represented by Kir4.1, a weak, inwardly rectifying K^+ channel expressed exclusively by glial cells (Kofuji and Newman (2004) [115]; Olsen and Sontheimer (2008) [121]; Brasko et al. (2017) [122]). Kir4.1 channels are open at resting membrane potential and contribute to its very negative value. Its voltage dependency is modulated by Mg^{2+} and intracellular polyamines, and it is reduced at depolarized membrane potentials. Kir4.1 is enriched at perisynaptic astrocytic processes and endfeet, i.e., processes contacting blood vessels, and is also present as a heterodimer with Kir5.1 (Hibino et al. (2004, 2010) [123,124]; Tan et al. (2008) [68]).

Other Kir channels include Kir2.1-3, Kir6.1, and Kir6.2. The latter compose the ATP-sensitive potassium channels (K_{ATP}) formed by four pore-forming Kir6.1 and Kir6.2 subunits and four sulfonylurea receptor subunits (Sun and Hu (2010) [125]). These channels participate in the maintenance of extracellular K^+ homeostasis, allowing the K^+ entry inside the cells and its release within capillaries or in adjacent regions of low activity. K_{ATP} channels are also reported to exert anti-inflammatory and neuroprotective effects (Hu et al. (2019) [126]; Chen et al. (2021) [127]).

Voltage-activated K^+ channels include K_V 1.1, 1.6, 3.4, and 4.3. Among these delayed-rectifying (KD) channels with higher permeability at depolarized membrane potentials and transient Ka are channels that mediate rapidly activating and inactivating currents (A-type currents).

The family of voltage-independent K_{2P} channels, also known as leak channels, includes 15 different members. Of these, TREK 1, TREK 2, and TWIK 1 channels have been reported in astrocytes. Their role in setting the negative membrane potential is supported by pharmacological inhibition and shRNA approaches (Zhou et al. (2009) [128]; Hwang et al. (2014) [129]). However, experimental evidence is missing in knock-out mouse models (Du et al. (2016) [130]).

Calcium-activated K^+ channels (K_{Ca}), which are also voltage-dependent and named according to their biophysical properties, include big conductance (BK; K_{Ca}1.1), intermediate conductance (IK; K_{Ca}3.1), and small conductance (SK; K_{Ca}2.1-3) channels. Opening of these channels requires both intracellular Ca^{2+} and membrane depolarization. Their localization is mainly reported in perivascular processes and endfeet where they participate in neurovascular coupling (Price et al. (2002) [131]; Filosa et al. (2006) [132]; Carmignoto and Gómez-Gonzalo (2010) [133]; Filosa and Iddings (2013) [134]).

2.2. Sodium Channels

Although voltage-dependent Na^+ channels are typical of excitable cells such as neurons, their expression is also reported in glial cells, including astrocytes. In situ studies revealed the expression of Na^+ channels such as Nav1.5, the cardiac tetrodotoxin (TTX) resistant channel type, and to a lesser extent Nav1.2, 1.3, and 1.6, with substantial regional differences (Pappalardo et al. (2016) [135]; Verkhratsky and Nedergaard (2018) [1]). The functional roles of these channels are unclear. It is important to remember that these channels modulate the activity of other cotransporters/pumps, such as Na^+/K^+ ATPase and glutamate and GABA transporters.

Nav1.5 and Nav1.6 may be important in NDs as their expression is increased in reactive astrocytes (Black et al. (2010) [136]; Zhu et al. (2016) [137]) and microglia (Pappalardo et al. (2016) [135]).

2.3. Calcium Permeable Channels

Calcium homeostasis is crucial for many astrocyte functions. Accordingly, astrocyte plasma membrane and intracellular organelles express several calcium channels and calcium-permeable ionotropic receptors. As mentioned above, astrocytes display complex and dynamic intracellular calcium transients that can be evoked by network activity or occur spontaneously. The spatial and temporal pattern of Ca^{2+} transients is very heterogeneous, showing frequent and localized events in small processes and PAPs, which are called Ca^{2+} microdomains (Grosche et al. (1999) [138]), or large events occurring in soma and major processes in response to sustained network activity (Panatier et al. (2011) [139]; Di Castro et al. (2011) [140]; Shigetomi et al. (2013a) [141]; Kanemaru et al. (2014) [142]; Srinivasan et al. (2015) [143]; Bindocci et al. (2017) [144]; Mariotti et al. (2018) [145]; Stobart et al. (2018) [146]; Arizono et al. (2020) [147]) or in pathological conditions (Shigetomi et al. (2019) [148]). Calcium transients may evoke the release of gliotransmitters (such as glutamate, D-serine, and ATP) that finely modulate synaptic functions (Araque et al. (2014) [21]; Bazargani and Attwell (2016) [22] and, ultimately, behavior (Lyon and Allen (2022) [24]; Nagai et al. (2021) [149]; Kofuji and Araque (2021) [25]).

The mechanisms that generate calcium transients are complex and involve both calcium release from intracellular stores and calcium influx from the plasma membrane (Lia et al. (2021) [28]; Ahmadpour et al. (2021) [29]). Intracellular calcium channels and receptors mediate the release of calcium from the endoplasmic reticulum (ER) or other calcium stores such as mitochondria. Calcium release from the ER is triggered by the activation of metabotropic receptors coupled to G protein (GPCRs) and the consequent production of inositol trisphosphate (IP3), which opens IP3 receptors expressed on ER membranes, leading to the final Ca^{2+} release in the cytoplasm. Indeed, astrocytes sense network activity through high-affinity metabotropic receptors of neurotransmitters, including glutamate, GABA, dopamine, and norepinephrine, as well as other messengers such as ATP and purines. Of note, astrocyte GPCRs are associated not only with Gq but also with Gi, as in the case of $GABA_B$ receptors, inducing IP3-mediated Ca^{2+} transients (Mariotti et al. (2015, 2018) [145,150]; Durkee et al. (2019) [151]; Caudal et al. (2020) [152]).

Calcium influx from the external space may be mediated by different ion channels, such as transient receptor potential (TRP) channels, Orai channels, mechanoreceptors (Piezo-1), and ionotropic receptors for different neurotransmitters (Lia et al. (2021) [28]; Ahmadpour et al. (2021) [29]). In addition, reverse operation of the Na^+-Ca^{2+} exchanger (NCX) can also mediate cytosolic Ca^{2+} events (Rose et al. (2020) [153]).

Orai and TRP channels type C (TRPC) take part in store-operated calcium entry (SOCE), a mechanism that involves the concerted action of these plasma membrane ion channels and the ER-Ca^{2+} sensor STIM1 (Figure 1). Under conditions of a reduced calcium content in the ER, STIM1 activates Orai or TRPC channels that allow calcium influx from the extracellular space (Verkhratsky and Parpura (201) [154]; Yoast et al. (2020) [155]). Indeed, astrocytes express functional Orai channels type 1 and 3 (Kwon et al. (2017) [156]) and TRPC receptors.

TRP channels are a group of ion channels present in different tissues that can be activated either by physical stimuli, such as cell membrane stretch, osmotic pressure, and temperature, or by molecules such as signaling lipids and others found in hot spices. TRP channels expressed by astrocytes include TRPA1, TRPC1, TRPC4-6, TRPV1, and TRPV4, and their functions are only partially understood (Verkhratsky et al. (2014) [157]; Verkhratsky and Nedergaard (2018) [1]; Ahmadpour et al. (2021) [29]). For instance, TRPA1 takes part in calcium activity in fine processes modulating synaptic inhibition in the striatum and long-term potentiation (LTP) in the hippocampus (Shigetomi et al. (2011, 2013b) [158,159]), while TRPV4 channels are involved in the control of local blood flow (Dunn et al. (2013) [160]) and brain ischemia (Sucha et al. (2022) [161]; Tureckova et al. (2023) [162]).

Other calcium-permeable channels are Piezo channels, transmembrane proteins that act as mechanoreceptors. Stretch-induced membrane tension, caused, for instance, by cell swelling, opens Piezo channels, which are permeable to calcium, potassium, and sodium (Lewis and Grandl (2015) [163]). The role of Piezo-1 channels in glial cells is starting to emerge (Benfenati et al. (2011) [164]; Blumenthal et al. (2014) [165]; Chi et al. (2022) [166]).

2.4. Anion Channels

Astrocytes express a large number of anion-permeable channels, including voltage-dependent chloride channels (ClC1-3), volume-regulated anion channels (VRAC), Maxi-Cl^- channels (MAC), and Ca^{2+}-dependent Cl^- channels like Bestrophin 1 (Best1), which are all involved in brain diseases (Verkhratsky and Nedergaard (2018) [1]; Elorza-Vidal et al. (2019) [167]).

The most studied voltage-dependent chloride channel in astrocytes is ClC-2 (Makara et al. (2003) [168]), which is widely expressed in all tissues. In astrocytes, the auxiliary subunit GlialCAM modulates ClC-2 electrophysiological properties and subcellular localization (Jeworutzki et al. (2012) [169]). ClC channels are thought to play a role in GABAergic signaling as they favor Cl^- exit from astrocytes at rest and especially during swelling (Verkhratsky and Nedergaard (2018) [1]; Elorza-Vidal et al. (2019) [167]).

VRAC channels are formed by leucine-rich repeats containing the 8A subunit (LRRC8A) and at least one other LRRC8 subunit (B-E). VRACs are also known as SWELL1 (Qiu et al. (2014) [170]; Voss et al. (2014) [171]). These channels play a crucial role in cell volume regulation after cell swelling by allowing the efflux of Cl^- and other osmolytes, including glutamate and taurine (Mongin (2016) [172]; Osei-Owusu et al. (2018) [173]; Elorza-Vidal et al. (2019) [167]). SWELL1 mediates glutamate release from astrocytes and contributes to the modulation of basal synaptic transmission and excitotoxicity after stroke (Yang et al. (2019) [174]). In addition, astrocytic glutamate release through VRACs evokes slow inward currents in nearby neurons (Gómez-Gonzalo et al. (2018) [175]), which may be involved in pathological conditions, synchronizing the principal cells and favoring seizure onset (Fellin et al. (2004) [176]; Gómez-Gonzalo et al. (2010) [177]). Recently, it was shown that astrocytes in the ventral tegmental area (VTA) release GABA through SWELL1, modulating dopaminergic neuron disinhibition in cocaine-induced rewards (Yang et al. (2023) [178]).

MAC channels are anion channels with very high single-channel conductance that is permeable to anions such as Cl^-, pyruvate, glutamate, and ATP (Lalo et al. (2014) [179]; Sabirov et al. (2021) [180]). The core of these channels was recently shown to be SLCO2A1, which is also known to be a prostaglandin transporter (Kanai et al. (1995) [181]; Sabirov et al. (2017) [182]). In astrocytes, MAC channels are gated by swelling or hypoxia, releasing ATP (Liu et al. (2006, 2008) [183,184]). Therefore, different pathological conditions can enhance the opening of these channels.

Best1 is a calcium-dependent anion channel largely expressed on astrocytes, particularly on PAPs (Woo et al. (2012) [185]; Park et al. (2013) [186]). Best1 is permeable to chloride and anions like glutamate and GABA. This channel was shown to modulate synaptic plasticity and GABA tonic currents, affecting network activity and also cognitive functions in pathological conditions such as AD (Lee et al. (2010) [187]; Yoon et al. (2011, 2012) [188,189]; Jo et al. (2014) [52]; Park et al. (2015) [190]; Kwak et al. (2020) [191]).

2.5. Ligand-Gated Ion Channels (Ionotropic Receptors)

Astrocytes express several ionotropic receptors, i.e., ligand-gated ion channels, whose functions are only partially understood. The main known ionotropic receptors reported in astrocytes are glutamatergic (iGluRs), GABAergic (GABARs), nicotinic cholinergic (nAChRs), purinergic (P2XRs), glycinergic, and serotonergic receptors (Verkhratsky and Nedergaard (2018) [1]).

iGluRs include α-amino-3-hydroxy-5-methyl4-isoxazolepropionic acid (AMPA), Kainate (KA), and NMDA receptors. Astrocytic AMPA receptors are expressed in the cortex, cerebellum, hippocampus, olfactory bulb, and spinal cord, with different GluA1-4 subunit compositions (Höft et al. (2014) [192]; Mölders et al. (2018) [193]). AMPARs in astrocytes have been reported to modulate glutamate transporters (López-Bayghen et al. (2003) [194]) and motor coordination (Saab et al. (2012) [195]). KARs in astrocytes were reported to be involved in a model of temporal lobe epilepsy (Vargas et al. (2013) [196]); however, their physiological role is unknown.

Although initially controversial, likely due to a lower expression compared to other iGluRs, the presence of NMDARs in astrocytes is now established. Functional NMDA receptors are reported in astrocytes of the cortex, spinal cord, hippocampus, and cerebellum (Verkhratsky and Kirchhoff (2007) [197]; Verkhratsky and Nedergaard (2018) [1]). In astrocytes, NMDARs are composed of GluN1, GluN2A-D, and GluN3 subunits, mainly in combinations that limit Mg^{2+} blocks. The role of NMDARs in astrocytes has been poorly explored (Skowrońska et al. (2019) [198]).

Astrocytes express both ionotropic $GABA_A$ and metabotropic $GABA_B$ receptors. $GABA_A$ receptors are pentameric ion channels that bind to γGABA and are permeable to Cl^- and $HCO_3{}^-$ ions. Depending on the intracellular Cl^- concentration, activation of $GABA_A$Rs mediates either hyperpolarization or depolarization. Indeed, their activation hyperpolarizes mature neurons, whereas it depolarizes glial cells and immature neurons (Labrakakis et al. (1998) [199]). Typically, $GABA_A$ receptors are composed of five subunits selected from a pool of 19 isoforms, including two α, two β, and a third type of subunit. The functional and pharmacological characteristics of $GABA_A$ receptors in astrocytes, which depend on their subunit composition, are only partially known (Müller et al. (1994) [200]; Fraser et al. (1995) [201]; Höft et al. (2014) 95]). Functionally, astrocytic $GABA_A$Rs are thought to modulate extracellular Cl^- concentrations, thus indirectly modulating GABAergic signaling (Egawa et al. (2013) [202]), intracellular pH, and K^+ influx (Ma et al. (2012) [203]). Notably, astrocytes synthesize neurosteroids (Puia and Belelli (2001) [204]; Belelli and Lambert (2005) [205]; Reddy (2010) [206]), potent modulators of GABA receptors with paracrine and possible autocrine effects, although these latter effects are largely unexplored.

Nicotinic acetylcholine receptors are pentameric ligand-gated ion channels composed of different subunits (α2-10 and β2-4) and expressed by neurons and astrocytes in different brain regions (Gotti et al. (2006) [207]; Shen and Yakel (2012) [208]; Zoli et al. (2018) [209]). Neuronal nAChRs are involved in gene expression, neurotransmitter release, and synaptic plasticity, thus affecting learning and memory (Gotti et al. (2006) [207]; Zoli et al. (2018) [209]; Koukouli and Changeux (2020) [210]). Astrocytes express mainly α7s or α4β2 receptors. Activation of α7-nAChRs triggers calcium transients in astrocytes and the release of gliotransmitters such as glutamate, GABA, ATP, and D-serine, which affect neuronal and synaptic activity (Pirttimaki et al. (2013) [56]; Wang et al. (2013) [211]; Papouin et al. (2017) [212]; Lezmy et al. (2021) [213]). Of note, nAChRs are also actively involved in neuroinflammation by acting on NfkB and STAT3 and could represent a target for novel therapeutic strategies (Wang et al. (2003) [214]).

Purinergic receptors are largely expressed in astrocytes and glial cells, where they play a major role in different pathological conditions (Fields and Burnstock (2006) [215]; Burnstock (2018) [216]; Huang et al. (2021b) [217]), being suitable targets for future therapies. Purinergic receptors include G-protein coupled (P2Y) and ionotropic (P2X) receptors, nonselective cation channels with high Ca^{2+} permeability. ATP can bind to $P2X_7R$, opening the channel and leading to the influx of Ca^{2+} and Na^+ and the efflux of K^+.

Purinergic signaling is actively involved in brain disease. Indeed, ATP and its breakdown products ADM, AMP, and adenosine are enriched in extracellular space in response to cellular damage and a tumor microenvironment (TME), acting as damage-associated molecular patterns (DAMPs), contributing to inflammatory responses (Di Virgilio et al. (2020) [218]; Huang et al. (2021b) [217]; Engel et al. (2022) [219]).

2.6. Other Channels

2.6.1. Aquaporins

Aquaporins (AQPs) are a class of transmembrane proteins forming a pore responsible for the bi-directional passage of water and small solutes across the cell membranes while preserving ion gradients. To date, 13 members of this family have been identified in mammals (AQP0-12) (Ishibashi et al. (2017) [220]). AQP4 is expressed mainly by astrocytes in small processes in the proximity of the subarachnoid area, ventricular spaces, and blood vessels (Nielsen et al. (1997) [221]; Aoyama et al. (2012) [222]). Accumulated evidence suggests that AQP4 is necessary not only for water homeostasis but also to clear small molecules and metabolites, including Aβ and Tau proteins, from interstitial space (Iliff et al. (2012, 2014) [223,224]; Kress et al. (2014) [225]). Indeed, impaired AQP expression/function is involved in different neurological conditions affecting the blood-brain barrier (BBB) or causing water accumulation and brain edema (Vella et al. (2015) [226]; Filippidis et al. (2016) [227]; Clément et al. (2020) [228]).

2.6.2. GAP Junctions and Hemichannels

Astrocytes are known to be interconnected in a network by gap junctions, large pores formed by the juxtaposition of two connexons expressed in adjacent astrocytes, that allow intercellular exchange of ions and molecules, including second messengers and metabolites (Pannasch and Rouach (2013) [229]; Giaume et al. (2013) [230]). Connexons are formed by six subunits named connexins (Cx), such as Cx43, Cx26, and Cx30, whose opening depends on intracellular calcium, pH, phosphorylation, or molecules like spermine and spermidine (Saez et al. (2003) [231]; Harris (2007) [232]; Houades et al. (2008) [233]; Decrock et al. (2015) [234]).

Connexins, as well as pannexins, can also form plasma membrane pores called hemichannels, which are not connected to an adjacent astrocyte. Hemichannels may release molecules such as ATP, glutamate, and metabolites into the extracellular space in response to changes in extracellular or intracellular calcium, pH, phosphorylation state, and pro-inflammatory cytokines (Giaume et al. (2013) [230]).

Connexins and hemichannels are considered possible new therapeutic targets for different brain diseases (Decrock et al. (2015) [234]; Charvériat et al. (2017) [235]; Xing et al. (2019) [111]), particularly for epilepsy (Mylvaganam et al. (2014) [236]; Li et al. (2019) [237]; Guo et al. (2022) [238]) and cancer (Sinyuk et al. (2018) [239]; Zhou et al. (2023) [240]).

3. Alzheimer's Disease

Alzheimer's disease is a common neurodegenerative disease affecting more than 55 million people worldwide (World Health Organization) with a constantly growing incidence. Although different hypotheses have been proposed to clarify the etiopathology of the disease, a complete picture of the underlying mechanisms is missing, as well as an efficient treatment to block AD progression.

The hallmark of AD is the presence of aggregated amyloid-β (Aβ) plaques and neurofibrillary tangles of phosphorylated tau (Henstridge et al. (2019) [241]). Reactive astrocytes and neuroinflammation are always present in AD (Chun and Lee (2018) [242]; Perez-Nievas and Serrano-Pozo (2018) [243]; Habib et al. (2020) [244]; Viejo et al. (2022) [245]), leading progressively to significant synapse and neuronal dysfunction and, ultimately, to their loss. Currently, neurotoxicity is thought to be linked to extracellular $A\beta_{42}$ oligomers present since disease onset rather than to the plaques that occur at later stages of the disease. Although the large majority of AD is idiopathic, in a small percentage of cases, AD is due to a hereditary component related to mutations in genes involved in Aβ production. In more detail, familial forms of AD are related to the genes encoding amyloid precursor protein (APP) or presenilin-1 and presenilin-2 (PS1 and PS2, respectively), both implicated in APP cleavage. These mutations are useful for obtaining genetic mouse models of the disease that recapitulate many of its features (Webster et al. (2014) [246]).

The involvement of astrocytes in AD pathogenesis is linked to their importance in oxidative stress and neuroinflammatory responses, as supported by several studies performed both in animal models and in humans (Pekny et al. (2016) [9]; Nanclares et al. (2021) [247]). In humans, higher blood levels of glial fibrillary acidic protein (GFAP), a marker of astrocyte reactivity, have been reported in AD patients compared to controls (Abdelhak et al. (2022) [248]). Moreover, a study revealed a positive correlation between cortical Aβ deposition and GFAP levels in symptomatic AD patients (Asken et al. (2020) [249]).

3.1. Potassium Channels

Several studies have shown alterations of K^+ channels in AD. Kir4.1 KO animal models, despite the exclusive Kir4.1 expression in macroglia, showed neuronal hyperexcitability associated with neurodegenerative diseases, including AD (Nwaobi et al. (2016) [40]). Accordingly, postmortem human AD brains showed reduced Kir4.1 levels (Wilcock et al. (2009) [41]). Comparing multiple AD mouse models, Wilcock and colleagues concluded that reduced levels of Kir4.1 mRNA and protein are associated with cerebral amyloid angiopathy (CAA), a condition that affects 70% of AD patients (Kalaria and Ballard (1999) [250]). Finally, as for AQP4, it was proposed that decreased levels of both AQP4 and Kir4.1 are a consequence of the diminished expression of anchoring protein dystrophin 1 (DP71), a protein highly affected by vascular amyloid deposition (Wilcock et al. (2009) [41]).

Conversely, Huffels et al. revealed that in the APP/PS1 AD mouse model, dentate gyrus astrocytes in the proximity of Aβ plaques showed increased expression of Kir4.1 but regular function, possibly to compensate for increased extracellular K^+ levels (Huffels et al. (2022) [42]).

K_{ATP} channels expressed in astrocytes and neurons seem to play a critical role during the development of AD. Griffith and colleagues demonstrated that the Kir6.2 subunit is upregulated in astrocytes in the hippocampus of 3 × Tg-AD mice and the postmortem brains of AD patients. Moreover, increased levels of the pore-forming Kir6.2

were found in the plasma membrane of reactive astrocytes of aged 3 × Tg-AD animals (Griffith et al. (2016) [43]). The altered expression levels on the astrocytes' plasma membranes may affect the resting potential of the cells, resulting in impaired metabolism and gliotransmission.

In vivo studies showed that inducing the opening of K_{ATP} with diazoxide in 3 × Tg-AD mice results in reduced levels of cortical and hippocampal Aβ oligomers and hyperphosphorylated tau, and improved cognitive performance (Liu et al. (2010) [44]). On the other hand, delivering the K_{ATP} channel blocker glibenclamide to the APP/PS1 hippocampus increases extracellular Aβ levels (Macauley et al. (2015) [45]).

The voltage-gated potassium $K_v3.4$ channel subunits are responsible for fast-inactivating K^+ currents. $K_v3.4$ is upregulated in AD human brains and 6-month-old Tg2576 mice, suggesting a possible role in the development of AD (Angulo et al. (2004) [46]; Pannaccione et al. (2007) [47]; Boscia et al. (2017) [48]). In vitro studies performed in primary rat astrocytes showed that Aβ oligomer treatment induces $K_v3.4$ expression. Interestingly, silencing $K_v3.4$ subunits by intracerebral infusion of selective siRNA was sufficient to reduce both GFAP and Aβ oligomer levels (Boscia et al. (2017) [48]).

Among the K^+ channels, the intermediate conductance calcium-activated potassium channel 3.1 ($K_{Ca}3.1$) is involved in the control of membrane potential by controlling the K^+ efflux in response to the inward flow of Ca^{2+} (Yi et al. (2017) [251]). $K_{Ca}3.1$ was found to be up-regulated in astrocytes in senescence-accelerated mouse prone 8 (SAMP8) mice and AD patients (Yi et al. (2016) [49]). Notably, blockage of $K_{Ca}3.1$ with the selective inhibitor TRAM-34 or by genetic ablation was sufficient to reduce astrogliosis and microglia activation in 7-month-old SAMP8 animals and APP-PS1 mice. Moreover, this neuroinflammation reduction was accompanied by an attenuation of memory deficits (Wei et al. (2016) [50]; Yi et al. (2016) [49]; Yu et al. (2018) [51]). Interestingly, in cultured astrocytes, TRAM-34 reduces the Ca^{2+} elevation induced by Aβ oligomers treatment (Yi et al. (2016) [49]). Indeed, it was reported that $K_{Ca}3.1$ is able to induce ER stress by increasing ER Ca^{2+} overload in an Orai1-dependent manner. Accordingly, $K_{Ca}3.1$ gene deletion or pharmacological inhibition prevents astrocyte activation by reducing Ca^{2+}-induced ER stress and activating the neuroprotective AKT/mTOR pathway (Yu et al. (2018) [51]).

3.2. Calcium Permeable Channels and Intracellular Calcium Activity

The astrocyte role in AD has been extensively studied in relationship with Ca^{2+} activity, with different outcomes present in different AD models. Initial experiments in APP-PS1 mice showed that amyloidosis is linked to an increased spontaneous somatic astrocyte Ca^{2+} activity (Kuchibhotla et al. (2009) [252]) that is P2Y1R-mediated (Delekate et al. (2014) [253]). On the other hand, a diminished sensory-evoked astrocyte response has been recently reported in a different mouse model that is also based on APP and PS1 mutations (Lines et al. (2022) [254]). A reduced astrocyte Ca^{2+} response to locomotion has also been reported in the neocortex of awake-behaving 15-month-old tg-ArcSwe mice (Åbjørsbråten et al. (2022) [255]). These studies were performed at a single time point in AD mouse models.

Conversely, a recent work performed longitudinally in the PS2APP mouse model of AD found that astrocyte Ca^{2+} activity evolves with disease progression, with a drastic reduction of both spontaneous and evoked activity at the onset of Aβ plaque deposition. The reduction of astrocyte Ca^{2+} activity leads to long-term memory impairments. It is noteworthy that by acting on SOCE through astrocyte-specific STIM1 overexpression, both spontaneous and evoked astrocyte Ca^{2+} activity together with synaptic plasticity were rescued (Lia et al. (2023b) [256]).

Although these studies point to a heterogeneous astrocyte response to different AD-related mutations, AD progression is certainly accompanied by alterations of astrocyte Ca^{2+} activity. Therefore, it is a potential target for therapeutic intervention.

TRPA1 channels mediate Ca^{2+} influx in astrocytes, contributing to free basal Ca^{2+} levels (Shigetomi et al. (2011, 2013b) [158,159]). In 2016, the first study on a mouse model based on APP and PS1 mutants investigated the role of TRPA1 in AD. TRPA1 is upregulated in the late stage of disease in APP/PS1 mice but not in the early phase. Accordingly, knockout (KO) of TRPA1 in APP/PS1 mice improves spatial learning and memory and decreases Aβ deposition (Lee et al. (2016) [35]). Interestingly, another model of AD based on APP and PS1 mutations (APP/PS1-21) showed that, concomitantly with Aβ production, astrocytes became hyperactive. This Ca^{2+} hyperactivity is mediated by TRPA1 channels and is linked to hippocampal CA1 neuronal hyperexcitability (Bosson et al. (2017) [257]). In line with this, chronic pharmacological inhibition of TRPA1 channels has positive effects on disease outcomes in the same AD model by normalizing spine density and maturation and reducing astrocyte Ca^{2+} hyperactivity (Paumier et al. (2022) [36]).

Piezo channels are mechanosensitive cation-conducting channels activated upon increased membrane tension by unknown chemical ligands. It is conceivable that in the context of AD, where Aβ plaque deposition leads to a change in the mechanical environment, Piezo channel function could be significantly altered. Both hippocampal and cortical astrocytes express Piezo1, which allows an increase of cytosolic Ca^{2+} in the presence of mechanical stimuli or upon Yoda1 perfusion, a Piezo1 agonist. Piezo1 activation elicits ATP release by astrocytes, although the mechanisms involved are still unclear (Chi et al. (2022) (46)).

The role of Piezo1 in the context of AD has been studied in the 5xFAD mouse model, where Yoda1 administration reduced Aβ accumulation and improved synaptic function together with learning and memory, in contrast to selective Piezo1 KO in microglia that exacerbates AD pathology (Hu et al. (2023) [37]). In the 5xFAD model, the authors clearly showed that the protective effect is mediated by microglia, but it has already been shown that astrocyte Piezo1 is upregulated around Aβ plaques in postmortem AD brains (Satoh et al. (2006) [258]; Velasco-Estevez et al. (2018) [259]). Therefore, the role of astrocyte Piezo1 in other AD mouse models could be of potential interest.

3.3. Anion Channels

The Best1 channel is highly expressed at astrocyte microdomains in the hippocampus (Woo et al. (2012) [185]), and its role in AD was studied by Jo and co-workers in 2014. The authors showed in 5xFAD and APP/PS1 models of AD an abnormal GABA release in the dentate gyrus mediated by astrocytic Best1. Of note, the authors showed a redistribution of Best1 in the hippocampus of the AD mice. Although Best1 is expressed similarly in control and AD mice, in the AD scenario it is found mainly at the level of the soma and proximal processes. The excess of released GABA activates both ionotropic ($GABA_A$) and metabotropic ($GABA_B$) receptors with a consequent presynaptic inhibition that affects neurotransmitter release, LTP, and cognitive functions (Jo et al. (2014) [52]).

3.4. Ionotropic Receptors

Calcium permeable α7nAChRs are emerging as key players in AD. Cholinergic signaling is deeply affected in AD, and the few medications currently used, like donepezil, rivastigmine, and galantamine, are all acetylcholinesterase inhibitors. AChRs are expressed by all brain cell populations, and the specific role of astrocyte α7nAChRs is under investigation in several laboratories. In particular, it was shown that α7nAChRs expressing astrocytes are increased in the hippocampus and entorhinal cortex of patients with AD but not in other forms of dementia (Teaktong et al. (2003) [53]; Yu et al. (2005) [54]. Other works revealed that astrocytic α7nAChRs are activated by both physiological and pathological concentrations of Aβ, leading to Ca^{2+} transients that favor or oppose synaptic plasticity, respectively (Wang et al. (2002) [55]); Pirttimaki et al. (2013) [56]; Gulisano et al. (2019) [57]).

Interestingly Tropea et al. showed that α7nAChRs KO mice develop an AD-like phenotype with aging, showing increased Aβ and phospho-tau and cognitive impairment, suggesting that the lack of a physiological target of Aβ, i.e., α7nAChRs, may lead to a compensatory Aβ overproduction that induces neurotoxicity and AD-like symptoms (Tropea et al. (2021) [58]). The specific role of the lack of astrocytic vs. neuronal α7nAChRs is, however, to be determined.

$P2X_7R$ expression is upregulated in microglia, both in AD post-mortem brains and AD mouse models (McLarnon et al. (2006) [59]; Martínez-Frailes et al. (2019) [60]), and in astrocytes from mice overexpressing human tau protein (MAPT P301S)(Jin et al. (2018) [61]) and APP/PS1 mice (Martin et al. (2019) [62]). In a recent study, deleting $P2X_7R$ in an APP/PS1 model of AD had a protective role, reducing the Aβ plaque load and improving spatial memory. The author related this phenotype to a reduced level of chemokine production (i.e., CCL3). Interestingly, $P2X_7R$ deletion does not affect cytokines production (IL-1β) or microglia activation (Martin et al. (2019) [62]). The authors proposed a new model where high levels of Aβ peptide induce ATP release from microglia and astrocytes (Sanz et al. (2009) [260]; Orellana Roca et al. (2011) [261]), which activates $P2X_7Rs$, which in turn triggers the neurodegenerative process, increasing chemokine production, and favoring the pathogenic recruitment of T cells (Martin et al. (2019) [62]). Further studies are needed to verify if the inhibition of chemokine release is sufficient to prevent AD-related impairments.

3.5. *Aquaporins*

The involvement of AQP4 in the development of AD has been reported in several studies. As mentioned above, aquaporins are important for the cerebrospinal fluid (CSF)–interstitial fluid (ISF) exchange. Indeed, it was shown that AQP4 KO mice have a 55% reduction in I^{125}-$A\beta_{40}$ washout after intrastriatal injection (Iliff et al. (2012) [224]). The partial clearance of Aβ40 suggests the existence of other pathways involved in Aβ transport and degradation in the brain (Storck et al. 2016 [262]; Gallwitz et al. 2022 [263]).

Moreover, APP/PS1 mice lacking AQP4 showed increased Aβ deposition, exacerbating AD-related cognitive deficits (Xu et al. (2015) [38]). These observations obtained in animal models were confirmed in patients with AD, where an impaired glia–lymphatic system and mislocalization of astrocytic AQP4 was found (Reeves et al. (2020) [39]). A couple of studies questioned what could be the reason for the AQP4 mislocalization. Evidence supports the idea that the mislocalization of AQP4 on astrocytes is associated with the downregulation of DP71 dystrophin (Wilcock et al. (2009) [41]). Of note, AQP4 mislocalization on astrocytic endfeet could participate in blood-brain barrier (BBB) dysfunction, which in turn contributes to neurodegeneration (Wilcock et al. (2009) [41]; Zlokovic (2011) [264]).

Studies support the idea that, as for Aβ, tau clearance also occurs through the glymphatic system. Indeed, inhibition of AQP4 leads to impaired CSF–ISF exchange and decreased tau clearance (Harrison et al. (2020) [265]). Similar results were found in AQP4 KO mice, in which a severe neurogenic fibrillary pathology is present (Iliff et al. (2014) [223]).

In conclusion, ion homeostasis in astrocytes is widely modified in AD (see Table 1 and Figure 2). Notably, the trend shows an increased excitability during the early phases of the disease and a decrease in the late phases, in particular for Ca^{2+} activity. Therefore, trying to revert this trend by targeting ion channels or receptors in astrocytes could help prevent neuronal loss and AD progression.

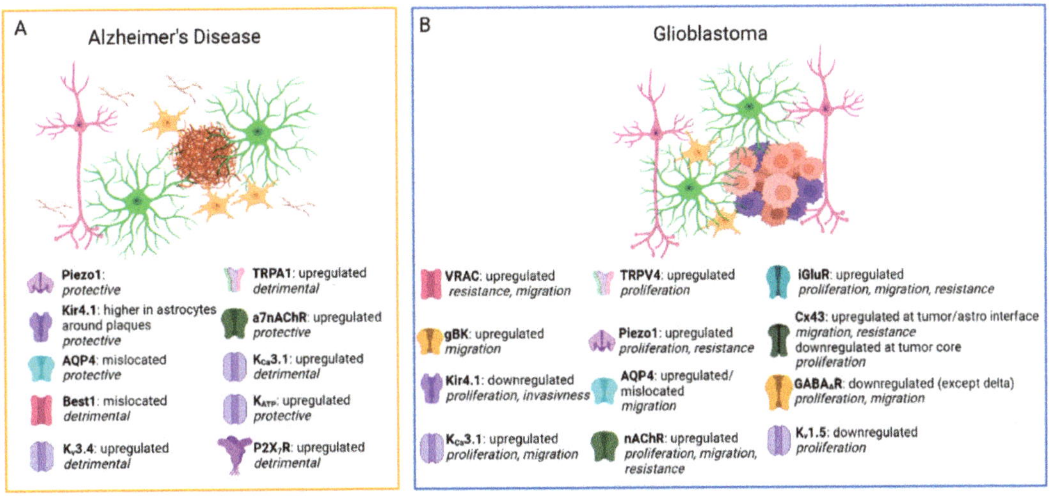

Figure 2. Ion permeable channels and receptors in astrocytes in AD and GBM. (**A**) Schematic view of an Aβ plaque surrounded by reactive astrocytes (green) and microglia (yellow), together with Aβ fibrils and neurons (purple). Ion channels and ionotropic receptors reported in Table 1 are highlighted below with information about their expression changes and the consequent protective or detrimental effects on brain functions. (**B**) Schematic view of GBM cells surrounded by reactive astrocytes (green) and microglia (yellow), together with neurons (purple). The different colors of the GBM cells highlight the GBM cell heterogeneity. Ion channels and ionotropic receptors reported in Table 2 are highlighted below with information about expression changes in GBM cells and their effects in promoting tumor proliferation, migration, or resistance to chemotherapy. Synaptic contacts between neurons are omitted for simplicity. Obtained from Biorender.

4. Glioblastoma

The astrocytic tumor known as glioblastoma (GBM, grade IV glioma, World Health Organization) is the most aggressive and common glioma in adults (Weller et al. (2015) [266]). GBM has very frequent recurrence and a poor prognosis (14–20 months) due to several factors, including high cellular heterogeneity, invasiveness, microvascular proliferation, therapeutic resistance, and recurrence after surgical removal (Wen et al. (2020) [267]). The existence of a functional network among GBM cells and the tumor microenvironment (TME) preserves and fosters its development (Osswald et al. (2015) [268]; Weil et al. (2017) [269]; Venkataramani et al. (2019) [95]). Indeed, GBM cells are interconnected (between them) through tumor microtubes (TMs) that join single cells via gap junctions (GJs), mainly formed by Cx43, and form an intracellular pathway (Osswald et al. (2015) [268]) with important roles both in tumor progression and in resistance to cytotoxic therapies (Weil et al. (2017) [269]; Li et al. (2020) [270]).

Furthermore, a tumor–neuron network exists: bona fide synapses were detected between presynaptic neurons and postsynaptic glioma cells, mostly at the TM level, and the crosstalk between neurons and specific tumor cells within the TME is a crucial step in cancer initiation and progression (Jung et al. (2020) [96]).

The release of glutamate near GM cells from neurons and from neighboring astrocytes, results in complex calcium signaling followed by de novo formation of TMs and increased invasion speed (Venkataramani et al. (2019, 2022 [95,271]). In addition to adjacent neurons and astrocytes, GBM cells themselves secrete glutamate through a cystine/glutamate antiporter (SLC7A11), producing hyper-excitability in the peritumoral zone that fosters GBM malignity (Lo et al. (2008) [272]).

In support of this hypothesis, it was shown that SLC7A11 upregulation in GBM cells correlates with tumor invasion and a worse outcome in GBM patients, as well as with the onset of tumor-related seizures in patients (Robert et al. (2015) [273]). Moreover, neuronal hyperactivity results in the release of factors, such as brain-derived neurotrophic factor (BDNF) and soluble neuroligin 3 (NLGN3) (Venkatesh et al. (2015) [274]), that add support to glioma progression and facilitate the synapsing of neurons onto glioma cells (Goethe et al. (2023) [275]).

GBM exhibits intra-tumoral heterogeneity, and a small population of neuronal and neural progenitor-like tumor cells has been identified as responsible for the aggressive behavior of the tumor and its resistance to treatments. Interestingly, these cells resemble migrating neurons during development (Elias et al. (2023) [276]).

Recently, Hausmann et al. showed that one of these subpopulations can generate rhythmic calcium oscillations that are important for driving tumor progression and invasiveness. This small fraction of tumoral cells having "pacemaker activity" express Ca^{2+} activated potassium channels ($K_{Ca}3.1$) that, by triggering calcium signaling, regulate cell proliferation and motility (Hausmann et al. (2023) [71]).

In GBM cells, the altered expression of specific ion channels with consequent ionic misbalance may contribute to tumor growth, progression, and resistance. Indeed, changes in ion channel activity can lead to the dysregulation of the cell cycle and cell growth, inhibition of apoptosis, and enhancement of cell migration and invasion, all of which contribute to malignant transformation (Litan and Langhans (2015) [277]).

Here, we review the current literature on ion channels and ionotropic receptors in GBM cells and nearby interacting populations.

4.1. Potassium Channels

These channels have a significant physiological function in GBM cells. In particular, the voltage-dependent large-conductance Ca^{2+}-activated BK channels (gBK in glioma) are involved in the aggressive growth and extensive migratory behavior of GBM cells. These channels are overexpressed in malignant gliomas (like GBM), and their expression level is positively correlated with the severity of the tumor (Molenaar (2011) [66]). These channels can contribute to alterations in the shape and volume of glioma cells during their migration into the TME (Wawrzkiewicz-Jałowiecka et al. (2020) [278]). Even though BK channel activation increases the movement of GBM cells by allowing K^+ efflux, preventing the same conductance from functioning when the cells are at rest does not stop GBM cells from invading. In fact, it was demonstrated that inhibiting BK channels can reduce GBM cell migration only if the intracellular calcium concentration is increased, which results in the BK channels being in the open state (Brandalise et al. (2020) [67]).

The ionic balance of GBM cells is also altered by the downregulation of the constitutively open Kir4.1 channel (Tan et al. (2008) [127]; Brandalise et al. (2020) [67]). The reduction of Kir4.1 expression in GBM leads to a change in cell phenotype, resulting in the increased formation of filopodia that promote cell invasion (Thuringer et al. (2017) [279]). Furthermore, it has been proposed that a small but steady outflow of potassium at the higher resting membrane potential of GBM cells is facilitated by the remaining portion of functional Kir4.1 channels. By doing so, these channels may cooperate with gBK channels in promoting tumor invasion. In fact, recent experiments have shown that when Kir4.1 and BK channels are simultaneously blocked, the migration of GBM cells is reduced (Brandalise et al. (2020) [67]). The depolarization of glioma cells caused by Kir4.1 loss is associated with increased proliferation, whereas introducing Kir4.1 via stable transfection with the resulting repolarization prompts the transition from G2/M to G0/G1 phase, thereby reducing proliferation. It is noteworthy that this reversal effect can be hindered by depolarization with high extracellular K^+ or by Ba^{2+} usage (Madadi et al. (2021) [69]). Altogether, these findings indicate that Kir4.1 targeting could be a potential treatment for glioblastoma.

In addition, several tumor types, including GBM, exhibit abnormally high levels of $K_{Ca}3.1$, which plays a major role in cellular activation, migration, and proliferation by regulating the membrane potential and Ca^{2+} signaling (Brown et al. (2018) [70]). Specifically, in migrating GBM cells, calcium signaling frequently manifests as oscillations in the intracellular Ca^{2+} concentration that are important for promoting critical processes in the migratory cycle (Brandalise et al. (2020) [67]). Considering this, it has been hypothesized that $K_{Ca}3.1$ channels may contribute to GBM cell migration by generating or modifying the pattern of these calcium oscillations (Catacuzzeno and Franciolini (2018) [72]). Studies have also shown that GBM cells exhibit an increase in the expression of IL-4 and $K_{Ca}3.1$ when exposed to high levels of radiation. This causes the mobilization of calcium within the cells and the transcription of genes that promote invasion (D'Alessandro et al. (2019) [280]). Given the correlation between $K_{Ca}3.1$ overexpression in glioma patients and their poor survival rate (Hausmann et al. (2023) [71]), it is reasonable to consider targeting $K_{Ca}3.1$ as a means of reducing glioma invasiveness and progression. In fact, both in vitro and in vivo experiments have shown that $K_{Ca}3.1$-silencing and using $K_{Ca}3.1$ inhibitors (such as TRAM-34) reduced tumor infiltration, glioma-associated microgliosis and astrogliosis, and increased survival time in mouse glioma models (Brown et al. (2018) [70]).

Moreover, K_V subtypes $K_V1.3$ and $K_V1.5$ have been demonstrated to play a specific role in the growth-related characteristics of normal glial cells, and it has been proposed that glioma subtypes may exhibit varied expression of these channels (Preußat et al. (2003) [73]). Specifically, there is an inverse correlation between the level of expression of $K_V1.5$ in astrocytomas and their grade of malignancy, with high expression observed in low-grade astrocytomas and low expression in glioblastoma. Consequently, there is a positive correlation between the high expression of $K_V1.5$ and a favorable outcome of patients with GBM.

No such association was observed for $K_V1.3$ expression (Preußat et al. (2003) [73]; Arvind et al. (2012) [74]). However, research has shown that $K_V1.3$ inhibition can regulate astrocytes and microglia reactivity in the context of glioma, leading to a decrease in tumor growth with a direct effect on the invasive properties of glioma cells. These results suggest that $K_V1.3$ channels could be promising targets for restoring glial cells functioning and reducing the damage caused by glioma to the surrounding brain tissue (Grimaldi et al. 2018). Other studies have demonstrated that $K_V1.3$ is expressed in different murine and human glioma cell lines and can be found in both the plasma membrane and mitochondria. In vitro experiments have shown that treatment with $K_V1.3$ inhibitors (such as clofazimine, PAPTP, or PCARBTP) can induce cell death in a significant portion of glioma cells. Nonetheless, modifying these drugs and/or the delivery method is necessary to enable the translation of these findings into clinical practice (Venturini et al. (2017) [76]).

In addition, KAaH1, a homologous Kv1 blocker from scorpion venom, has been discovered to have an impact on $K_V1.3$ by preventing the migration and adhesion of U87 cells. As a result, it may serve as a potential therapeutic approach for treating GBM (Aissaoui et al. (2018) [281]).

4.2. *Calcium Permeable Channels*

4.2.1. TRP Channels

Emerging and underexplored targets for GBM therapy are the TRP channels. Changes in the expression of TRP channels have been linked to cancer growth and development. Indeed, they can be useful as diagnostic and/or predictive markers for a variety of tumor types, including glioma (Chinigò et al. (2021) [282]). Changes in calcium homeostasis due to the abnormal function of these channels in tumors are associated with uncontrolled proliferation and resistance to cell death (Huang et al. (2021a) [78]). Among the TRP channels, TRPV4 (transient receptor potential vanilloid 4) expression is considerably higher in malignant glioma compared to both normal brain tissue and low-grade glioma. Furthermore, there is a negative correlation between the expression of TRPV4 and the prognosis of glioma patients (Yang et al. (2020) [77]). All of this suggests that TRPV4 could be an attractive therapeutic target and biomarker for GBM. Notably, a team of researchers investigated the impact of cannabidiol (CBD) on glioma and showed that TRPV4-mediated calcium influx triggered mitophagy, leading to the hypothesis that this might be the primary cause of glioma cell death in response to CBD treatment. Furthermore, the same group provided in vitro and in vivo evidence that a combination of temozolomide (TMZ) and CBD resulted in a powerful antitumoral effect (Huang et al. (2021a) [78]).

4.2.2. Piezo Channels

Other ion channels important in GBM "biology" are the mechanosensitive Piezo ion channels. The stretch-induced membrane tension caused by cell swelling opens the Piezo channels, which then allow the permeation of calcium, potassium, and sodium (Lewis and Grandl (2015) [163]). Piezo1 specifically is overexpressed in aggressive human gliomas, and its expression is inversely correlated with the patient's prognosis. The mechanical microenvironment provided by ECM stiffening in tumor tissue activates Piezo1, whose activity promotes focal adhesion and activates the integrin/FAK signaling pathway. Piezo1-mediated signaling also regulates cell proliferation and controls the expression of genes involved in ECM remodeling, which further modulates the tissue stiffness. As a result, the harsher environment increases the expression of Piezo1, which enhances the mechanosensory capacity of cancer cells and promotes glioma aggressiveness. These processes create a feed-forward circuit between Piezo1-dependent mechano-transduction and abnormal tissue mechanics in gliomas, exacerbating the disease (Chen et al. (2018) [79]).

4.3. *Anion Channels*

One channel that is highly expressed in GBM cells is the anion channel VRAC, which mediates the flow of chloride triggered by cell swelling. The primary function of VRAC is to restore a normal cell volume, which may be altered due to various pathological conditions. Additionally, VRAC promotes cell shape and volume changes that are required for cell proliferation and migration (Caramia et al. (2019) [63]). In GBM, VRAC arguably plays a major role. This channel can be activated in vivo by severe hypoxia, a condition that occurs in GBM tissues. In this state, following mechanical stress of the plasma membrane, it has been observed that VRAC activation promotes cell migration and resistance to cell death, both of which increase the severity of GBM (Caramia et al. (2019) [63]). Also, another study revealed that lowering the expression of LRRC8A leads to a decrease in the growth of GBM cells and enhances their sensitivity to the chemotherapeutic drugs TMZ and carmustine, which are commonly used in clinical settings (Rubino et al. (2018) [283]).

Conversely, a different investigation showed that pharmacological VRAC inhibition and LRRC8A knockdown do not have any impact on GBM cell proliferation or migration, suggesting that VRAC may not be necessary for tumor development (Liu and Stauber (2019) [64]). Further investigation is needed to elucidate the roles that this ion channel may play in cell migration and invasion.

Moreover, VRAC can transport anticancer drugs like cisplatin and carboplatin into the cell, and the selectivity of the substrate, as well as the pharmacology of VRAC, depends on its subunit composition. Notably, one of the subunits of VRAC, the LRRC8D subunit, plays an important pharmacological role, supporting the transport of these anticancer drugs (Planells-Cases et al. (2015) [65]).

4.4. *Ligand Gated Ion Channels*

4.4.1. $GABA_A$ Receptors

In GBM cells, all of the different isoforms of $GABA_A$ receptor subunits are present and generally down-regulated compared to grade II and III gliomas, except for the up-regulation of the δ subunit (Smits et al. (2012) [83]). $GABA_AR$ expression in GBM cells is triggered by contacts with neurons (Synowitz et al. (2001) [284]), fostering the hypothesis of communication between the two cell types. Interestingly, in the neurons surrounding the tumor, a loss of expression of the $\alpha 1$ subunit, leading to a reduction of GABAergic input, was reported (Tantillo et al. (2023) [84]). In the peritumoral area, the expression of specific potassium chloride co-transporters (NKCC1 and KCC2) is dysregulated. NKCC1 and KCC2 levels are increased and decreased, respectively, which results in a high intracellular chloride concentration that, when $GABA_ARs$ are activated, induces functional excitation (Pallud et al. (2014) [285];

Campbell et al. (2015) [286]). Similarly, GBM cells could be depolarized by high concentrations of GABA present in the TME (Blanchart et al. (2017) [85]), and the increased intracellular calcium may activate several pathways, also involving fibroblast growth factor, which reduces tumor proliferation and growth (Babateen et al. (2015) [287]; Huberfeld and Vecht (2016) [288]; Blanchart et al. (2017) [85]). GABA does not affect the initial phase of tumorigenesis but it limits growth and disease progression (Blanchart et al. (2017) [85]). It has also been shown that GABA activity promotes tumor cell quiescence, which ends when $GABA_AR$ is blocked. Since GBM produces high levels of GABA, lowering overall GABA levels by tumor resection can possibly result in rapid recurrence by residual GBM cells (Blanchart et al. (2017) [85]). The importance of GABA signaling is supported by the fact that the loss of functional $GABA_AR$ correlates with the grade of the glioma and the clinical outcome (Smits et al. (2012) [83]).

Since GABA signaling limits tumor growth, it is conceivable that enhancing $GABA_AR$ function may represent a valuable approach to GBM therapy. Exogenous (benzodiazepines) and endogenous (neurosteroids, NS) substances are able to positively modulate $GABA_AR$ function (Puia et al. (2012) [289]). It was shown that progesterone and its 5α-reduced metabolite allopregnanolone (Allo) favor the progression of GBM cell lines at nanomolar concentrations (Zamora-Sánchez et al. (2017) [88]) while having opposite effects at high micromolar concentrations (Zamora-Sánchez et al. (2022) [87]; Feng et al. (2022) [89]). Because GBM cells express steroidogenic enzymes and produce NS (Zamora-Sánchez et al. (2017) [88]), it is possible that these endogenous substances, depending on their metabolism, play significant roles in the maintenance and progression of GBM (Pinacho-Garcia et al. (2019) [290]). Indeed, Allo has a therapeutic effect on GBM cell lines, enhancing TMZ inhibition of cell migration and TMZ-induced apoptosis (Feng et al. (2022) [89]).

On the other hand, we should keep in mind that GABA, by acting as an excitatory neurotransmitter, may bolster the oncogenic effects of the neuronal hyperactivity previously detailed (Venkatesh et al. (2015) [274]). In fact, it has been observed that while GABA appears to inhibit GBM progression, it also stimulates peritumoral neurons to release glutamate, leading to epilepsy (Radin and Tsirka (2020) [86]).

4.4.2. Glutamate Ionotropic Receptors

The GBM microenvironment is characterized by increased glutamatergic signaling, which fuels tumor progression and induces hyperexcitability of the peritumoral neurons. This heightened activity, in turn, may contribute to the genesis of seizures or epilepsy and to widespread neurodegeneration (Venkataramani et al. (2019) [95]; Jung et al. (2020) [96]). GBM cells express iGluRs such as NMDA (Nandakumar et al. (2019) [98]), KAR and AMPAR (in parts, Ca^{2+} permeable AMPAR) (Maas et al. (2001) [291]; Venkataramani et al. (2019) [95]; Venkatesh et al. (2019) [99]). GBM iGluRs may be stimulated either via neuron–tumor synapses or by glutamate released in an autocrine fashion by tumor cells through the cystine–glutamate antiporter (xCT) (Takano et al. (2001) [292]) or by neighboring reactive astrocytes (Sin et al. (2013) [107]).

The importance of neuronal activity for GBM progression is widely accepted (Corsi et al. (2019) [293]). Specifically, neurons and GBM cells communicate directly through bona fide synapses and also indirectly via paracrine signals such as NLGN3 or BDNF (Venkatesh et al. (2015, 2017) [274,294]). Aside from the direct unidirectional synaptic contacts between neurons and tumor cells, in GBM, there are indirect presynaptic contacts resembling the tripartite synapses between neurons and astrocytes. Glutamatergic signals may induce calcium transients through calcium-permeable AMPARs, activating downstream Akt-PKB pathways that promote GBM growth and invasion (Ishiuchi et al. (2002) [295]; Piao et al. (200) [296]; Venkatesh et al. (2019) [99]). NMDARs are highly expressed in GBM cells, and their activation also contributes to increased intracellular Ca^{2+} levels (Nandakumar et al. (2019) [98]) and the phosphorylation of transcription factors, contributing to GBM survival and migration (Längle et al. 2019 [297]).

A cross-talk between these calcium-permeable receptors (NMDAR and AMPAR) in the glutamate-rich microenvironment of GBM probably has an important role in tumor progression (Nandakumar et al. (2019) [98]). Both AMPAR and NMDAR activation influence extracellular matrix proteins (Piao et al. (2009) [296]; Ramaswamy et al. (2014) [298]).

Finally, the involvement of KAR in GBM has been poorly investigated (Lange et al. (2021) [299]). High expression of GluK4 was found in GBM cell lines (Stepulak et al. (2009) [300]), but more studies are needed to shed light on the role of KARs in GBM progression and glioma-associated epilepsy since glutamate released from GBM cells could activate KARs along with AMPARs and NMDARs (Lyons et al. (2007) [97]).

Since GBM, like several cancers of different tissues, relies on glutamate signaling to survive and proliferate (Yu et al. (2017) [301]), pharmacological targeting of iGluRs or the systems responsible for glutamate release may slow the disease progression while simultaneously limiting neurodegeneration and seizure onset. Considering the interdependency between neuronal hyperexcitability and GBM progression, a GBM therapy with iGluRs antagonists acting on both sides (tumor cells and neurons) is particularly suitable. Indeed, an antiepileptic drug with AMPAR antagonistic properties has been successfully employed in clinical trials for GBM in association with radiochemotherapy (Grossman et al. (2009) [100]; Salmaggi et al. (2021) [302]). Because of the biological relevance of calcium-permeable AMPARs in tumor progression, specific inhibitors of these receptors could represent a promising therapeutic strategy for GBM (Venkataramani et al. (2021) [105]).

NMDAR antagonists, such as memantine, are useful in the management of GBM (Albayrak et al. (2021) [103]), especially considering their neuroprotective activity. For instance, MP1 and MP2 are two NMDA antagonists derived from memantine that increase autophagy in the human U87MG glioblastoma cell line and reduce its proliferative activity (Cacciatore et al. (2017) [101]). Of note, Blyufer et al. highlighted the potential of riluzole, a drug that inhibits glutamate release, as a potential therapeutic tool for GBM (Yamada et al. (2020) [102]; Blyufer et al. (2021) [104]).

4.4.3. Nicotinic ACh Receptors

GBM cells are capable of synthesizing and releasing Ach, which, acting in an autocrine or paracrine fashion, affects proliferation, survival, and tumor invasion (Thompson and Sontheimer (2019) [90]). The expression of most nAchR subunits in GBM is very low, except for the muscle-type $\alpha1$ and $\beta1$ subunits and the neuronal $\alpha7$ and $\alpha1$ subunits (Thompson and Sontheimer (2019) [90]; Pucci et al. (2021, 2022) [91,92]).

AChR activation increases cell invasion by enhancing the activity of matrix metalloproteinase-9 (MMP-9) through a Ca^{2+}-dependent mechanism (Thompson and Sontheimer (2019) [90]). A recent study (Pucci et al. (2021) [92]) found that $\alpha9$ and $\alpha5$ subunit mRNAs are highly expressed in GBM and that chronic treatment with selective agonists increases the proliferation of tumor cell lines (U87MG and GBM5). By silencing the expression of the $\alpha7$ and $\alpha9$ subunits or by applying selective $\alpha7$ and $\alpha9$ AChR antagonists, this effect was prevented, thus indicating that the presence of both subunits was required (Pucci et al. (2021, 2022) [91,92]). Stimulation of $\alpha9$ and $\alpha7$ nAChRs can activate various intracellular signaling pathways and regulate gene expression through non-conventional metabotropic channel signaling. In GBM cell lines, the activation of $\alpha7$nAChRs and $\alpha9$nAChRs has been shown to inhibit cell apoptosis via the EGFR/Akt pathway and the promotion of cell proliferation via the EGFR/ERK pathway (Khalil et al. (2013) [303]; Pucci et al. 2022 (2022) [91]).

Considering the importance of nAChRs in GBM progression in the last few years, non-selective and selective antagonists of nAchR have been developed and tested on GBM cell lines. For example, Atracurium Besylate (an nAChR antagonist) (Spina et al. (2016) [93]) and StN2/4/8 ($\alpha7$ and $\alpha9/\alpha10$ nAChR antagonists) (Pucci et al. (2022) [91]; Bavo et al. (2023) [94]) have been shown to have significant inhibitory effects on GBM cell line vitality.

4.5. Other Channels

4.5.1. Aquaporins

AQPs play a role in enhancing tumor mobility through several mechanisms: cell volume regulation, cell-cell and cell-matrix adhesion, actin cytoskeleton interaction, control of enzymes and molecules involved in extracellular matrix degradation, and ion channel and transporter co-localization. In the case of GBM, the overexpression of AQP-1, -4, and -9 is consistent with their suggested functions in facilitating the mobility, growth, and survival of glioma cells (Varricchio et al. (2021) [80]). AQP4, the primary aquaporin expressed in the CNS, plays a crucial role in maintaining water and potassium homeostasis. In glioblastomas and other types of astrocytomas, AQP4 expression is elevated and redistributed. In fact, in high-grade brain tumors, AQP4 shows a loss of its polarized distribution at the tips of astroglial pedicels, which can contribute to increased tumor cell migration (Vandebroek and Yasui (2020) [81]). In this regard, a research group suggested that high concentrations of T3 thyroid hormone could be effective in decreasing AQP4 in GBM tumor cells, leading to improved outcomes in terms of reducing the migration and growth of brain tumor cells (Costa et al. (2019) [82]).

4.5.2. Connexin 43

Connexin 43 is an important building block of GJs and hemichannels, and its expression, which is very high in astrocytes, is reduced when they undergo malignant transformation (Caltabiano et al. (2010) [106]). Indeed, the amount of Cx43 protein inversely correlates with the level of malignancy of astrocytomas, and its presence is low in GBM (Sin et al. (2016) [107]; Gielen et al. (2013) [112]). Hence, this relationship between Cx43 expression and glioma severity suggests that Cx43 should have a tumor suppressor ability, restraining the proliferation of glioma cells.

On the other hand, several pieces of evidence have pointed out the capability of Cx43 to enhance the mobility and invasiveness of GBM cells, facilitating their migration from the tumor core into the surrounding tissues (Sin et al. (2016) [107]; Dong et al. (2017) [110]). The ambivalence of Cx43 effects, ranging from tumor suppressor to cell migration booster, can be partly explained by the heterogeneous expression of Cx43 in the GBM mass. In fact, the cells in the core with low Cx43 expression proliferate, while those with high Cx43 levels are expected to migrate (Sin et al. (2016) [107]; McCutcheon and Spray (2022) [109]). The presence of GJs between GBM cells might inhibit migration, whereas GJs between GBM cells and astrocytes, or between astrocytes themselves, promote tumor progression. Indeed, Cx43 expression is enhanced significantly in the glioma-associated astrocytes of the peritumoral zone that have a decisive role in granting the dissemination of tumoral cells (Sin et al. (2016) [107]). Cx43 can foster invasiveness by activating different pathways (i.e., by interacting with different cytoskeletal proteins) and also by promoting the transfer of oncogenic signaling molecules from tumor cells to neighboring astrocytes or among astrocytes (Sin et al. (2016) [107]; Uzu et al. (2018) [108]; McCutcheon and Spray (2022) [109]).

Cx43 is also involved in tumor progression because it increases GBM resistance to apoptosis and has an impact on cell homeostasis through paracrine hemichannels activity (Sin et al. (2016) [107]).

Moreover, Cx43 and GJ/hemichannels may contribute to glioma-associated epileptic activity in the peritumoral zone induced by changes in the tumor microenvironment. In particular, Cx43 may have a role in the regulation of neurotransmitters such as glutamate and ATP (Dong et al. (2017) [110]).

Furthermore, the expression of Cx43 in GBM cells is associated with the development of resistance to TMZ treatment (Gielen et al. (2013) [112]). It has been shown that Cx43-dependent resistance involves both GJ-dependent and GJ-independent mechanisms that influence the invasion and migration of tumor cells. In response to TMZ treatment, Cx43 modulates mitochondrial apoptotic pathways by regulating Bax and Bcl-2 levels and by influencing the release of cytochrome C from mitochondria (Gielen et al. (2013) [112]; Dong et al. (2017) [110]; Xing et al. 2019 [111]).

Considering the different roles played by Cx43 in GBM progression, it will be extremely important in the development of novel therapeutic strategies for GBM treatment (Xing et al. 2019 [111]) to selectively target Cx43 of specific cell types (tumoral or non tumoral), carefully considering the stage of glioma genesis.

In conclusion, the published research in the field converges toward a complex view of "GBM biology." The influence of ion channels and ionotropic receptors on tumor development, invasiveness, and resistance to therapies is widely recognized, but on the other hand, is extremely complex. The channels/receptors that are differently expressed in GBM and the results of these changes are a composite (see Table 2 and Figure 2) because they affect not only the single tumoral cell but also the multicellular network of the TM that drives tumor growth and resistance to therapy. Since a major reason for the incurability of gliomas by local treatment is the diffuse infiltration of the brain, determined also by changes in channels/receptors, they are important pharmacological targets for new GBM drugs and as powerful prognostic biomarkers. The development of new therapeutic approaches that target ion channels should always consider not only the effect on single cellular entities but also on the whole "GBM network".

5. Conclusions and Perspectives

When considering the poor results obtained by neurocentric drug research for different NDs such as AD, fifteen years ago, the great scientist Ben Barres made the following statement: "Quite possibly saving astrocytes from dying in neurological disease would be a far more effective strategy than trying to save neurons, glia already know how to save neurons, whereas neuroscientists still have no clue" (Barres (2008) [304]). Scientific advancements in the last fifteen years confirmed this assertion. Indeed, it is now clear that astrocytes contribute to brain disorders with several intermingled mechanisms that depend on different factors such as subcellular localization, disease stage, comorbidity, and others. Technical advances over the last years helped to reveal the morphological, functional, and molecular complexity of glial cells (Endo et al. (2022) [305]) and their plethora of functions, including modulation of behavior and cognition in health and disease. Multiphoton or supersolution microscopy, genetic tools, and transcriptomics allow for a deeper understanding of the molecular fingerprint of reactive astrocytes in brain diseases. In addition, astrocyte small processes, which are in strict contact with synapses, have emerged as the most interesting and yet elusive cellular district.

Finally, human astrocytes are even more complex and heterogeneous than in rodents (Oberheim et al. (2009) [306]; Vasile et al. (2017) [307]), suggesting their great relevance in human brain physiology and pathology. Accordingly, expanding our knowledge on astrocyte mechanisms at different stages of brain disorders is a required step to design new preventive or therapeutic treatments oreand finally prove that acting on glia is an efficient strategy to save brain functions.

Author Contributions: A.L. and A.D.S. wrote the section dedicated to AD; L.V. and G.P. wrote the section of GBM; A.L., M.T., G.P. and G.L. wrote the abstract, introduction and conclusions. G.L. and A.L. wrote the section on physiology. All authors have read and agreed to the published version of the manuscript.

Funding: This research was funded by Italian Ministry of University and Research (MUR), grant PRIN2020 n.2020AMLXHH to G.L.

Institutional Review Board Statement: Not applicable.

Informed Consent Statement: Not applicable.

Data Availability Statement: Not applicable.

Acknowledgments: We thank Giorgio Carmignoto for critical reading of the manuscript and helpful discussion and Isabella Martusciello for proofreading the text.

Conflicts of Interest: The authors declare no conflict of interest.

References

1. Verkhratsky, A.; Nedergaard, M. Physiology of Astroglia. *Physiol. Rev.* **2018**, *98*, 239–389. [CrossRef] [PubMed]
2. Khakh, B.S.; Deneen, B. The Emerging Nature of Astrocyte Diversity. *Annu. Rev. Neurosci.* **2019**, *42*, 187–207. [CrossRef] [PubMed]
3. Khakh, B.S.; Sofroniew, M.V. Diversity of astrocyte functions and phenotypes in neural circuits. *Nat. Neurosci.* **2015**, *18*, 942–952. [CrossRef]
4. Allen, N.J.; Eroglu, C. Cell Biology of Astrocyte-Synapse Interactions. *Neuron* **2017**, *96*, 697–708. [CrossRef] [PubMed]
5. Allaman, I.; Bélanger, M.; Magistretti, P.J. Astrocyte–neuron metabolic relationships: For better and for worse. *Trends Neurosci.* **2011**, *34*, 76–87. [CrossRef] [PubMed]
6. Allen, N.J.; Lyons, D.A. Glia as architects of central nervous system formation and function. *Science* **2018**, *362*, 181–185. [CrossRef] [PubMed]
7. Lia, A.; Di Spiezio, A.; Speggiorin, M.; Zonta, M. Two decades of astrocytes in neurovascular coupling. *Front. Netw. Physiol.* **2023**, *3*, 1162757. [CrossRef]
8. Semyanov, A.; Verkhratsky, A. Astrocytic processes: From tripartite synapses to the active milieu. *Trends Neurosci.* **2021**, *44*, 781–792. [CrossRef]
9. Pekny, M.; Pekna, M.; Messing, A.; Steinhäuser, C.; Lee, J.-M.; Parpura, V.; Hol, E.M.; Sofroniew, M.V.; Verkhratsky, A. Astrocytes: A central element in neurological diseases. *Acta Neuropathol.* **2016**, *131*, 323–345. [CrossRef]
10. Liddelow, S.A.; Barres, B.A. Reactive Astrocytes: Production, Function, and Therapeutic Potential. *Immunity* **2017**, *46*, 957–967. [CrossRef]
11. Sofroniew, M.V. Astrocyte Reactivity: Subtypes, States, and Functions in CNS Innate Immunity. *Trends Immunol.* **2020**, *41*, 758–770. [CrossRef] [PubMed]
12. Escartin, C.; Galea, E.; Lakatos, A.; O'callaghan, J.P.; Petzold, G.C.; Serrano-Pozo, A.; Steinhäuser, C.; Volterra, A.; Carmignoto, G.; Agarwal, A.; et al. Reactive astrocyte nomenclature, definitions, and future directions. *Nat. Neurosci.* **2021**, *24*, 312–325. [CrossRef] [PubMed]
13. Guttenplan, K.A.; Weigel, M.K.; Prakash, P.; Wijewardhane, P.R.; Hasel, P.; Rufen-Blanchette, U.; Münch, A.E.; Blum, J.A.; Fine, J.; Neal, M.C.; et al. Neurotoxic reactive astrocytes induce cell death via saturated lipids. *Nature* **2021**, *599*, 102–107. [CrossRef]
14. Brandebura, A.N.; Paumier, A.; Onur, T.S.; Allen, N.J. Astrocyte contribution to dysfunction, risk and progression in neurodegenerative disorders. *Nat. Rev. Neurosci.* **2023**, *24*, 23–39. [CrossRef] [PubMed]
15. Patani, R.; Hardingham, G.E.; Liddelow, S.A. Functional roles of reactive astrocytes in neuroinflammation and neurodegeneration. *Nat. Rev. Neurol.* **2023**, *19*, 395–409. [CrossRef]
16. Brandao, M.; Simon, T.; Critchley, G.; Giamas, G. Astrocytes, the rising stars of the glioblastoma microenvironment. *Glia* **2019**, *67*, 779–790. [CrossRef] [PubMed]
17. Yang, Y.; Schubert, M.C.; Kuner, T.; Wick, W.; Winkler, F.; Venkataramani, V. Brain Tumor Networks in Diffuse Glioma. *Neurotherapeutics* **2022**, *19*, 1832–1843. [CrossRef]
18. Kuffler, S.W.; Nicholls, J.G.; Orkand, R.K.; Sawyer, J.E.R.; Hennebry, J.E.; Revill, A.; Brown, A.M.; Du, Y.; Ma, B.; Kiyoshi, C.M.; et al. Physiological properties of glial cells in the central nervous system of amphibia. *J. Neurophysiol.* **1966**, *29*, 768–787. [CrossRef]
19. Orkand, R.K.; Nicholls, J.G.; Kuffler, S.W.; Verkhratsky, A.; Nedergaard, M.; Sawyer, J.E.R.; Hennebry, J.E.; Revill, A.; Brown, A.M.; Wanke, E.; et al. Effect of nerve impulses on the membrane potential of glial cells in the central nervous system of amphibia. *J. Neurophysiol.* **1966**, *29*, 788–806. [CrossRef]
20. Kuffler, S.W. The Ferrier Lecture–Neuroglial cells: Physiological properties and a potassium mediated effect of neuronal activity on the glial membrane potential. *Proc. R. Soc. Lond. Ser. B Biol. Sci.* **1967**, *168*, 1–21. [CrossRef]
21. Araque, A.; Carmignoto, G.; Haydon, P.G.; Oliet, S.H.; Robitaille, R.; Volterra, A. Gliotransmitters Travel in Time and Space. *Neuron* **2014**, *81*, 728–739. [CrossRef] [PubMed]
22. Bazargani, N.; Attwell, D. Astrocyte calcium signaling: The third wave. *Nat. Neurosci.* **2016**, *19*, 182–189. [CrossRef] [PubMed]
23. Santello, M.; Toni, N.; Volterra, A. Astrocyte function from information processing to cognition and cognitive impairment. *Nat. Neurosci.* **2019**, *22*, 154–166. [CrossRef] [PubMed]
24. Lyon, K.A.; Allen, N.J. From Synapses to Circuits, Astrocytes Regulate Behavior. *Front. Neural Circuits* **2022**, *15*, 786293. [CrossRef] [PubMed]
25. Kofuji, P.; Araque, A. Astrocytes and Behavior. *Annu. Rev. Neurosci.* **2021**, *44*, 49–67. [CrossRef]
26. Ventura, R.; Harris, K.M. Three-Dimensional Relationships between Hippocampal Synapses and Astrocytes. *J. Neurosci.* **1999**, *19*, 6897–6906. [CrossRef]
27. Bushong, E.A.; Martone, M.E.; Jones, Y.Z.; Ellisman, M.H. Protoplasmic Astrocytes in CA1 Stratum Radiatum Occupy Separate Anatomical Domains. *J. Neurosci.* **2002**, *22*, 183–192. [CrossRef]
28. Lia, A.; Henriques, V.J.; Zonta, M.; Chiavegato, A.; Carmignoto, G.; Gómez-Gonzalo, M.; Losi, G. Calcium Signals in Astrocyte Microdomains, a Decade of Great Advances. *Front. Cell. Neurosci.* **2021**, *15*, 673433. [CrossRef]
29. Ahmadpour, N.; Kantroo, M.; Stobart, J.L. Extracellular Calcium Influx Pathways in Astrocyte Calcium Microdomain Physiology. *Biomolecules* **2021**, *11*, 1467. [CrossRef]
30. Armbruster, M.; Naskar, S.; Garcia, J.P.; Sommer, M.; Kim, E.; Adam, Y.; Haydon, P.G.; Boyden, E.S.; Cohen, A.E.; Dulla, C.G. Neuronal activity drives pathway-specific depolarization of peripheral astrocyte processes. *Nat. Neurosci.* **2022**, *25*, 607–616. [CrossRef]
31. McNeill, J.; Rudyk, C.; Hildebrand, M.E.; Salmaso, N. Ion Channels and Electrophysiological Properties of Astrocytes: Implications for Emergent Stimulation Technologies. *Front. Cell. Neurosci.* **2021**, *15*, 644126. [CrossRef] [PubMed]

32. Osborn, L.M.; Kamphuis, W.; Wadman, W.J.; Hol, E.M. Astrogliosis: An integral player in the pathogenesis of Alzheimer's disease. *Prog. Neurobiol.* **2016**, *144*, 121–141. [CrossRef] [PubMed]
33. Losi, G.; Cammarota, M.; Carmignoto, G. The Role of Astroglia in the Epileptic Brain. *Front. Pharmacol.* **2012**, *3*, 132. [CrossRef] [PubMed]
34. Khakh, B.S.; Goldman, S.A. Astrocytic contributions to Huntington's disease pathophysiology. *Ann. N. Y. Acad. Sci.* **2023**, *1522*, 42–59. [CrossRef] [PubMed]
35. Lee, K.-I.; Lee, H.-T.; Lin, H.-C.; Tsay, H.-J.; Tsai, F.-C.; Shyue, S.-K.; Lee, T.-S. Role of transient receptor potential ankyrin 1 channels in Alzheimer's disease. *J. Neuroinflamm.* **2016**, *13*, 92. [CrossRef]
36. Paumier, A.; Boisseau, S.; Jacquier-Sarlin, M.; Pernet-Gallay, K.; Buisson, A.; Albrieux, M. Astrocyte–neuron interplay is critical for Alzheimer's disease pathogenesis and is rescued by TRPA1 channel blockade. *Brain* **2022**, *145*, 388–405. [CrossRef]
37. Hu, J.; Chen, Q.; Zhu, H.; Hou, L.; Liu, W.; Yang, Q.; Shen, H.; Chai, G.; Zhang, B.; Chen, S.; et al. Microglial Piezo1 senses Aβ fibril stiffness to restrict Alzheimer's disease. *Neuron* **2023**, *111*, 15–29.e8. [CrossRef]
38. Xu, Z.; Xiao, N.; Chen, Y.; Huang, H.; Marshall, C.; Gao, J.; Cai, Z.; Wu, T.; Hu, G.; Xiao, M. Deletion of aquaporin-4 in APP/PS1 mice exacerbates brain Aβ accumulation and memory deficits. *Mol. Neurodegener.* **2015**, *10*, 58. [CrossRef]
39. Reeves, B.C.; Karimy, J.K.; Kundishora, A.J.; Mestre, H.; Cerci, H.M.; Matouk, C.; Alper, S.L.; Lundgaard, I.; Nedergaard, M.; Kahle, K.T. Glymphatic System Impairment in Alzheimer's Disease and Idiopathic Normal Pressure Hydrocephalus. *Trends Mol. Med.* **2020**, *26*, 285–295. [CrossRef]
40. Nwaobi, S.E.; Cuddapah, V.A.; Patterson, K.C.; Randolph, A.C.; Olsen, M.L. The role of glial-specific Kir4.1 in normal and pathological states of the CNS. *Acta Neuropathol.* **2016**, *132*, 1–21. [CrossRef]
41. Wilcock, D.; Vitek, M.; Colton, C. Vascular amyloid alters astrocytic water and potassium channels in mouse models and humans with Alzheimer's disease. *Neuroscience* **2009**, *159*, 1055–1069. [CrossRef] [PubMed]
42. Huffels, C.F.M.; Osborn, L.M.; Hulshof, L.A.; Kooijman, L.; Henning, L.; Steinhäuser, C.; Hol, E.M. Amyloid-β plaques affect astrocyte Kir4.1 protein expression but not function in the dentate gyrus of APP/PS1 mice. *Glia* **2022**, *70*, 748–767. [CrossRef] [PubMed]
43. Griffith, C.M.; Xie, M.-X.; Qiu, W.-Y.; Sharp, A.A.; Ma, C.; Pan, A.; Yan, X.-X.; Patrylo, P.R. Aberrant expression of the pore-forming KATP channel subunit Kir6.2 in hippocampal reactive astrocytes in the 3xTg-AD mouse model and human Alzheimer's disease. *Neuroscience* **2016**, *336*, 81–101. [CrossRef] [PubMed]
44. Liu, D.; Pitta, M.; Lee, J.-H.; Ray, B.; Lahiri, D.K.; Furukawa, K.; Mughal, M.; Jiang, H.; Villarreal, J.; Cutler, R.G.; et al. The KATP Channel Activator Diazoxide Ameliorates Amyloid-β and Tau Pathologies and Improves Memory in the 3xTgAD Mouse Model of Alzheimer's Disease. *J. Alzheimer's Dis.* **2010**, *22*, 443–457. [CrossRef]
45. Macauley, S.L.; Stanley, M.; Caesar, E.E.; Yamada, S.A.; Raichle, M.E.; Perez, R.; Mahan, T.E.; Sutphen, C.L.; Holtzman, D.M. Hyperglycemia modulates extracellular amyloid-β concentrations and neuronal activity in vivo. *J. Clin. Investig.* **2015**, *125*, 2463–2467. [CrossRef]
46. Angulo, E.; Noe, V.; Casado, V.; Mallol, J.; Gomez-Isla, T.; Lluis, C.; Ferrer, I.; Ciudad, C.J.; Franco, R. Up-regulation of the Kv3.4 potassium channel subunit in early stages of Alzheimer's disease. *J. Neurochem.* **2004**, *91*, 547–557. [CrossRef]
47. Pannaccione, A.; Boscia, F.; Scorziello, A.; Adornetto, A.; Castaldo, P.; Sirabella, R.; Taglialatela, M.; Di Renzo, G.F.; Annunziato, L. Up-Regulation and Increased Activity of $K_V3.4$ Channels and Their Accessory Subunit MinK-Related Peptide 2 Induced by Amyloid Peptide Are Involved in Apoptotic Neuronal Death. *Mol. Pharmacol.* **2007**, *72*, 665–673. [CrossRef]
48. Boscia, F.; Pannaccione, A.; Ciccone, R.; Casamassa, A.; Franco, C.; Piccialli, I.; de Rosa, V.; Vinciguerra, A.; Di Renzo, G.; Annunziato, L. The expression and activity of K V 3.4 channel subunits are precociously upregulated in astrocytes exposed to Aβ oligomers and in astrocytes of Alzheimer's disease Tg2576 mice. *Neurobiol. Aging* **2017**, *54*, 187–198. [CrossRef]
49. Yi, M.; Yu, P.; Lu, Q.; Geller, H.M.; Yu, Z.; Chen, H. KCa3.1 constitutes a pharmacological target for astrogliosis associated with Alzheimer's disease. *Mol. Cell. Neurosci.* **2016**, *76*, 21–32. [CrossRef]
50. Wei, T.; Yi, M.; Gu, W.; Hou, L.; Lu, Q.; Yu, Z.; Chen, H. The Potassium Channel KCa3.1 Represents a Valid Pharmacological Target for Astrogliosis-Induced Neuronal Impairment in a Mouse Model of Alzheimer's Disease. *Front. Pharmacol.* **2016**, *7*, 528. [CrossRef]
51. Yu, Z.; Dou, F.; Wang, Y.; Hou, L.; Chen, H. Ca^{2+}-dependent endoplasmic reticulum stress correlation with astrogliosis involves upregulation of KCa3.1 and inhibition of AKT/mTOR signaling. *J. Neuroinflamm.* **2018**, *15*, 316. [CrossRef] [PubMed]
52. Jo, S.; Yarishkin, O.; Hwang, Y.J.; Chun, Y.E.; Park, M.; Woo, D.H.; Bae, J.Y.; Kim, T.; Lee, J.; Chun, H.; et al. GABA from reactive astrocytes impairs memory in mouse models of Alzheimer's disease. *Nat. Med.* **2014**, *20*, 886–896. [CrossRef] [PubMed]
53. Teaktong, T.; Graham, A.; Court, J.; Perry, R.; Jaros, E.; Johnson, M.; Hall, R.; Perry, E. Alzheimer's disease is associated with a selective increase in ?7 nicotinic acetylcholine receptor immunoreactivity in astrocytes. *Glia* **2003**, *41*, 207–211. [CrossRef] [PubMed]
54. Yu, W.-F.; Guan, Z.-Z.; Bogdanovic, N.; Nordberg, A. High selective expression of α7 nicotinic receptors on astrocytes in the brains of patients with sporadic Alzheimer's disease and patients carrying Swedish APP 670/671 mutation: A possible association with neuritic plaques. *Exp. Neurol.* **2005**, *192*, 215–225. [CrossRef] [PubMed]
55. Wang, H.-Y.; Lee, D.H.S.; Davis, C.B.; Shank, R.P. Amyloid Peptide Aβ1-42 Binds Selectively and with Picomolar Affinity to α7 Nicotinic Acetylcholine Receptors. *J. Neurochem.* **2002**, *75*, 1155–1161. [CrossRef] [PubMed]
56. Pirttimaki, T.M.; Codadu, N.K.; Awni, A.; Pratik, P.; Nagel, D.A.; Hill, E.J.; Dineley, K.T.; Parri, H.R. α7 Nicotinic Receptor-Mediated Astrocytic Gliotransmitter Release: Aβ Effects in a Preclinical Alzheimer's Mouse Model. *PLoS ONE* **2013**, *8*, e81828. [CrossRef]

57. Gulisano, W.; Melone, M.; Ripoli, C.; Tropea, M.R.; Puma, D.D.L.; Giunta, S.; Cocco, S.; Marcotulli, D.; Origlia, N.; Palmeri, A.; et al. Neuromodulatory Action of Picomolar Extracellular Aβ42 Oligomers on Presynaptic and Postsynaptic Mechanisms Underlying Synaptic Function and Memory. *J. Neurosci.* **2019**, *39*, 5986–6000. [CrossRef]
58. Tropea, M.R.; Puma, D.D.L.; Melone, M.; Gulisano, W.; Arancio, O.; Grassi, C.; Conti, F.; Puzzo, D. Genetic deletion of α7 nicotinic acetylcholine receptors induces an age-dependent Alzheimer's disease-like pathology. *Prog. Neurobiol.* **2021**, *206*, 102154. [CrossRef]
59. McLarnon, J.G.; Ryu, M.J.K.; Walker, D.G.; Choi, B.H.B. Upregulated Expression of Purinergic $P2X_7$Receptor in Alzheimer Disease and Amyloid-β Peptide-Treated Microglia and in Peptide-Injected Rat Hippocampus. *J. Neuropathol. Exp. Neurol.* **2006**, *65*, 1090–1097. [CrossRef]
60. Martínez-Frailes, C.; Di Lauro, C.; Bianchi, C.; de Diego-García, L.; Sebastián-Serrano, A.; Boscá, L.; Díaz-Hernández, M. Amyloid Peptide Induced Neuroinflammation Increases the P2X7 Receptor Expression in Microglial Cells, Impacting on Its Functionality. *Front. Cell. Neurosci.* **2019**, *13*, 143. [CrossRef]
61. Jin, H.; Han, J.; Resing, D.; Liu, H.; Yue, X.; Miller, R.L.; Schoch, K.M.; Miller, T.M.; Perlmutter, J.S.; Egan, T.M.; et al. Synthesis and in vitro characterization of a P2X7 radioligand [123I]TZ6019 and its response to neuroinflammation in a mouse model of Alzheimer disease. *Eur. J. Pharmacol.* **2018**, *820*, 8–17. [CrossRef] [PubMed]
62. Martin, E.; Amar, M.; Dalle, C.; Youssef, I.; Boucher, C.; Le Duigou, C.; Brückner, M.; Prigent, A.; Sazdovitch, V.; Halle, A.; et al. New role of P2X7 receptor in an Alzheimer's disease mouse model. *Mol. Psychiatry* **2018**, *24*, 108–125. [CrossRef] [PubMed]
63. Caramia, M.; Sforna, L.; Franciolini, F.; Catacuzzeno, L. The Volume-Regulated Anion Channel in Glioblastoma. *Cancers* **2019**, *11*, 307. [CrossRef] [PubMed]
64. Liu, T.; Stauber, T. The Volume-Regulated Anion Channel LRRC8/VRAC Is Dispensable for Cell Proliferation and Migration. *Int. J. Mol. Sci.* **2019**, *20*, 2663. [CrossRef] [PubMed]
65. Planells-Cases, R.; Lutter, D.; Guyader, C.; Gerhards, N.M.; Ullrich, F.; Elger, D.A.; Kucukosmanoglu, A.; Xu, G.; Voss, F.K.; Reincke, S.M.; et al. Subunit composition of VRAC channels determines substrate specificity and cellular resistance to P t-based anti-cancer drugs. *EMBO J.* **2015**, *34*, 2993–3008. [CrossRef]
66. Molenaar, R.J. Ion Channels in Glioblastoma. *ISRN Neurol.* **2011**, *2011*, e590249. [CrossRef]
67. Brandalise, F.; Ratto, D.; Leone, R.; Olivero, F.; Roda, E.; Locatelli, C.A.; Bottone, M.G.; Rossi, P. Deeper and Deeper on the Role of BK and Kir4.1 Channels in Glioblastoma Invasiveness: A Novel Summative Mechanism? *Front. Neurosci.* **2020**, *14*, 595664. [CrossRef]
68. Tan, G.; Sun, S.-Q.; Yuan, D.-L. Expression of Kir 4.1 in human astrocytic tumors: Correlation with pathologic grade. *Biochem. Biophys. Res. Commun.* **2008**, *367*, 743–747. [CrossRef]
69. Madadi, A.; Wolfart, J.; Lange, F.; Brehme, H.; Linnebacher, M.; Bräuer, A.U.; Büttner, A.; Freiman, T.; Henker, C.; Einsle, A.; et al. Correlation between Kir4.1 expression and barium-sensitive currents in rat and human glioma cell lines. *Neurosci. Lett.* **2021**, *741*, 135481. [CrossRef]
70. Brown, B.M.; Pressley, B.; Wulff, H. KCa3.1 Channel Modulators as Potential Therapeutic Compounds for Glioblastoma. *Curr. Neuropharmacol.* **2018**, *16*, 618–626. [CrossRef]
71. Hausmann, D.; Hoffmann, D.C.; Venkataramani, V.; Jung, E.; Horschitz, S.; Tetzlaff, S.K.; Jabali, A.; Hai, L.; Kessler, T.; Azoñn, D.D.; et al. Autonomous rhythmic activity in glioma networks drives brain tumour growth. *Nature* **2022**, *613*, 179–186. [CrossRef] [PubMed]
72. Catacuzzeno, L.; Franciolini, F. Role of KCa3.1 Channels in Modulating Ca^{2+} Oscillations during Glioblastoma Cell Migration and Invasion. *Int. J. Mol. Sci.* **2018**, *19*, 2970. [CrossRef] [PubMed]
73. Preußat, K.; Beetz, C.; Schrey, M.; Kraft, R.; Wölfl, S.; Kalff, R.; Patt, S. Expression of voltage-gated potassium channels Kv1.3 and Kv1.5 in human gliomas. *Neurosci. Lett.* **2003**, *346*, 33–36. [CrossRef] [PubMed]
74. Arvind, S.; Arivazhagan, A.; Santosh, V.; Chandramouli, B.A. Differential expression of a novel voltage gated potassium channel–Kv 1.5 in astrocytomas and its impact on prognosis in glioblastoma. *Br. J. Neurosurg.* **2012**, *26*, 16–20. [CrossRef] [PubMed]
75. Grimaldi, A.; D'alessandro, G.; Di Castro, M.A.; Lauro, C.; Singh, V.; Pagani, F.; Sforna, L.; Grassi, F.; Di Angelantonio, S.; Catacuzzeno, L.; et al. Kv1.3 activity perturbs the homeostatic properties of astrocytes in glioma. *Sci. Rep.* **2018**, *8*, 7654. [CrossRef]
76. Venturini, E.; Leanza, L.; Azzolini, M.; Kadow, S.; Mattarei, A.; Weller, M.; Tabatabai, G.; Edwards, M.J.; Zoratti, M.; Paradisi, C.; et al. Targeting the Potassium Channel Kv1.3 Kills Glioblastoma Cells. *Neurosignals* **2017**, *25*, 26–38. [CrossRef]
77. Yang, W.; Wu, P.-F.; Ma, J.-X.; Liao, M.-J.; Xu, L.-S.; Yi, L. TRPV4 activates the Cdc42/N-wasp pathway to promote glioblastoma invasion by altering cellular protrusions. *Sci. Rep.* **2020**, *10*, 14151. [CrossRef]
78. Huang, T.; Xu, T.; Wang, Y.; Zhou, Y.; Yu, D.; Wang, Z.; He, L.; Chen, Z.; Zhang, Y.; Davidson, D.; et al. Cannabidiol inhibits human glioma by induction of lethal mitophagy through activating TRPV4. *Autophagy* **2021**, *17*, 3592–3606. [CrossRef]
79. Chen, X.; Wanggou, S.; Bodalia, A.; Zhu, M.; Dong, W.; Fan, J.J.; Yin, W.C.; Min, H.-K.; Hu, M.; Draghici, D.; et al. A Feedforward Mechanism Mediated by Mechanosensitive Ion Channel PIEZO1 and Tissue Mechanics Promotes Glioma Aggression. *Neuron* **2018**, *100*, 799–815.e7. [CrossRef]
80. Varricchio, A.; Ramesh, S.A.; Yool, A.J. Novel Ion Channel Targets and Drug Delivery Tools for Controlling Glioblastoma Cell Invasiveness. *Int. J. Mol. Sci.* **2021**, *22*, 11909. [CrossRef]
81. Vandebroek, A.; Yasui, M. Regulation of AQP4 in the Central Nervous System. *Int. J. Mol. Sci.* **2020**, *21*, 1603. [CrossRef] [PubMed]

82. Costa, L.E.S.; Clementino-Neto, J.; Mendes, C.B.; Franzon, N.H.; Costa, E.d.O.; Moura-Neto, V.; Ximenes-Da-Silva, A. Evidence of Aquaporin 4 Regulation by Thyroid Hormone during Mouse Brain Development and in Cultured Human Glioblastoma Multiforme Cells. *Front. Neurosci.* **2019**, *13*, 317. [CrossRef] [PubMed]
83. Smits, A.; Jin, Z.; Elsir, T.; Pedder, H.; Nistér, M.; Alafuzoff, I.; Dimberg, A.; Edqvist, P.-H.; Pontén, F.; Aronica, E.; et al. GABA-A Channel Subunit Expression in Human Glioma Correlates with Tumor Histology and Clinical Outcome. *PLoS ONE* **2012**, *7*, e37041. [CrossRef]
84. Tantillo, E.; Scalera, M.; De Santis, E.; Meneghetti, N.; Cerri, C.; Menicagli, M.; Mazzoni, A.; Costa, M.; Mazzanti, C.M.; Vannini, E.; et al. Molecular changes underlying decay of sensory responses and enhanced seizure propensity in peritumoral neurons. *Neuro-Oncology* **2023**, *25*, 1463–1473. [CrossRef] [PubMed]
85. Blanchart, A.; Fernando, R.; Häring, M.; Assaife-Lopes, N.; Romanov, R.A.; Andäng, M.; Harkany, T.; Ernfors, P. Endogenous GABAA receptor activity suppresses glioma growth. *Oncogene* **2017**, *36*, 777–786. [CrossRef]
86. Radin, D.P.; Tsirka, S.E. Interactions between Tumor Cells, Neurons, and Microglia in the Glioma Microenvironment. *Int. J. Mol. Sci.* **2020**, *21*, 8476. [CrossRef]
87. Zamora-Sánchez, C.J.; Bello-Alvarez, C.; Rodríguez-Dorantes, M.; Camacho-Arroyo, I. Allopregnanolone Promotes Migration and Invasion of Human Glioblastoma Cells through the Protein Tyrosine Kinase c-Src Activation. *Int. J. Mol. Sci.* **2022**, *23*, 4996. [CrossRef]
88. Zamora-Sánchez, C.J.; Hansberg-Pastor, V.; Salido-Guadarrama, I.; Rodríguez-Dorantes, M.; Camacho-Arroyo, I. Allopregnanolone promotes proliferation and differential gene expression in human glioblastoma cells. *Steroids* **2017**, *119*, 36–42. [CrossRef]
89. Feng, Y.-H.; Lim, S.-W.; Lin, H.-Y.; Wang, S.-A.; Hsu, S.-P.; Kao, T.-J.; Ko, C.-Y.; Hsu, T.-I. Allopregnanolone suppresses glioblastoma survival through decreasing DPYSL3 and S100A11 expression. *J. Steroid Biochem. Mol. Biol.* **2022**, *219*, 106067. [CrossRef]
90. Thompson, E.G.; Sontheimer, H. Acetylcholine Receptor Activation as a Modulator of Glioblastoma Invasion. *Cells* **2019**, *8*, 1203. [CrossRef]
91. Pucci, S.; Bolchi, C.; Bavo, F.; Pallavicini, M.; De Palma, C.; Renzi, M.; Fucile, S.; Benfante, R.; Di Lascio, S.; Lattuada, D.; et al. Evidence of a dual mechanism of action underlying the anti-proliferative and cytotoxic effects of ammonium-alkyloxy-stilbene-based α7- and α9-nicotinic ligands on glioblastoma cells. *Pharmacol. Res.* **2022**, *175*, 105959. [CrossRef]
92. Pucci, S.; Zoli, M.; Clementi, F.; Gotti, C. α9-Containing Nicotinic Receptors in Cancer. *Front. Cell. Neurosci.* **2021**, *15*, 805123. [CrossRef]
93. Spina, R.; Voss, D.M.; Asnaghi, L.; Sloan, A.; Bar, E.E. Atracurium Besylate and other neuromuscular blocking agents promote astroglial differentiation and deplete glioblastoma stem cells. *Oncotarget* **2016**, *7*, 459–472. [CrossRef] [PubMed]
94. Bavo, F.; Pallavicini, M.; Pucci, S.; Appiani, R.; Giraudo, A.; Oh, H.; Kneisley, D.L.; Eaton, B.; Lucero, L.; Gotti, C.; et al. Subnanomolar Affinity and Selective Antagonism at α7 Nicotinic Receptor by Combined Modifications of 2-Triethylammonium Ethyl Ether of 4-Stilbenol (MG624). *J. Med. Chem.* **2023**, *66*, 306–332. [CrossRef] [PubMed]
95. Venkataramani, V.; Tanev, D.I.; Strahle, C.; Studier-Fischer, A.; Fankhauser, L.; Kessler, T.; Körber, C.; Kardorff, M.; Ratliff, M.; Xie, R.; et al. Glutamatergic synaptic input to glioma cells drives brain tumour progression. *Nature* **2019**, *573*, 532–538. [CrossRef] [PubMed]
96. Jung, E.; Alfonso, J.; Monyer, H.; Wick, W.; Winkler, F. Neuronal signatures in cancer. *Int. J. Cancer* **2020**, *147*, 3281–3291. [CrossRef]
97. Lyons, S.A.; Chung, W.J.; Weaver, A.K.; Ogunrinu, T.; Sontheimer, H. Autocrine Glutamate Signaling Promotes Glioma Cell Invasion. *Cancer Res.* **2007**, *67*, 9463–9471. [CrossRef]
98. Nandakumar, D.; Ramaswamy, P.; Prasad, C.; Srinivas, D.; Goswami, K. Glioblastoma invasion and NMDA receptors: A novel prospect. *Imaging* **2019**, *106*, 250–260. [CrossRef]
99. Venkatesh, H.S.; Morishita, W.; Geraghty, A.C.; Silverbush, D.; Gillespie, S.M.; Arzt, M.; Tam, L.T.; Espenel, C.; Ponnuswami, A.; Ni, L.; et al. Electrical and synaptic integration of glioma into neural circuits. *Nature* **2019**, *573*, 539–545. [CrossRef]
100. Grossman, S.A.; Ye, X.; Chamberlain, M.; Mikkelsen, T.; Batchelor, T.; Desideri, S.; Piantadosi, S.; Fisher, J.; Fine, H.A. Talampanel With Standard Radiation and Temozolomide in Patients With Newly Diagnosed Glioblastoma: A Multicenter Phase II Trial. *J. Clin. Oncol.* **2009**, *27*, 4155–4161. [CrossRef]
101. Cacciatore, I.; Fornasari, E.; Marinelli, L.; Eusepi, P.; Ciulla, M.; Ozdemir, O.; Tatar, A.; Turkez, H.; Di Stefano, A. Memantine-derived drugs as potential antitumor agents for the treatment of glioblastoma. *Eur. J. Pharm. Sci.* **2017**, *109*, 402–411. [CrossRef] [PubMed]
102. Yamada, T.; Tsuji, S.; Nakamura, S.; Egashira, Y.; Shimazawa, M.; Nakayama, N.; Yano, H.; Iwama, T.; Hara, H. Riluzole enhances the antitumor effects of temozolomide via suppression of MGMT expression in glioblastoma. *J. Neurosurg.* **2020**, *134*, 701–710. [CrossRef] [PubMed]
103. Albayrak, G.; Konac, E.; Dere, U.A.; Emmez, H. Targeting cancer cell metabolism with metformin, dichloroacetate and memantine in glioblastoma (gbm). *Turk. Neurosurg.* **2021**, *31*, 233–237. [CrossRef] [PubMed]
104. Blyufer, A.; Lhamo, S.; Tam, C.; Tariq, I.; Thavornwatanayong, T.; Mahajan, S.S. Riluzole: A neuroprotective drug with potential as a novel anti-cancer agent (Review). *Int. J. Oncol.* **2021**, *59*, 95. [CrossRef]
105. Venkataramani, V.; Tanev, D.I.; Kuner, T.; Wick, W.; Winkler, F. Synaptic input to brain tumors: Clinical implications. *Neuro-Oncology* **2021**, *23*, 23–33. [CrossRef]
106. Caltabiano, R.; Torrisi, A.; Condorelli, D.; Albanese, V.; Lanzafame, S. High levels of connexin 43 mRNA in high grade astrocytomas. Study of 32 cases with in situ hybridization. *Acta Histochem.* **2010**, *112*, 529–535. [CrossRef]
107. Sin, W.C.; Aftab, Q.; Bechberger, J.F.; Leung, J.H.; Chen, H.; Naus, C.C. Astrocytes promote glioma invasion via the gap junction protein connexin43. *Oncogene* **2016**, *35*, 1504–1516. [CrossRef]

108. Uzu, M.; Sin, W.C.; Shimizu, A.; Sato, H. Conflicting Roles of Connexin43 in Tumor Invasion and Growth in the Central Nervous System. *Int. J. Mol. Sci.* **2018**, *19*, 1159. [CrossRef]
109. McCutcheon, S.; Spray, D.C. Glioblastoma–Astrocyte Connexin 43 Gap Junctions Promote Tumor Invasion. *Mol. Cancer Res.* **2022**, *20*, 319–331. [CrossRef]
110. Dong, H.; Zhou, X.-W.; Wang, X.; Yang, Y.; Luo, J.-W.; Liu, Y.-H.; Mao, Q. Complex role of connexin 43 in astrocytic tumors and possible promotion of glioma-associated epileptic discharge (Review). *Mol. Med. Rep.* **2017**, *16*, 7890–7900. [CrossRef]
111. Xing, L.; Yang, T.; Cui, S.; Chen, G. Connexin Hemichannels in Astrocytes: Role in CNS Disorders. *Front. Mol. Neurosci.* **2019**, *12*, 23. [CrossRef] [PubMed]
112. Gielen, P.R.; Aftab, Q.; Ma, N.; Chen, V.C.; Hong, X.; Lozinsky, S.; Naus, C.C.; Sin, W.C. Connexin43 confers Temozolomide resistance in human glioma cells by modulating the mitochondrial apoptosis pathway. *Neuropharmacology* **2013**, *75*, 539–548. [CrossRef] [PubMed]
113. Ransom, B.R.; Goldring, S.; Sawyer, J.E.R.; Hennebry, J.E.; Revill, A.; Brown, A.M.; Kohn, A.; Metz, C.; Tommerdahl, M.A.; Whitsel, B.L.; et al. Ionic determinants of membrane potential of cells presumed to be glia in cerebral cortex of cat. *J. Neurophysiol.* **1973**, *36*, 855–868. [CrossRef] [PubMed]
114. Seifert, G.; Henneberger, C.; Steinhäuser, C. Diversity of astrocyte potassium channels: An update. *Brain Res. Bull.* **2018**, *136*, 26–36. [CrossRef] [PubMed]
115. Kofuji, P.; Newman, E.A. Potassium buffering in the central nervous system. *Neuroscience* **2004**, *129*, 1043–1056. [CrossRef]
116. Shih, P.-Y.; Savtchenko, L.P.; Kamasawa, N.; Dembitskaya, Y.; McHugh, T.J.; Rusakov, D.A.; Shigemoto, R.; Semyanov, A. Retrograde Synaptic Signaling Mediated by K+ Efflux through Postsynaptic NMDA Receptors. *Cell Rep.* **2013**, *5*, 941–951. [CrossRef]
117. Tyurikova, O.; Shih, P.; Dembitskaya, Y.; Savtchenko, L.P.; McHugh, T.J.; Rusakov, D.A.; Semyanov, A. K^+ efflux through postsynaptic NMDA receptors suppresses local astrocytic glutamate uptake. *Glia* **2022**, *70*, 961–974. [CrossRef]
118. Kaila, K.; Lamsa, K.; Smirnov, S.; Taira, T.; Voipio, J. Long-Lasting GABA-Mediated Depolarization Evoked by High-Frequency Stimulation in Pyramidal Neurons of Rat Hippocampal Slice Is Attributable to a Network-Driven, Bicarbonate-Dependent K^+Transient. *J. Neurosci.* **1997**, *17*, 7662–7672. [CrossRef]
119. Voipio, J.; Kaila, K. GABAergic excitation and K+-mediated volume transmission in the hippocampus. *Prog. Brain Res.* **2000**, *125*, 329–338. [CrossRef]
120. Viitanen, T.; Ruusuvuori, E.; Kaila, K.; Voipio, J. The K^+-Cl^- cotransporter KCC2 promotes GABAergic excitation in the mature rat hippocampus. *J. Physiol.* **2010**, *588*, 1527–1540. [CrossRef]
121. Olsen, M.L.; Sontheimer, H. Functional implications for Kir4.1 channels in glial biology: From K^+ buffering to cell differentiation. *J. Neurochem.* **2008**, *107*, 589–601. [CrossRef] [PubMed]
122. Brasko, C.; Hawkins, V.; De La Rocha, I.C.; Butt, A.M. Expression of Kir4.1 and Kir5.1 inwardly rectifying potassium channels in oligodendrocytes, the myelinating cells of the CNS. *Anat. Embryol.* **2017**, *222*, 41–59. [CrossRef] [PubMed]
123. Hibino, H.; Fujita, A.; Iwai, K.; Yamada, M.; Kurachi, Y. Differential Assembly of Inwardly Rectifying K+ Channel Subunits, Kir4.1 and Kir5.1, in Brain Astrocytes. *J. Biol. Chem.* **2004**, *279*, 44065–44073. [CrossRef]
124. Hibino, H.; Inanobe, A.; Furutani, K.; Murakami, S.; Findlay, I.; Kurachi, Y.; Yamamura, H.; Suzuki, Y.; Yamamura, H.; Asai, K.; et al. Inwardly Rectifying Potassium Channels: Their Structure, Function, and Physiological Roles. *Physiol. Rev.* **2010**, *90*, 291–366. [CrossRef] [PubMed]
125. Sun, X.-L.; Hu, G. ATP-sensitive potassium channels: A promising target for protecting neurovascular unit function in stroke. *Clin. Exp. Pharmacol. Physiol.* **2010**, *37*, 243–252. [CrossRef] [PubMed]
126. Hu, Z.-L.; Sun, T.; Lu, M.; Ding, J.-H.; Du, R.-H.; Hu, G. Kir6.1/K-ATP channel on astrocytes protects against dopaminergic neurodegeneration in the MPTP mouse model of Parkinson's disease via promoting mitophagy. *Brain Behav. Immun.* **2019**, *81*, 509–522. [CrossRef]
127. Chen, M.-M.; Hu, Z.-L.; Ding, J.-H.; Du, R.-H.; Hu, G. Astrocytic Kir6.1 deletion aggravates neurodegeneration in the lipopolysaccharide-induced mouse model of Parkinson's disease via astrocyte-neuron cross talk through complement C3-C3R signaling. *Brain Behav. Immun.* **2021**, *95*, 310–320. [CrossRef]
128. Zhou, M.; Xu, G.; Xie, M.; Zhang, X.; Schools, G.P.; Ma, L.; Kimelberg, H.K.; Chen, H. TWIK-1 and TREK-1 Are Potassium Channels Contributing Significantly to Astrocyte Passive Conductance in Rat Hippocampal Slices. *J. Neurosci.* **2009**, *29*, 8551–8564. [CrossRef]
129. Hwang, E.M.; Kim, E.; Yarishkin, O.; Woo, D.H.; Han, K.-S.; Park, N.; Bae, Y.; Woo, J.; Kim, D.; Park, M.; et al. A disulphide-linked heterodimer of TWIK-1 and TREK-1 mediates passive conductance in astrocytes. *Nat. Commun.* **2014**, *5*, 3227. [CrossRef]
130. Du, Y.; Kiyoshi, C.M.; Wang, Q.; Wang, W.; Ma, B.; Alford, C.C.; Zhong, S.; Wan, Q.; Chen, H.; Lloyd, E.E.; et al. Genetic Deletion of TREK-1 or TWIK-1/TREK-1 Potassium Channels does not Alter the Basic Electrophysiological Properties of Mature Hippocampal Astrocytes In Situ. *Front. Cell. Neurosci.* **2016**, *10*, 13. [CrossRef]
131. Price, D.L.; Ludwig, J.W.; Mi, H.; Schwarz, T.L.; Ellisman, M.H. Distribution of rSlo Ca^{2+}-activated K+ channels in rat astrocyte perivascular endfeet. *Brain Res.* **2002**, *956*, 183–193. [CrossRef]
132. Filosa, J.A.; Bonev, A.D.; Straub, S.V.; Meredith, A.L.; Wilkerson, M.K.; Aldrich, R.W.; Nelson, M.T. Local potassium signaling couples neuronal activity to vasodilation in the brain. *Nat. Neurosci.* **2006**, *9*, 1397–1403. [CrossRef]
133. Carmignoto, G.; Gómez-Gonzalo, M. The contribution of astrocyte signalling to neurovascular coupling. *Brain Res. Rev.* **2010**, *63*, 138–148. [CrossRef]

134. Filosa, J.A.; Iddings, J.A.; Shih, E.K.; Robinson, M.B.; Sun, S.; Li, H.; Chen, J.; Qian, Q.; Merchant, S.; Medow, M.S.; et al. Astrocyte regulation of cerebral vascular tone. *Am. J. Physiol. Circ. Physiol.* **2013**, *305*, H609–H619. [CrossRef] [PubMed]
135. Pappalardo, L.W.; Black, J.A.; Waxman, S.G. Sodium channels in astroglia and microglia. *Glia* **2016**, *64*, 1628–1645. [CrossRef] [PubMed]
136. Black, J.A.; Newcombe, J.; Waxman, S.G. Astrocytes within multiple sclerosis lesions upregulate sodium channel Nav1.5. *Brain* **2010**, *133*, 835–846. [CrossRef] [PubMed]
137. Zhu, H.; Zhao, Y.; Wu, H.; Jiang, N.; Wang, Z.; Lin, W.; Jin, J.; Ji, Y. Remarkable alterations of Nav1.6 in reactive astrogliosis during epileptogenesis. *Sci. Rep.* **2016**, *6*, 38108. [CrossRef] [PubMed]
138. Grosche, J.; Matyash, V.; Möller, T.; Verkhratsky, A.; Reichenbach, A.; Kettenmann, H. Microdomains for neuron–glia interaction: Parallel fiber signaling to Bergmann glial cells. *Nat. Neurosci.* **1999**, *2*, 139–143. [CrossRef] [PubMed]
139. Panatier, A.; Vallée, J.; Haber, M.; Murai, K.K.; Lacaille, J.-C.; Robitaille, R. Astrocytes Are Endogenous Regulators of Basal Transmission at Central Synapses. *Cell* **2011**, *146*, 785–798. [CrossRef]
140. Di Castro, M.A.; Chuquet, J.; Liaudet, N.; Bhaukaurally, K.; Santello, M.; Bouvier, D.; Tiret, P.; Volterra, A. Local Ca^{2+} detection and modulation of synaptic release by astrocytes. *Nat. Neurosci.* **2011**, *14*, 1276–1284. [CrossRef]
141. Shigetomi, E.; Bushong, E.A.; Haustein, M.D.; Tong, X.; Jackson-Weaver, O.; Kracun, S.; Xu, J.; Sofroniew, M.V.; Ellisman, M.H.; Khakh, B.S. Imaging calcium microdomains within entire astrocyte territories and endfeet with GCaMPs expressed using adeno-associated viruses. *J. Gen. Physiol.* **2013**, *141*, 633–647. [CrossRef] [PubMed]
142. Kanemaru, K.; Sekiya, H.; Xu, M.; Satoh, K.; Kitajima, N.; Yoshida, K.; Okubo, Y.; Sasaki, T.; Moritoh, S.; Hasuwa, H.; et al. In Vivo Visualization of Subtle, Transient, and Local Activity of Astrocytes Using an Ultrasensitive Ca^{2+} Indicator. *Cell Rep.* **2014**, *8*, 311–318. [CrossRef] [PubMed]
143. Srinivasan, R.; Huang, B.S.; Venugopal, S.; Johnston, A.D.; Chai, H.; Zeng, H.; Golshani, P.; Khakh, B.S. Ca^{2+} signaling in astrocytes from Ip3r2−/− mice in brain slices and during startle responses in vivo. *Nat. Neurosci.* **2015**, *18*, 708–717. [CrossRef] [PubMed]
144. Bindocci, E.; Savtchouk, I.; Liaudet, N.; Becker, D.; Carriero, G.; Volterra, A. Three-dimensional Ca^{2+} imaging advances understanding of astrocyte biology. *Science* **2017**, *356*, eaai8185. [CrossRef] [PubMed]
145. Mariotti, L.; Losi, G.; Lia, A.; Melone, M.; Chiavegato, A.; Gómez-Gonzalo, M.; Sessolo, M.; Bovetti, S.; Forli, A.; Zonta, M.; et al. Interneuron-specific signaling evokes distinctive somatostatin-mediated responses in adult cortical astrocytes. *Nat. Commun.* **2018**, *9*, 82. [CrossRef] [PubMed]
146. Stobart, J.L.; Ferrari, K.D.; Barrett, M.J.P.; Glück, C.; Stobart, M.J.; Zuend, M.; Weber, B. Cortical Circuit Activity Evokes Rapid Astrocyte Calcium Signals on a Similar Timescale to Neurons. *Neuron* **2018**, *98*, 726–735.e4. [CrossRef]
147. Arizono, M.; Inavalli, V.V.G.K.; Panatier, A.; Pfeiffer, T.; Angibaud, J.; Levet, F.; Ter Veer, M.J.T.; Stobart, J.; Bellocchio, L.; Mikoshiba, K.; et al. Structural basis of astrocytic Ca^{2+} signals at tripartite synapses. *Nat. Commun.* **2020**, *11*, 1906. [CrossRef]
148. Shigetomi, E.; Saito, K.; Sano, F.; Koizumi, S. Aberrant Calcium Signals in Reactive Astrocytes: A Key Process in Neurological Disorders. *Int. J. Mol. Sci.* **2019**, *20*, 996. [CrossRef]
149. Nagai, J.; Yu, X.; Papouin, T.; Cheong, E.; Freeman, M.R.; Monk, K.R.; Hastings, M.H.; Haydon, P.G.; Rowitch, D.; Shaham, S.; et al. Behaviorally consequential astrocytic regulation of neural circuits. *Neuron* **2021**, *109*, 576–596. [CrossRef]
150. Mariotti, L.; Losi, G.; Sessolo, M.; Marcon, I.; Carmignoto, G. The inhibitory neurotransmitter GABA evokes long-lasting Ca^{2+} oscillations in cortical astrocytes. *Glia* **2015**, *64*, 363–373. [CrossRef]
151. Durkee, C.A.; Covelo, A.; Lines, J.; Kofuji, P.; Aguilar, J.; Araque, A. $G_{i/o}$ protein-coupled receptors inhibit neurons but activate astrocytes and stimulate gliotransmission. *Glia* **2019**, *67*, 1076–1093. [CrossRef]
152. Caudal, L.C.; Gobbo, D.; Scheller, A.; Kirchhoff, F. The Paradox of Astroglial Ca^{2+} Signals at the Interface of Excitation and Inhibition. *Front. Cell. Neurosci.* **2020**, *14*, 609947. [CrossRef] [PubMed]
153. Rose, C.R.; Ziemens, D.; Verkhratsky, A. On the special role of NCX in astrocytes: Translating Na^{+}-transients into intracellular Ca^{2+} signals. *Cell Calcium* **2020**, *86*, 102154. [CrossRef]
154. Verkhratsky, A.; Parpura, V. Store-operated calcium entry in neuroglia. *Neurosci. Bull.* **2013**, *30*, 125–133. [CrossRef] [PubMed]
155. Yoast, R.E.; Emrich, S.M.; Trebak, M. The anatomy of native CRAC channel(s). *Curr. Opin. Physiol.* **2020**, *17*, 89–95. [CrossRef]
156. Kwon, J.; An, H.; Sa, M.; Won, J.; Shin, J.; Lee, C.J. Orai1 and Orai3 in Combination with Stim1 Mediate the Majority of Store-operated Calcium Entry in Astrocytes. *Exp. Neurobiol.* **2017**, *26*, 42–54. [CrossRef]
157. Verkhratsky, A.; Reyes, R.C.; Parpura, V. TRP Channels Coordinate Ion Signalling in Astroglia. *Ergeb. Physiol.* **2014**, *166*, 1–22. [CrossRef]
158. Shigetomi, E.; Jackson-Weaver, O.; Huckstepp, R.T.; O'Dell, T.J.; Khakh, B.S. TRPA1 Channels Are Regulators of Astrocyte Basal Calcium Levels and Long-Term Potentiation via Constitutive D-Serine Release. *J. Neurosci.* **2013**, *33*, 10143–10153. [CrossRef]
159. Shigetomi, E.; Tong, X.; Kwan, K.Y.; Corey, D.P.; Khakh, B.S. TRPA1 channels regulate astrocyte resting calcium and inhibitory synapse efficacy through GAT-3. *Nat. Neurosci.* **2011**, *15*, 70–80. [CrossRef]
160. Dunn, K.M.; Hill-Eubanks, D.C.; Liedtke, W.B.; Nelson, M.T. TRPV4 channels stimulate Ca^{2+} -induced Ca^{2+} release in astrocytic endfeet and amplify neurovascular coupling responses. *Proc. Natl. Acad. Sci. USA* **2013**, *110*, 6157–6162. [CrossRef]
161. Sucha, P.; Hermanova, Z.; Chmelova, M.; Kirdajova, D.; Garcia, S.C.; Marchetti, V.; Vorisek, I.; Tureckova, J.; Shany, E.; Jirak, D.; et al. The absence of AQP4/TRPV4 complex substantially reduces acute cytotoxic edema following ischemic injury. *Front. Cell. Neurosci.* **2022**, *16*, 1054919. [CrossRef] [PubMed]
162. Tureckova, J.; Hermanova, Z.; Marchetti, V.; Anderova, M. Astrocytic TRPV4 Channels and Their Role in Brain Ischemia. *Int. J. Mol. Sci.* **2023**, *24*, 7101. [CrossRef]

163. Lewis, A.H.; Grandl, J. Mechanical sensitivity of Piezo1 ion channels can be tuned by cellular membrane tension. *eLife* **2015**, *4*, e12088. [CrossRef] [PubMed]
164. Benfenati, V.; Caprini, M.; Dovizio, M.; Mylonakou, M.N.; Ferroni, S.; Ottersen, O.P.; Amiry-Moghaddam, M. An aquaporin-4/transient receptor potential vanilloid 4 (AQP4/TRPV4) complex is essential for cell-volume control in astrocytes. *Proc. Natl. Acad. Sci. USA* **2011**, *108*, 2563–2568. [CrossRef] [PubMed]
165. Blumenthal, N.R.; Hermanson, O.; Heimrich, B.; Shastri, V.P. Stochastic nanoroughness modulates neuron–astrocyte interactions and function via mechanosensing cation channels. *Proc. Natl. Acad. Sci. USA* **2014**, *111*, 16124–16129. [CrossRef] [PubMed]
166. Chi, S.; Cui, Y.; Wang, H.; Jiang, J.; Zhang, T.; Sun, S.; Zhou, Z.; Zhong, Y.; Xiao, B. Astrocytic Piezo1-mediated mechanotransduction determines adult neurogenesis and cognitive functions. *Neuron* **2022**, *110*, 2984–2999.e8. [CrossRef]
167. Elorza-Vidal, X.; Gaitán-Peñas, H.; Estévez, R. Chloride Channels in Astrocytes: Structure, Roles in Brain Homeostasis and Implications in Disease. *Int. J. Mol. Sci.* **2019**, *20*, 1034. [CrossRef]
168. Makara, J.K.; Rappert, A.; Matthias, K.; Steinhäuser, C.; Spät, A.; Kettenmann, H. Astrocytes from mouse brain slices express ClC-2-mediated Cl− currents regulated during development and after injury. *Mol. Cell. Neurosci.* **2003**, *23*, 521–530. [CrossRef]
169. Jeworutzki, E.; López-Hernández, T.; Capdevila-Nortes, X.; Sirisi, S.; Bengtsson, L.; Montolio, M.; Zifarelli, G.; Arnedo, T.; Müller, C.S.; Schulte, U.; et al. GlialCAM, a Protein Defective in a Leukodystrophy, Serves as a ClC-2 Cl^- Channel Auxiliary Subunit. *Neuron* **2012**, *73*, 951–961. [CrossRef]
170. Qiu, Z.; Dubin, A.E.; Mathur, J.; Tu, B.; Reddy, K.; Miraglia, L.J.; Reinhardt, J.; Orth, A.P.; Patapoutian, A. SWELL1, a Plasma Membrane Protein, Is an Essential Component of Volume-Regulated Anion Channel. *Cell* **2014**, *157*, 447–458. [CrossRef]
171. Voss, F.K.; Ullrich, F.; Münch, J.; Lazarow, K.; Lutter, D.; Mah, N.; Andrade-Navarro, M.A.; von Kries, J.P.; Stauber, T.; Jentsch, T.J. Identification of LRRC8 Heteromers as an Essential Component of the Volume-Regulated Anion Channel VRAC. *Science* **2014**, *344*, 634–638. [CrossRef]
172. Mongin, A.A. Volume-regulated anion channel—A frenemy within the brain. *Pflügers Archiv-Eur. J. Physiol.* **2016**, *468*, 421–441. [CrossRef]
173. Osei-Owusu, J.; Yang, J.; Vitery, M.D.C.; Qiu, Z. Molecular Biology and Physiology of Volume-Regulated Anion Channel (VRAC). *Curr. Top Membr.* **2018**, *81*, 177–203. [PubMed]
174. Yang, J.; Vitery, M.d.C.; Chen, J.; Osei-Owusu, J.; Chu, J.; Qiu, Z. Glutamate-Releasing SWELL1 Channel in Astrocytes Modulates Synaptic Transmission and Promotes Brain Damage in Stroke. *Neuron* **2019**, *102*, 813–827.e6. [CrossRef] [PubMed]
175. Gómez-Gonzalo, M.; Zehnder, T.; Requie, L.M.; Bezzi, P.; Carmignoto, G. Insights into the release mechanism of astrocytic glutamate evoking in neurons NMDA receptor-mediated slow depolarizing inward currents. *Glia* **2018**, *66*, 2188–2199. [CrossRef]
176. Fellin, T.; Pascual, O.; Gobbo, S.; Pozzan, T.; Haydon, P.G.; Carmignoto, G. Neuronal Synchrony Mediated by Astrocytic Glutamate through Activation of Extrasynaptic NMDA Receptors. *Neuron* **2004**, *43*, 729–743. [CrossRef]
177. Gómez-Gonzalo, M.; Losi, G.; Chiavegato, A.; Zonta, M.; Cammarota, M.; Brondi, M.; Vetri, F.; Uva, L.; Pozzan, T.; de Curtis, M.; et al. An Excitatory Loop with Astrocytes Contributes to Drive Neurons to Seizure Threshold. *PLOS Biol.* **2010**, *8*, e1000352. [CrossRef] [PubMed]
178. Yang, J.; Chen, J.; Liu, Y.; Chen, K.H.; Baraban, J.M.; Qiu, Z. Ventral tegmental area astrocytes modulate cocaine reward by tonically releasing GABA. *Neuron* **2023**, *111*, 1104–1117.e6. [CrossRef]
179. Lalo, U.; Palygin, O.; Rasooli-Nejad, S.; Andrew, J.; Haydon, P.G.; Pankratov, Y. Exocytosis of ATP From Astrocytes Modulates Phasic and Tonic Inhibition in the Neocortex. *PLoS Biol.* **2014**, *12*, e1001747. [CrossRef]
180. Sabirov, R.Z.; Islam, R.; Okada, T.; Merzlyak, P.G.; Kurbannazarova, R.S.; Tsiferova, N.A.; Okada, Y. The ATP-Releasing Maxi-Cl Channel: Its Identity, Molecular Partners, and Physiological/Pathophysiological Implications. *Life* **2021**, *11*, 509. [CrossRef]
181. Kanai, N.; Lu, R.; Satriano, J.A.; Bao, Y.; Wolkoff, A.W.; Schuster, V.L. Identification and Characterization of a Prostaglandin Transporter. *Science* **1995**, *268*, 866–869. [CrossRef] [PubMed]
182. Sabirov, R.Z.; Merzlyak, P.G.; Okada, T.; Islam, R.; Uramoto, H.; Mori, T.; Makino, Y.; Matsuura, H.; Xie, Y.; Okada, Y. The organic anion transporter SLCO2A1 constitutes the core component of the Maxi-Cl channel. *EMBO J.* **2017**, *36*, 3309–3324. [CrossRef] [PubMed]
183. Liu, H.-T.; Sabirov, R.Z.; Okada, Y. Oxygen-glucose deprivation induces ATP release via maxi-anion channels in astrocytes. *Purinergic Signal.* **2008**, *4*, 147–154. [CrossRef] [PubMed]
184. Liu, H.; Tashmukhamedov, B.A.; Inoue, H.; Okada, Y.; Sabirov, R.Z. Roles of two types of anion channels in glutamate release from mouse astrocytes under ischemic or osmotic stress. *Glia* **2006**, *54*, 343–357. [CrossRef]
185. Woo, D.H.; Han, K.-S.; Shim, J.W.; Yoon, B.-E.; Kim, E.; Bae, J.Y.; Oh, S.-J.; Hwang, E.M.; Marmorstein, A.D.; Bae, Y.C.; et al. TREK-1 and Best1 Channels Mediate Fast and Slow Glutamate Release in Astrocytes upon GPCR Activation. *Cell* **2012**, *151*, 25–40. [CrossRef]
186. Park, H.; Han, K.-S.; Oh, S.-J.; Jo, S.; Woo, J.; Yoon, B.-E.; Lee, C.J. High glutamate permeability and distal localization of Best1 channel in CA1 hippocampal astrocyte. *Mol. Brain* **2013**, *6*, 54. [CrossRef]
187. Lee, S.; Yoon, B.-E.; Berglund, K.; Oh, S.-J.; Park, H.; Shin, H.-S.; Augustine, G.J.; Lee, C.J. Channel-Mediated Tonic GABA Release from Glia. *Science* **2010**, *330*, 790–796. [CrossRef]
188. Yoon, B.-E.; Jo, S.; Woo, J.; Lee, J.-H.; Kim, T.; Kim, D.; Lee, C.J. The amount of astrocytic GABA positively correlates with the degree of tonic inhibition in hippocampal CA1 and cerebellum. *Mol. Brain* **2011**, *4*, 42. [CrossRef]
189. Yoon, B.-E.; Woo, J.; Lee, C.J. Astrocytes as GABA-ergic and GABA-ceptive Cells. *Neurochem. Res.* **2012**, *37*, 2474–2479. [CrossRef]
190. Park, H.; Han, K.-S.; Seo, J.; Lee, J.; Dravid, S.M.; Woo, J.; Chun, H.; Cho, S.; Bae, J.Y.; An, H.; et al. Channel-mediated astrocytic glutamate modulates hippocampal synaptic plasticity by activating postsynaptic NMDA receptors. *Mol. Brain* **2015**, *8*, 7. [CrossRef]

191. Kwak, H.; Koh, W.; Kim, S.; Song, K.; Shin, J.-I.; Lee, J.M.; Lee, E.H.; Bae, J.Y.; Ha, G.E.; Oh, J.-E.; et al. Astrocytes Control Sensory Acuity via Tonic Inhibition in the Thalamus. *Neuron* **2020**, *108*, 691–706.e10. [CrossRef] [PubMed]
192. Höft, S.; Griemsmann, S.; Seifert, G.; Steinhäuser, C. Heterogeneity in expression of functional ionotropic glutamate and GABA receptors in astrocytes across brain regions: Insights from the thalamus. *Philos. Trans. R. Soc. B Biol. Sci.* **2014**, *369*, 20130602. [CrossRef]
193. Mölders, A.; Koch, A.; Menke, R.; Klöcker, N. Heterogeneity of the astrocytic AMPA-receptor transcriptome. *Glia* **2018**, *66*, 2604–2616. [CrossRef]
194. López-Bayghen, E.; Espinoza-Rojo, M.; Ortega, A. Glutamate down-regulates GLAST expression through AMPA receptors in Bergmann glial cells. *Mol. Brain Res.* **2003**, *115*, 1–9. [CrossRef]
195. Saab, A.S.; Neumeyer, A.; Jahn, H.M.; Cupido, A.; Šimek, A.A.M.; Boele, H.-J.; Scheller, A.; Le Meur, K.; Götz, M.; Monyer, H.; et al. Bergmann Glial AMPA Receptors Are Required for Fine Motor Coordination. *Science* **2012**, *337*, 749–753. [CrossRef]
196. Vargas, J.R.; Takahashi, D.K.; Thomson, K.E.; Wilcox, K.S. The Expression of Kainate Receptor Subunits in Hippocampal Astrocytes After Experimentally Induced Status Epilepticus. *J. Neuropathol. Exp. Neurol.* **2013**, *72*, 919–932. [CrossRef] [PubMed]
197. Verkhratsky, A.; Kirchhoff, F. NMDA Receptors in Glia. *Neuroscientist* **2007**, *13*, 28–37. [CrossRef] [PubMed]
198. Skowrońska, K.; Obara-Michlewska, M.; Zielińska, M.; Albrecht, J. NMDA Receptors in Astrocytes: In Search for Roles in Neurotransmission and Astrocytic Homeostasis. *Int. J. Mol. Sci.* **2019**, *20*, 309. [CrossRef]
199. Labrakakis, C.; Patt, S.; Hartmann, J.; Kettenmann, H. Functional $GABA_A$ receptors on human glioma cells. *Eur. J. Neurosci.* **1998**, *10*, 231–238. [CrossRef]
200. Muller, T.; Fritschy, J.; Grosche, J.; Pratt, G.; Mohler, H.; Kettenmann, H. Developmental regulation of voltage-gated K^+ channel and GABAA receptor expression in Bergmann glial cells. *J. Neurosci.* **1994**, *14*, 2503–2514. [CrossRef]
201. Fraser, D.; Duffy, S.; Angelides, K.; Perez-Velazquez, J.; Kettenmann, H.; MacVicar, B. GABAA/benzodiazepine receptors in acutely isolated hippocampal astrocytes. *J. Neurosci.* **1995**, *15*, 2720–2732. [CrossRef]
202. Egawa, K.; Yamada, J.; Furukawa, T.; Yanagawa, Y.; Fukuda, A. Cl^- homeodynamics in gap junction-coupled astrocytic networks on activation of GABAergic synapses. *J. Physiol.* **2013**, *591*, 3901–3917. [CrossRef] [PubMed]
203. Ma, B.-F.; Xie, M.-J.; Zhou, M. Bicarbonate efflux via GABAA receptors depolarizes membrane potential and inhibits two-pore domain potassium channels of astrocytes in rat hippocampal slices. *Glia* **2012**, *60*, 1761–1772. [CrossRef]
204. Puia, G.; Belelli, D. Neurosteroids on our minds. *Trends Pharmacol. Sci.* **2001**, *22*, 266–267. [CrossRef] [PubMed]
205. Belelli, D.; Lambert, J.J. Neurosteroids: Endogenous regulators of the GABAA receptor. *Nat. Rev. Neurosci.* **2005**, *6*, 565–575. [CrossRef] [PubMed]
206. Reddy, D.S. Neurosteroids. *Progress Brain Res.* **2010**, *186*, 113–137. [CrossRef]
207. Gotti, C.; Zoli, M.; Clementi, F. Brain nicotinic acetylcholine receptors: Native subtypes and their relevance. *Trends Pharmacol. Sci.* **2006**, *27*, 482–491. [CrossRef]
208. Shen, J.-X.; Yakel, J.L. Functional $\alpha 7$ Nicotinic ACh Receptors on Astrocytes in Rat Hippocampal CA1 Slices. *J. Mol. Neurosci.* **2012**, *48*, 14–21. [CrossRef]
209. Zoli, M.; Pucci, S.; Vilella, A.; Gotti, C. Neuronal and Extraneuronal Nicotinic Acetylcholine Receptors. *Curr. Neuropharmacol.* **2018**, *16*, 338–349. [CrossRef]
210. Koukouli, F.; Changeux, J.-P. Do Nicotinic Receptors Modulate High-Order Cognitive Processing? *Trends Neurosci.* **2020**, *43*, 550–564. [CrossRef]
211. Wang, X.; Lippi, G.; Carlson, D.M.; Berg, D.K. Activation of $\alpha 7$-containing nicotinic receptors on astrocytes triggers AMPA receptor recruitment to glutamatergic synapses. *J. Neurochem.* **2013**, *127*, 632–643. [CrossRef] [PubMed]
212. Papouin, T.; Dunphy, J.M.; Tolman, M.; Dineley, K.T.; Haydon, P.G. Septal Cholinergic Neuromodulation Tunes the Astrocyte-Dependent Gating of Hippocampal NMDA Receptors to Wakefulness. *Neuron* **2017**, *94*, 840–854.e7. [CrossRef] [PubMed]
213. Lezmy, J.; Arancibia-Cárcamo, I.L.; Quintela-López, T.; Sherman, D.L.; Brophy, P.J.; Attwell, D. Astrocyte Ca^{2+}-evoked ATP release regulates myelinated axon excitability and conduction speed. *Science* **2021**, *374*, eabh2858. [CrossRef]
214. Wang, H.; Yu, M.; Ochani, M.; Amella, C.A.; Tanovic, M.; Susarla, S.; Li, J.H.; Wang, H.; Yang, N.; Ulloa, L.; et al. Nicotinic acetylcholine receptor $\alpha 7$ subunit is an essential regulator of inflammation. *Nature* **2003**, *421*, 384–388. [CrossRef] [PubMed]
215. Fields, R.D.; Burnstock, G. Purinergic signalling in neuron–glia interactions. *Nat. Rev. Neurosci.* **2006**, *7*, 423–436. [CrossRef]
216. Burnstock, G. Purine and purinergic receptors. *Brain Neurosci. Adv.* **2018**, *2*, 2398212818817494. [CrossRef] [PubMed]
217. Huang, Z.; Xie, N.; Illes, P.; Di Virgilio, F.; Ulrich, H.; Semyanov, A.; Verkhratsky, A.; Sperlagh, B.; Yu, S.-G.; Huang, C.; et al. From purines to purinergic signalling: Molecular functions and human diseases. *Signal Transduct. Target. Ther.* **2021**, *6*, 162. [CrossRef]
218. Di Virgilio, F.; Sarti, A.C.; Coutinho-Silva, R. Purinergic signaling, DAMPs, and inflammation. *Am. J. Physiol. Physiol.* **2020**, *318*, C832–C835. [CrossRef]
219. Engel, T.; Jiménez-Mateos, E.M.; Diaz-Hernandez, M. Purinergic Signalling and Inflammation-Related Diseases. *Cells* **2022**, *11*, 3748. [CrossRef]
220. Ishibashi, K.; Morishita, Y.; Tanaka, Y. The Evolutionary Aspects of Aquaporin Family. *Adv. Exp. Med. Biol.* **2017**, *969*, 35–50. [CrossRef]
221. Nielsen, S.; Nagelhus, E.A.; Amiry-Moghaddam, M.; Bourque, C.; Agre, P.; Ottersen, O.P. Specialized Membrane Domains for Water Transport in Glial Cells: High-Resolution Immunogold Cytochemistry of Aquaporin-4 in Rat Brain. *J. Neurosci.* **1997**, *17*, 171–180. [CrossRef] [PubMed]
222. Aoyama, M.; Kakita, H.; Kato, S.; Tomita, M.; Asai, K. Region-specific expression of a water channel protein, aquaporin 4, on brain astrocytes. *J. Neurosci. Res.* **2012**, *90*, 2272–2280. [CrossRef] [PubMed]

223. Iliff, J.J.; Chen, M.J.; Plog, B.A.; Zeppenfeld, D.M.; Soltero, M.; Yang, L.; Singh, I.; Deane, R.; Nedergaard, M. Impairment of Glymphatic Pathway Function Promotes Tau Pathology after Traumatic Brain Injury. *J. Neurosci.* **2014**, *34*, 16180–16193. [CrossRef] [PubMed]
224. Iliff, J.J.; Wang, M.; Liao, Y.; Plogg, B.A.; Peng, W.; Gundersen, G.A.; Benveniste, H.; Vates, G.E.; Deane, R.; Goldman, S.A.; et al. A Paravascular Pathway Facilitates CSF Flow Through the Brain Parenchyma and the Clearance of Interstitial Solutes, Including Amyloid β. *Sci. Transl. Med.* **2012**, *4*, 147ra111. [CrossRef]
225. Kress, B.T.; Iliff, J.J.; Xia, M.; Wang, M.; Wei, H.S.; Zeppenfeld, D.; Xie, L.; Kang, H.; Xu, Q.; Liew, J.A.; et al. Impairment of paravascular clearance pathways in the aging brain. *Ann. Neurol.* **2014**, *76*, 845–861. [CrossRef]
226. Vella, J.; Zammit, C.; Di Giovanni, G.; Muscat, R.; Valentino, M. The central role of aquaporins in the pathophysiology of ischemic stroke. *Front. Cell. Neurosci.* **2015**, *9*, 108. [CrossRef]
227. Filippidis, A.S.; Carozza, R.B.; Rekate, H.L. Aquaporins in Brain Edema and Neuropathological Conditions. *Int. J. Mol. Sci.* **2016**, *18*, 55. [CrossRef]
228. Clément, T.; Rodriguez-Grande, B.; Badaut, J. Aquaporins in brain edema. *J. Neurosci. Res.* **2018**, *98*, 9–18. [CrossRef]
229. Pannasch, U.; Rouach, N. Emerging role for astroglial networks in information processing: From synapse to behavior. *Trends Neurosci.* **2013**, *36*, 405–417. [CrossRef]
230. Giaume, C.; Leybaert, L.; Naus, C.C.; Sáez, J.C. Connexin and pannexin hemichannels in brain glial cells: Properties, pharmacology, and roles. *Front. Pharmacol.* **2013**, *4*, 88. [CrossRef]
231. Saez, J.C.; Berthoud, V.M.; Branes, M.C.; Martinez, A.D.; Beyer, E.C. Plasma membrane channels formed by connexins: Their regulation and functions. *Physiol. Rev.* **2003**, *83*, 1359–1400. [CrossRef] [PubMed]
232. Harris, A.L. Connexin channel permeability to cytoplasmic molecules. *Prog. Biophys. Mol. Biol.* **2007**, *94*, 120–143. [CrossRef]
233. Houades, V.; Koulakoff, A.; Ezan, P.; Seif, I.; Giaume, C. Gap Junction-Mediated Astrocytic Networks in the Mouse Barrel Cortex. *J. Neurosci.* **2008**, *28*, 5207–5217. [CrossRef]
234. Decrock, E.; De Bock, M.; Wang, N.; Bultynck, G.; Giaume, C.; Naus, C.C.; Green, C.R.; Leybaert, L. Connexin and pannexin signaling pathways, an architectural blueprint for CNS physiology and pathology? *Cell. Mol. Life Sci.* **2015**, *72*, 2823–2851. [CrossRef] [PubMed]
235. Charvériat, M.; Naus, C.C.; Leybaert, L.; Sáez, J.C.; Giaume, C. Connexin-Dependent Neuroglial Networking as a New Therapeutic Target. *Front. Cell. Neurosci.* **2017**, *11*, 174. [CrossRef] [PubMed]
236. Mylvaganam, S.; Ramani, M.; Krawczyk, M.; Carlen, P.L. Roles of gap junctions, connexins, and pannexins in epilepsy. *Front. Physiol.* **2014**, *5*, 172. [CrossRef]
237. Li, Q.; Li, Q.-Q.; Jia, J.-N.; Liu, Z.-Q.; Zhou, H.-H.; Mao, X.-Y. Targeting gap junction in epilepsy: Perspectives and challenges. *Biomed. Pharmacother.* **2019**, *109*, 57–65. [CrossRef]
238. Guo, A.; Zhang, H.; Li, H.; Chiu, A.; García-Rodríguez, C.; Lagos, C.F.; Sáez, J.C.; Lau, C.G. Inhibition of connexin hemichannels alleviates neuroinflammation and hyperexcitability in temporal lobe epilepsy. *Proc. Natl. Acad. Sci. USA* **2022**, *119*, e2213162119. [CrossRef]
239. Sinyuk, M.; Mulkearns-Hubert, E.E.; Reizes, O.; Lathia, J. Cancer Connectors: Connexins, Gap Junctions, and Communication. *Front. Oncol.* **2018**, *8*, 646. [CrossRef]
240. Zhou, M.; Zheng, M.; Zhou, X.; Tian, S.; Yang, X.; Ning, Y.; Li, Y.; Zhang, S. The roles of connexins and gap junctions in the progression of cancer. *Cell Commun. Signal.* **2023**, *21*, 8. [CrossRef]
241. Henstridge, C.M.; Hyman, B.T.; Spires-Jones, T.L. Beyond the neuron–cellular interactions early in Alzheimer disease pathogenesis. *Nat. Rev. Neurosci.* **2019**, *20*, 94–108. [CrossRef] [PubMed]
242. Chun, H.; Lee, C.J. Reactive astrocytes in Alzheimer's disease: A double-edged sword. *Neurosci. Res.* **2018**, *126*, 44–52. [CrossRef] [PubMed]
243. Perez-Nievas, B.G.; Serrano-Pozo, A. Deciphering the Astrocyte Reaction in Alzheimer's Disease. *Front. Aging Neurosci.* **2018**, *10*, 114. [CrossRef] [PubMed]
244. Habib, N.; McCabe, C.; Medina, S.; Varshavsky, M.; Kitsberg, D.; Dvir-Szternfeld, R.; Green, G.; Dionne, D.; Nguyen, L.; Marshall, J.L.; et al. Disease-associated astrocytes in Alzheimer's disease and aging. *Nat. Neurosci.* **2020**, *23*, 701–706. [CrossRef] [PubMed]
245. Viejo, L.; Noori, A.; Merrill, E.; Das, S.; Hyman, B.T.; Serrano-Pozo, A. Systematic review of human post-mortem immunohistochemical studies and bioinformatics analyses unveil the complexity of astrocyte reaction in Alzheimer's disease. *Neuropathol. Appl. Neurobiol.* **2022**, *48*, e12753. [CrossRef] [PubMed]
246. Webster, S.J.; Bachstetter, A.D.; Nelson, P.T.; Schmitt, F.A.; Van Eldik, L.J. Using mice to model Alzheimer's dementia: An overview of the clinical disease and the preclinical behavioral changes in 10 mouse models. *Front. Genet.* **2014**, *5*, 88. [CrossRef]
247. Nanclares, C.; Baraibar, A.M.; Araque, A.; Kofuji, P. Dysregulation of Astrocyte–Neuronal Communication in Alzheimer's Disease. *Int. J. Mol. Sci.* **2021**, *22*, 7887. [CrossRef]
248. Abdelhak, A.; Foschi, M.; Abu-Rumeileh, S.; Yue, J.K.; D'anna, L.; Huss, A.; Oeckl, P.; Ludolph, A.C.; Kuhle, J.; Petzold, A.; et al. Blood GFAP as an emerging biomarker in brain and spinal cord disorders. *Nat. Rev. Neurol.* **2022**, *18*, 158–172. [CrossRef]
249. Asken, B.M.; Elahi, F.M.; La Joie, R.; Strom, A.; Staffaroni, A.M.; Lindbergh, C.A.; Apple, A.C.; You, M.; Weiner-Light, S.; Brathaban, N.; et al. Plasma Glial Fibrillary Acidic Protein Levels Differ Along the Spectra of Amyloid Burden and Clinical Disease Stage1. *J. Alzheimer's Dis.* **2020**, *78*, 265–276. [CrossRef]
250. Kalaria, R.N.; Ballard, C. Overlap Between Pathology of Alzheimer Disease and Vascular Dementia. *Alzheimer Dis. Assoc. Disord.* **1999**, *13*, S115–S123. [CrossRef]

251. Yi, M.; Wei, T.; Wang, Y.; Lu, Q.; Chen, G.; Gao, X.; Geller, H.M.; Chen, H.; Yu, Z. The potassium channel KCa3.1 constitutes a pharmacological target for astrogliosis associated with ischemia stroke. *J. Neuroinflamm.* **2017**, *14*, 203. [CrossRef]
252. Kuchibhotla, K.V.; Lattarulo, C.R.; Hyman, B.T.; Bacskai, B.J. Synchronous Hyperactivity and Intercellular Calcium Waves in Astrocytes in Alzheimer Mice. *Science* **2009**, *323*, 1211–1215. [CrossRef] [PubMed]
253. Delekate, A.; Füchtemeier, M.; Schumacher, T.; Ulbrich, C.; Foddis, M.; Petzold, G.C. Metabotropic P2Y1 receptor signalling mediates astrocytic hyperactivity in vivo in an Alzheimer's disease mouse model. *Nat. Commun.* **2014**, *5*, 5422. [CrossRef]
254. Lines, J.; Baraibar, A.M.; Fang, C.; Martin, E.D.; Aguilar, J.; Lee, M.K.; Araque, A.; Kofuji, P. Astrocyte-neuronal network interplay is disrupted in Alzheimer's disease mice. *Glia* **2022**, *70*, 368–378. [CrossRef]
255. Åbjørsbråten, K.S.; Skaaraas, G.H.S.; Cunen, C.; Bjørnstad, D.M.; Binder, K.M.G.; Bojarskaite, L.; Jensen, V.; Nilsson, L.N.; Rao, S.B.; Tang, W.; et al. Impaired astrocytic Ca^{2+} signaling in awake-behaving Alzheimer's disease transgenic mice. *eLife* **2022**, *11*, e75055. [CrossRef] [PubMed]
256. Lia, A.; Sansevero, G.; Chiavegato, A.; Sbrissa, M.; Pendin, D.; Mariotti, L.; Pozzan, T.; Berardi, N.; Carmignoto, G.; Fasolato, C.; et al. Rescue of astrocyte activity by the calcium sensor STIM1 restores long-term synaptic plasticity in female mice modelling Alzheimer's disease. *Nat. Commun.* **2023**, *14*, 1590. [CrossRef] [PubMed]
257. Bosson, A.; Paumier, A.; Boisseau, S.; Jacquier-Sarlin, M.; Buisson, A.; Albrieux, M. TRPA1 channels promote astrocytic Ca^{2+} hyperactivity and synaptic dysfunction mediated by oligomeric forms of amyloid-β peptide. *Mol. Neurodegener.* **2017**, *12*, 53. [CrossRef]
258. Satoh, K.; Hata, M.; Takahara, S.; Tsuzaki, H.; Yokota, H.; Akatsu, H.; Yamamoto, T.; Kosaka, K.; Yamada, T. A novel membrane protein, encoded by the gene covering KIAA0233, is transcriptionally induced in senile plaque-associated astrocytes. *Brain Res.* **2006**, *1108*, 19–27. [CrossRef]
259. Velasco-Estevez, M.; Mampay, M.; Boutin, H.; Chaney, A.; Warn, P.; Sharp, A.; Burgess, E.; Moeendarbary, E.; Dev, K.K.; Sheridan, G.K. Infection Augments Expression of Mechanosensing Piezo1 Channels in Amyloid Plaque-Reactive Astrocytes. *Front. Aging Neurosci.* **2018**, *10*, 332. [CrossRef] [PubMed]
260. Sanz, J.M.; Chiozzi, P.; Ferrari, D.; Colaianna, M.; Idzko, M.; Falzoni, S.; Fellin, R.; Trabace, L.; Di Virgilio, F. Activation of Microglia by Amyloid β Requires P2X7 Receptor Expression. *J. Immunol.* **2009**, *182*, 4378–4385. [CrossRef]
261. Orellana, J.A.; Shoji, K.F.; Abudara, V.; Ezan, P.; Amigou, E.; Sáez, P.J.; Jiang, J.X.; Naus, C.C.; Sáez, J.C.; Giaume, C. Amyloid beta-Induced Death in Neurons Involves Glial and Neuronal Hemichannels. *J. Neurosci.* **2011**, *31*, 4962–4977. [CrossRef]
262. Storck, S.; Meister, S.; Nahrath, J.; Meißner, J.N.; Schubert, N.; Di Spiezio, A.; Baches, S.; Vandenbroucke, R.; Bouter, Y.; Prikulis, I.; et al. Endothelial LRP1 transports amyloid-β1–42 across the blood-brain barrier. *J. Clin. Investig.* **2016**, *126*, 123–136. [CrossRef] [PubMed]
263. Gallwitz, L.; Schmidt, L.; Marques, A.R.; Tholey, A.; Cassidy, L.; Ulku, I.; Multhaup, G.; Di Spiezio, A.; Saftig, P. Cathepsin D: Analysis of its potential role as an amyloid beta degrading protease. *Neurobiol. Dis.* **2022**, *175*, 105919. [CrossRef] [PubMed]
264. Zlokovic, B.V. Neurovascular pathways to neurodegeneration in Alzheimer's disease and other disorders. *Nat. Rev. Neurosci.* **2011**, *12*, 723–738. [CrossRef] [PubMed]
265. Harrison, I.F.; Ismail, O.; Machhada, A.; Colgan, N.; Ohene, Y.; Nahavandi, P.; Ahmed, Z.; Fisher, A.; Meftah, S.; Murray, T.K.; et al. Impaired glymphatic function and clearance of tau in an Alzheimer's disease model. *Brain* **2020**, *143*, 2576–2593. [CrossRef]
266. Weller, M.; Wick, W.; Aldape, K.; Brada, M.; Berger, M.; Pfister, S.M.; Nishikawa, R.; Rosenthal, M.; Wen, P.Y.; Stupp, R.; et al. Glioma. *Nat. Rev. Dis. Prim.* **2015**, *1*, 15017. [CrossRef]
267. Wen, P.Y.; Weller, M.; Lee, E.Q.; Alexander, B.M.; Barnholtz-Sloan, J.S.; Barthel, F.P.; Batchelor, T.T.; Bindra, R.S.; Chang, S.M.; Chiocca, E.A.; et al. Glioblastoma in adults: A Society for Neuro-Oncology (SNO) and European Society of Neuro-Oncology (EANO) consensus review on current management and future directions. *Neuro Oncol.* **2020**, *22*, 1073–1113. [CrossRef]
268. Osswald, M.; Jung, E.; Sahm, F.; Solecki, G.; Venkataramani, V.; Blaes, J.; Weil, S.; Horstmann, H.; Wiestler, B.; Syed, M.; et al. Brain tumour cells interconnect to a functional and resistant network. *Nature* **2015**, *528*, 93–98. [CrossRef]
269. Weil, S.; Osswald, M.; Solecki, G.; Grosch, J.; Jung, E.; Lemke, D.; Ratliff, M.; Hänggi, D.; Wick, W.; Winkler, F. Tumor microtubes convey resistance to surgical lesions and chemotherapy in gliomas. *Neuro-Oncol.* **2017**, *19*, 1316–1326. [CrossRef]
270. Li, X.; Spelat, R.; Bartolini, A.; Cesselli, D.; Ius, T.; Skrap, M.; Caponnetto, F.; Manini, I.; Yang, Y.; Torre, V. Mechanisms of malignancy in glioblastoma cells are linked to MCU upregulation and higher intracellular calcium level. *J. Cell Sci.* **2020**, *133*, jcs237503. [CrossRef]
271. Venkataramani, V.; Schneider, M.; Giordano, F.A.; Kuner, T.; Wick, W.; Herrlinger, U.; Winkler, F. Disconnecting multicellular networks in brain tumours. *Nat. Rev. Cancer* **2022**, *22*, 481–491. [CrossRef]
272. Lo, M.; Wang, Y.-Z.; Gout, P.W. The xc−cystine/glutamate antiporter: A potential target for therapy of cancer and other diseases. *J. Cell. Physiol.* **2008**, *215*, 593–602. [CrossRef] [PubMed]
273. Robert, S.M.; Buckingham, S.C.; Campbell, S.L.; Robel, S.; Holt, K.T.; Ogunrinu-Babarinde, T.; Warren, P.P.; White, D.M.; Reid, M.A.; Eschbacher, J.M.; et al. SLC7A11 expression is associated with seizures and predicts poor survival in patients with malignant glioma. *Sci. Transl. Med.* **2015**, *7*, 289ra86. [CrossRef]
274. Venkatesh, H.S.; Johung, T.B.; Caretti, V.; Noll, A.; Tang, Y.; Nagaraja, S.; Gibson, E.M.; Mount, C.W.; Polepalli, J.; Mitra, S.S.; et al. Neuronal Activity Promotes Glioma Growth through Neuroligin-3 Secretion. *Cell* **2015**, *161*, 803–816. [CrossRef]
275. Goethe, E.A.; Deneen, B.; Noebels, J.; Rao, G. The Role of Hyperexcitability in Gliomagenesis. *Int. J. Mol. Sci.* **2023**, *24*, 749. [CrossRef]

276. Elias, A.F.; Lin, B.C.; Piggott, B.J. Ion Channels in Gliomas—From Molecular Basis to Treatment. *Int. J. Mol. Sci.* **2023**, *24*, 2530. [CrossRef] [PubMed]
277. Litan, A.; Langhans, S.A. Cancer as a channelopathy: Ion channels and pumps in tumor development and progression. *Front. Cell. Neurosci.* **2015**, *9*, 86. [CrossRef]
278. Wawrzkiewicz-Jałowiecka, A.; Trybek, P.; Dworakowska, B.; Machura, Ł. Multifractal Properties of BK Channel Currents in Human Glioblastoma Cells. *J. Phys. Chem. B* **2020**, *124*, 2382–2391. [CrossRef] [PubMed]
279. Thuringer, D.; Chanteloup, G.; Boucher, J.; Pernet, N.; Boudesco, C.; Jego, G.; Chatelier, A.; Bois, P.; Gobbo, J.; Cronier, L.; et al. Modulation of the inwardly rectifying potassium channel Kir4.1 by the pro-invasive miR-5096 in glioblastoma cells. *Oncotarget* **2017**, *8*, 37681–37693. [CrossRef]
280. D'alessandro, G.; Monaco, L.; Catacuzzeno, L.; Antonangeli, F.; Santoro, A.; Esposito, V.; Franciolini, F.; Wulff, H.; Limatola, C. Radiation Increases Functional KCa3.1 Expression and Invasiveness in Glioblastoma. *Cancers* **2019**, *11*, 279. [CrossRef]
281. Aissaoui, D.; Mlayah-Bellalouna, S.; Jebali, J.; Abdelkafi-Koubaa, Z.; Souid, S.; Moslah, W.; Othman, H.; Luis, J.; ElAyeb, M.; Marrakchi, N.; et al. Functional role of Kv1.1 and Kv1.3 channels in the neoplastic progression steps of three cancer cell lines, elucidated by scorpion peptides. *Int. J. Biol. Macromol.* **2018**, *111*, 1146–1155. [CrossRef]
282. Chinigò, G.; Castel, H.; Chever, O.; Gkika, D. TRP Channels in Brain Tumors. *Front. Cell Dev. Biol.* **2021**, *9*, 617801. [CrossRef]
283. Rubino, S.; Bach, M.D.; Schober, A.L.; Lambert, I.H.; Mongin, A.A. Downregulation of Leucine-Rich Repeat-Containing 8A Limits Proliferation and Increases Sensitivity of Glioblastoma to Temozolomide and Carmustine. *Front. Oncol.* **2018**, *8*, 142. [CrossRef]
284. Synowitz, M.; Ahmann, P.; Matyash, M.; Kuhn, S.A.; Hofmann, B.; Zimmer, C.; Kirchhoff, F.; Kiwit, J.C.W.; Kettenmann, H. $GABA_A$-receptor expression in glioma cells is triggered by contact with neuronal cells. *Eur. J. Neurosci.* **2001**, *14*, 1294–1302. [CrossRef] [PubMed]
285. Pallud, J.; Le Van Quyen, M.; Bielle, F.; Pellegrino, C.; Varlet, P.; Labussiere, M.; Cresto, N.; Dieme, M.-J.; Baulac, M.; Duyckaerts, C.; et al. Cortical GABAergic excitation contributes to epileptic activities around human glioma. *Sci. Transl. Med.* **2014**, *6*, 244ra89. [CrossRef] [PubMed]
286. Campbell, S.L.; Robel, S.; Cuddapah, V.A.; Robert, S.; Buckingham, S.C.; Kahle, K.T.; Sontheimer, H. GABAergic disinhibition and impaired KCC2 cotransporter activity underlie tumor-associated epilepsy. *Glia* **2015**, *63*, 23–36. [CrossRef] [PubMed]
287. Babateen, O.; Jin, Z.; Bhandage, A.; Korol, S.V.; Westermark, B.; Nilsson, K.F.; Uhrbom, L.; Smits, A.; Birnir, B. Etomidate, propofol and diazepam potentiate GABA-evoked GABAA currents in a cell line derived from human glioblastoma. *Eur. J. Pharmacol.* **2015**, *748*, 101–107. [CrossRef]
288. Huberfeld, G.; Vecht, C.J. Seizures and gliomas—Towards a single therapeutic approach. *Nat. Rev. Neurol.* **2016**, *12*, 204–216. [CrossRef]
289. Puia, G.; Gullo, F.; Dossi, E.; Lecchi, M.; Wanke, E. Novel modulatory effects of neurosteroids and benzodiazepines on excitatory and inhibitory neurons excitability: A multi-electrode array recording study. *Front. Neural Circuits* **2012**, *6*, 94. [CrossRef]
290. Pinacho-Garcia, L.M.; Valdez, R.A.; Navarrete, A.; Cabeza, M.; Segovia, J.; Romano, M.C. The effect of finasteride and dutasteride on the synthesis of neurosteroids by glioblastoma cells. *Steroids* **2019**, *155*, 108556. [CrossRef]
291. Maas, S.; Patt, S.; Schrey, M.; Rich, A. Underediting of glutamate receptor GluR-B mRNA in malignant gliomas. *Proc. Natl. Acad. Sci. USA* **2001**, *98*, 14687–14692. [CrossRef] [PubMed]
292. Takano, T.; Lin, J.H.-C.; Arcuino, G.; Gao, Q.; Yang, J.; Nedergaard, M. Glutamate release promotes growth of malignant gliomas. *Nat. Med.* **2001**, *7*, 1010–1015. [CrossRef] [PubMed]
293. Corsi, L.; Mescola, A.; Alessandrini, A. Glutamate Receptors and Glioblastoma Multiforme: An Old "Route" for New Perspectives. *Int. J. Mol. Sci.* **2019**, *20*, 1796. [CrossRef] [PubMed]
294. Venkatesh, H.S.; Tam, L.T.; Woo, P.J.; Lennon, J.; Nagaraja, S.; Gillespie, S.M.; Ni, J.; Duveau, D.Y.; Morris, P.J.; Zhao, J.J.; et al. Targeting neuronal activity-regulated neuroligin-3 dependency in high-grade glioma. *Nature* **2017**, *549*, 533–537. [CrossRef]
295. Ishiuchi, S.; Tsuzuki, K.; Yoshida, Y.; Yamada, N.; Hagimura, N.; Okado, H.; Miwa, A.; Kurihara, H.; Nakazato, Y.; Tamura, M.; et al. Blockage of Ca^{2+}-permeable AMPA receptors suppresses migration and induces apoptosis in human glioblastoma cells. *Nat. Med.* **2002**, *8*, 971–978. [CrossRef]
296. Piao, Y.; Lu, L.; de Groot, J. AMPA receptors promote perivascular glioma invasion via β1 integrin–dependent adhesion to the extracellular matrix. *Neuro-Oncol.* **2009**, *11*, 260–273. [CrossRef]
297. Längle, M.; Lutz, H.; Hehlgans, S.; Rödel, F.; Rau, K.; Laube, B. NMDA Receptor-Mediated Signaling Pathways Enhance Radiation Resistance, Survival and Migration in Glioblastoma Cells—A Potential Target for Adjuvant Radiotherapy. *Cancers* **2019**, *11*, 503. [CrossRef]
298. Ramaswamy, P.; Devi, N.A.; Fathima, K.H.; Nanjaiah, N.D. Activation of NMDA receptor of glutamate influences MMP-2 activity and proliferation of glioma cells. *Neurol. Sci.* **2014**, *35*, 823–829. [CrossRef]
299. Lange, F.; Hörnschemeyer, J.; Kirschstein, T. Glutamatergic Mechanisms in Glioblastoma and Tumor-Associated Epilepsy. *Cells* **2021**, *10*, 1226. [CrossRef]
300. Stepulak, A.; Luksch, H.; Gebhardt, C.; Uckermann, O.; Marzahn, J.; Sifringer, M.; Rzeski, W.; Staufner, C.; Brocke, K.S.; Turski, L.; et al. Expression of glutamate receptor subunits in human cancers. *Histochem. Cell Biol.* **2009**, *132*, 435–445. [CrossRef]
301. Yu, L.J.; Wall, B.A.; Wangari-Talbot, J.; Chen, S. Metabotropic glutamate receptors in cancer. *Neuropharmacology* **2017**, *115*, 193–202. [CrossRef] [PubMed]

302. Salmaggi, A.; Corno, C.; Maschio, M.; Donzelli, S.; D'urso, A.; Perego, P.; Ciusani, E. Synergistic Effect of Perampanel and Temozolomide in Human Glioma Cell Lines. *J. Pers. Med.* **2021**, *11*, 390. [CrossRef] [PubMed]
303. Khalil, A.A.; Jameson, M.J.; Broaddus, W.C.; Lin, P.S.; Chung, T.D. Nicotine enhances proliferation, migration, and radioresistance of human malignant glioma cells through EGFR activation. *Brain Tumor Pathol.* **2013**, *30*, 73–83. [CrossRef] [PubMed]
304. Barres, B.A. The Mystery and Magic of Glia: A Perspective on Their Roles in Health and Disease. *Neuron* **2008**, *60*, 430–440. [CrossRef]
305. Endo, F.; Kasai, A.; Soto, J.S.; Yu, X.; Qu, Z.; Hashimoto, H.; Gradinaru, V.; Kawaguchi, R.; Khakh, B.S. Molecular basis of astrocyte diversity and morphology across the CNS in health and disease. *Science* **2022**, *378*, eadc9020. [CrossRef]
306. Oberheim, N.A.; Takano, T.; Han, X.; He, W.; Lin, J.H.C.; Wang, F.; Xu, Q.; Wyatt, J.D.; Pilcher, W.; Ojemann, J.; et al. Uniquely Hominid Features of Adult Human Astrocytes. *J. Neurosci.* **2009**, *29*, 3276–3287. [CrossRef]
307. Vasile, F.; Dossi, E.; Rouach, N. Human astrocytes: Structure and functions in the healthy brain. *Anat. Embryol.* **2017**, *222*, 2017–2029. [CrossRef]

life

MDPI

Article

Nucleoporin Nup358 Downregulation Tunes the Neuronal Excitability in Mouse Cortical Neurons

Vladimir A. Martínez-Rojas [1,†], Francesca Pischedda [2], Isabel Romero-Maldonado [3], Bouchra Khalaf [2,‡], Giovanni Piccoli [2], Paolo Macchi [2,*] and Carlo Musio [1,*]

[1] Institute of Biophysics—IBF, National Research Council—CNR, Via Sommarive 18, 38123 Trento, Italy; vlalx.mr@gmail.com

[2] Department of Cellular, Computational and Integrative Biology—CIBIO, University of Trento, Via Sommarive 9, 38123 Trento, Italy; francesca.pischedda@unitn.it (F.P.); bouchra.khalaf@gmail.com (B.K.); giovanni.piccoli@unitn.it (G.P.)

[3] Institute of Cellular Physiology, Universidad Autónoma de Mexico—UNAM, Ciudad Universitaria, Mexico City 04510, Mexico; isabel.romero@comunidad.unam.mx

* Correspondence: paolo.macchi@unitn.it (P.M.); carlo.musio@cnr.it (C.M.)

† Current address: Department of Physiology, Biophysics and Neurosciences, CINVESTAV, Mexico City 07360, Mexico.

‡ Current address: Sunnybrook Research Institute, Toronto, ON M4N 3M5, Canada.

Abstract: Nucleoporins (NUPs) are proteins that comprise the nuclear pore complexes (NPCs). The NPC spans the nuclear envelope of a cell and provides a channel through which RNA and proteins move between the nucleus and the cytoplasm and vice versa. NUP and NPC disruptions have a great impact on the pathophysiology of neurodegenerative diseases (NDDs). Although the downregulation of Nup358 leads to a reduction in the scaffold protein ankyrin-G at the axon initial segment (AIS) of mature neurons, the function of Nup358 in the cytoplasm of neurons remains elusive. To investigate whether Nup358 plays any role in neuronal activity, we downregulated Nup358 in non-pathological mouse cortical neurons and measured their active and passive bioelectrical properties. We identified that Nup358 downregulation is able to produce significant modifications of cell-membrane excitability via voltage-gated sodium channel kinetics. Our findings suggest that Nup358 contributes to neuronal excitability through a functional stabilization of the electrical properties of the neuronal membrane. Hypotheses will be discussed regarding the alteration of this active regulation as putatively occurring in the pathophysiology of NDDs.

Keywords: nucleoporins; ion channels; membrane excitability; voltage-gated sodium channels; neuronal activity; neurodegenerative disease; patch-clamp

Citation: Martínez-Rojas, V.A.; Pischedda, F.; Romero-Maldonado, I.; Khalaf, B.; Piccoli, G.; Macchi, P.; Musio, C. Nucleoporin Nup358 Downregulation Tunes the Neuronal Excitability in Mouse Cortical Neurons. *Life* **2023**, *13*, 1791. https://doi.org/10.3390/life13091791

Academic Editor: Tzer-Bin Lin

Received: 16 June 2023
Revised: 17 August 2023
Accepted: 21 August 2023
Published: 22 August 2023

1. Introduction

The nuclear pore complex (NPC) is a large multiprotein complex of nucleoporin (NUP) proteins that mediate nucleocytoplasmic transport, genome organization, and gene expression [1,2]. An important body of studies indicates that disruptions of the NPC contribute to the pathogenesis of neurodegenerative diseases (NDDs) by triggering pathophysiological intracellular cascade effects [3,4]. NUPs were first reported to localize at the nuclear rim, but emerging evidence has revealed that some individual nucleoporins are stably expressed in the cytoplasm and/or the nucleoplasm and contribute to numerous physiological events [5–8]. Nup358, also known as Ran-binding protein 2 (RanBP2), is by far the largest nucleoporin [9] of the NPC [10]. In neurons, Nup358 is indispensable for the process of axon specification and cell polarity during neuronal differentiation in culture [11]. The conditional loss of Nup358 in Thy1$^+$-motoneurons in mice causes the development of amyotrophic lateral sclerosis-like motor traits [12]. Furthermore, the loss of Nup358 in motoneurons triggers the dysregulation of neuronal–glial and chemokine signaling in sciatic nerves and somas of spinal motoneurons from mice without Nup358 [13]. Altogether, those reports underline the relevance of Nup358 in neuronal homeostasis during the development and maturation of neurons.

Recently, it has been shown that Nup358, through its N-terminal domain, is associated with the axon initial segment (AIS) in neurons in an ankyrin-G-dependent fashion [14]. Another study also confirmed that Nup358 is enriched at the AIS [15]. Stimulation through chemical depolarization of neurons results in an evident change in Nup358 subcellular distribution and/or expression, as demonstrated by (1) a reduction in the Nup358 signal at the nuclear rim, (2) more diffuse staining in the entire neuronal cell, and (3) a decrease in the total Nup358 protein expression level. Although evidence on the regulation of electrical activity exerted by nucleoporins has been reported in non-neuronal systems [16,17], the functional impact of the Nup-mediated electrophysiological activity in neuronal systems is still unexplored.

Hypothesizing that Nup358 modulates neuronal excitability, this work hence aims to determine the functional effects of Nup358 downregulation in neurons. Using whole-cell patch-clamp on dissociated cortical neurons, we show that Nup358 regulates neuron excitability by facilitating the conductance of voltage-gated sodium channels.

On the basis of functional interpretations of our results, and according to the structure/function relationship concept, we can postulate that (1) Nup358 promotes the stabilization of the membrane proteins that control neuronal firing and (2) impairments in Nup358 function might be involved in the pathophysiological mechanisms of NDDs.

2. Materials and Methods

2.1. Animals and Cell Culture

Mice (c57/bl6 strain) were kept in a normal light/dark cycle (12 h light:12 h dark) and had free access to water and food. Cortical neurons were prepared from E15 female and male embryos using a previously described method [14]. Briefly, we euthanized the pregnant mice with a carbon dioxide (CO_2) overdose. We isolated cortices under a stereomicroscope. After mechanical dissociation, cells were counted and plated on 12 mm coverslips coated with poly-D-lysine (Sigma, St. Louis, MO, USA) according to the density of 150–200 cells/mm^2. The cells were grown in neuronal complete medium (Neurobasal 1×, B-27 supplement 1×, 0.5 mM L-glutamine, 10 μg/mL gentamicin; B-27 supplement and Neurobasal medium were purchased at Life Technologies, Carlsbad, CA, USA) at 37 °C and 5% CO_2 until full maturation, i.e., 14 days in vitro (DIV).

2.2. Constructs

Scrambled negative control shRNA (shCTRL, Origene TR30021) and a pool of four mouse gene-specific Nup358 shRNA constructs (shNup358, Origene TL501860) in pGFP-C-shLenti Vector (Origene, Rockville, MD, USA) plasmids were used for the knockdown experiments as previously described [14]. Briefly, mouse cortical neurons were transfected on DIV 5 with shRNA constructs using Lipofectamine 2000 reagent (Invitrogen, Waltham, MA, USA) following the manufacturer's instructions and processed on DIV 14.

2.3. Electrophysiology

2.3.1. Recordings

Patch clamp recordings, in the configuration of the whole cell, were conducted in accordance with previous reports [18–20]. The neurons in the culture were placed into a recording chamber and underwent constant perfusion (flow rate, 3 mL per min) with standard extracellular solution consisting of (in mM) 140 NaCl, 4 KCl, 10 HEPES, 2.0 $MgCl_2$, 2.0 $CaCl_2$, and 10 glucose, pH 7.4 and 290 mOsm. The micropipettes were obtained from a temperature-controlled PIP6 pipette puller (HEKA, Reutlingen, Germany) and borosilicate glass capillaries (Harvard Apparatus, Cambridge, MA, USA). The resistance of the electrodes ranged between 3 and 5 MΩ upon filling them with an intracellular solution composed of (in mM) 130 K-gluconate, 10 KCl, 0.1 $CaCl_2$, 2.0 $MgCl_2$, 1.1 EGTA,

10 HEPES, 2.0 Mg-ATP, and 0.2 Na-GTP, pH 7.3 (280 mOsm). The cells selected for recording were identified subsequent to the 470 nm stimulation using a fiber-coupled LED (M89L01, Thorlabs, Newton, MA, USA), and the microelectrodes were positioned, for each neuron, in a distal somatic area through a water-immersion 40× objective. Once the whole-cell configuration was established, the series resistance and membrane capacitance were electronically compensated. Experimental signals were obtained using an ELC-03XS amplifier (npi, Tamm, Germany), filtered at 2 kHz with a low-pass filter, and sampled at 10 kHz with an INT-20X interface (npi electronic, Tamm, Germany). Analogic signals were recorded using an ELC-03XS amplifier (npi, Germany) and digitized with an INT-20X interface (npi electronic, Tamm, Germany). Data were acquired with WinWCP V5.2.7 software (©John Dempster, University of Strathclyde, Glasgow, UK) and low-pass-filtered at 2 kHz. The sampling rate employed was 10 kHz. Seal stability was tracked online, incorporating the following criteria: access resistance (<20 MΩ < 20% drift), stable RMP, and holding currents <100 pA. After a brief period of stabilization, the RMP was measured in a bridge balance configuration. The ion currents and action potentials were obtained from neurons displaying an RMP that was stable and lower than −50 mV. Thereafter, the membrane's passive properties were explored through hyperpolarizing current square pulses from 0 to −150 pA in −30 pA decreasing pulses. Still in a current-clamp configuration, a succession of depolarizing pulses required to trigger an action potential (AP) were injected from an imposed membrane potential of −70 mV, and unitary APs were evoked using 10 ms suprathreshold steps. In order to evaluate the AP frequency, we delivered long pulses (1000 milliseconds) of depolarizing current in 50 pA delta increments. Voltage-dependent ion conductances were elicited by 10 mV steps from −80 to +40 mV (500 ms) and a holding potential of −70 mV in a voltage-clamp configuration. In order to remove capacitive transients and leakage currents, both of no interest, from the macroscopic current providing only the ionic current under exam, P/4 (or P/N according to WinWCP V5.2.7 software term) leak subtraction protocol was applied online, with which capacitive and leakage currents, ideally linear and not voltage-dependent, were subtracted from the signal of interest.

2.3.2. Ionic Current Dissection

The experiments designed for the isolation of sodium current were performed as previously described [20], Cs+-based intracellular solution was composed of (in mM) 130 CsCl, 10 NaCl, 2 $MgCl_2$, 0.1 $CaCl_2$, 1.1 EGTA, 10 HEPES, 3 MgATP, 0.3 GTP, and pH 7.2 adjusted with CsOH, and 4-amino pyridine, 4-AP (20 μM) and tetraethylammonium, TEA (200 μM) were also added to the bath solution. The voltage dependency of activation was computed through a stepwise procedure starting from a holding potential of −70 mV and with increasing voltage steps of +5 mV across a range of potentials from −80 to +40 mV. The voltage dependency of inactivation was explored using a two-step procedure, with an initial conditioning pulse lasting 40 ms delivered from a holding potential of −70 mV to a series of potentials from −80 to 0 mV in 10 mV increments, succeeded by a test pulse to +10 mV. To analyze the recovery from fast inactivation, we employed a multiple-pulse procedure executed with a conditioning depolarization from a holding potential of −70 to −15 mV (10 ms); then, a subsequent test pulse at the equal depolarizing pulse value was delivered beforehand to evoke the Na^+ current fraction. The time between conditioning and test pulses (Δ) was modified throughout 2–18 ms, in 2 ms augmentations, to analyze the rate of recovery [21].

2.4. Data Analysis

Mean ± SEM values are provided for the data, with n representing the sample size of cells recorded. The AP waveform kinetics were calculated by differentiating the spike voltage with respect to time (dV/dt). Plane phase graphs were created by plotting the first derivative of the membrane potential (in mV/ms) against the membrane voltage. The threshold was established based on the membrane potential, where dV/dt exhibited an abrupt increase. The current density was estimated by dividing the peak current values by the total membrane surface area obtained from whole-cell capacitance determination. Conductance (G) was calculated through the equation $G(V_m) = I(V_m)/(V_m - E_{rev})$, where $I(V_m)$ is the current measured at the used membrane potential and V_m, and E_{rev} is the computed Nernst potential. The maximum conductivity (G_{max}) of every cell was used for normalization. To determine a measure of gating conductance, peak currents were adjusted using a Boltzmann distribution with the form $G = G_0 + (G_{max} - G_0)/1 + \exp[(V_{1/2} - V)/k]$, where G refers to the conductance that varies with voltage, while G_0 represents the voltage-independent baseline conductance. $V_{1/2}$ is the voltage of half-maximal activation, while k denotes the slope factor employed to calculate the activation kinetics [20]. The inactivation time constants were obtained by plotting the peak current amplitude during each test pulse normalized to the current amplitude of the first test pulse and graphed as a function of the pre-pulse potential. Last, the Boltzmann equation was used to fit the values of the voltage dependency of inactivation. The time constant of the recovery from inactivation was obtained by plotting the ratio $I_{test}/I_{conditioning}$ against the interval time (Δt) and the data were fitted with a single exponential equation. In order to perform data analysis, curve fitting, and plotting, we used the WinWCP, OriginPro 8.1 (OriginLab, Northampton, MA, USA), and Prism 8 (GraphPad, Boston, MA, USA) software packages, respectively. The Kolmogorov–Smirnov test was employed to test data normality. An unpaired Student's t was employed when comparing two groups; in all analyses, a threshold for statistical significance was set at $p < 0.05$.

3. Results

We first downregulated Nup358 using RNAi as previously described [14]. Neurons were transfected either with an shNup358 or with a non-targeted shCTRL construct as a negative control. The efficiency of the shRNAs in eliminating the expression of Nup358 was evaluated via immunostaining (Figure 1). Transfected neurons showed a significant reduction in Nup358 signal compared to shCTRL transfected neurons.

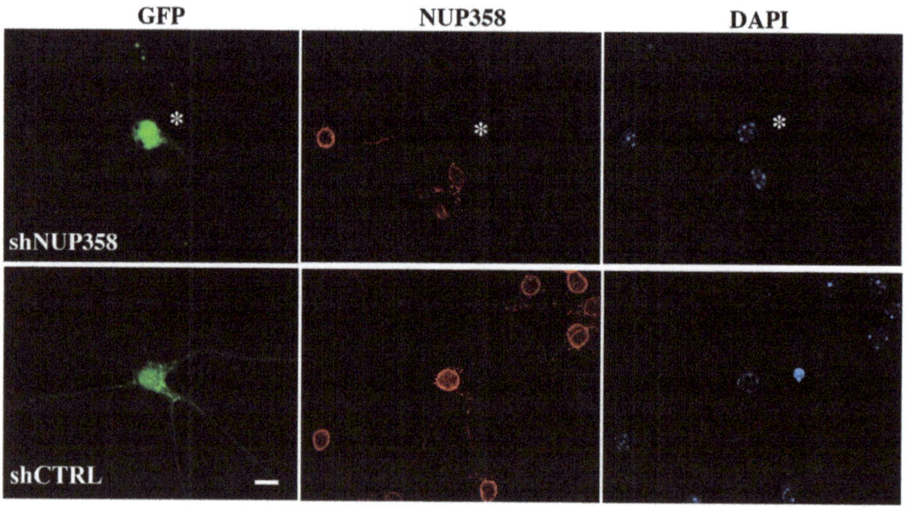

Figure 1. Downregulation of Nup358 expression in cortical neurons. Cortical neurons transfected at 5 DIV, with shRNA constructs co-expressing GFP to downregulate Nup358 (shNup358). As a control, neurons were transfected with a non-targeted shRNA construct (shCTRL). Immunostaining was conducted for Nup358 (red), and the nuclei are labeled with DAPI (blue). Asterisks are used to mark the transfected neuronal cells (green) with a substantial decrease in the Nup358 signal compared to the control untransfected cells. Scale bar: 10 μm.

We subsequently examined the effect of Nup358 downregulation on the spontaneous activity of the cultured mouse cortical neurons at 12 days in vitro (DIV12). The transfected neurons were identified through the presence of the reporter protein, GFP, and only those exhibiting positive expression were chosen for whole-cell patch-clamp recordings. At their respective physiological resting membrane potential (RMP), we tracked the intrinsic neuronal activity and found no significant differences in the RMP between the neurons with induced downregulation of Nup358 (56.85 ± 1.52 mV) and those treated with scrambled shRNA (57.18 ± 1.17 mV) (Figure 2A). While we did not detect differences in the amplitude, threshold, or the one-half width of the action potential (Figure 2B), the frequency of membrane potentials was significantly higher in the shNup358-treated neurons (0.42 ± 0.10 Hz) than in the control neurons (0.10 ± 0.03 Hz; here and throughout the text, values are mean ± SEM).

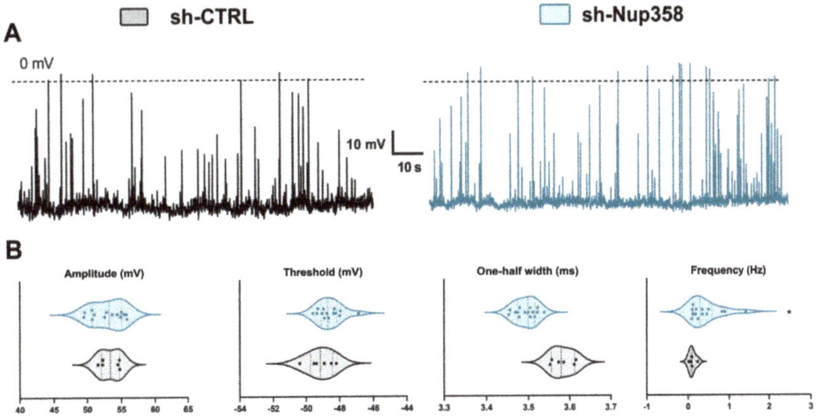

Figure 2. The effect of Nup358 downregulation on the spontaneous activity of cultured mouse cortical neurons. (**A**) Tracks of membrane potential changes recorded in gap-free mode in the control scrambled-treated (**left**) and shNup358-treated (**right**) neurons. (**B**) Different properties of the resting membrane potentials are computed from the traces shown in (**A**). Each dot in the graphs represents an individual cell, and the dashed line in the violin plots is the median; the left and right dashed lines are the upper and lower quartiles, respectively. $n = 14$. * $p < 0.05$, Student's unpaired t-test.

Next, we assessed whether Nup358 downregulation alters the active neuronal membrane properties. We injected neurons with 1000 ms long hyperpolarizing/depolarizing current pulses of increasing amplitude and recorded variations of membrane potentials from the membrane resting potential.

The injection of a 150 pA current evoked a tonic firing in neurons from both experimental groups (Figure 3A); however, the number of spikes was significantly higher in the shNup358-treated neurons. We performed raster plot and frequency analyses to estimate the shNup358-mediated rise in the action potential tonic firing.

Compared to the control group, the downregulation of Nup358 resulted in an increase in the number of spikes per individual neuron and a leftward shift in the InterSpike Interval (ISI) temporal window, indicating prompter excitability of the neurons (Figure 3B,C). These results demonstrate that Nup358 plays a critical role in regulating the resting membrane potential and the frequency of action potential firing of neurons.

Further, we analyzed whether Nup358 downregulation impacts the biophysical properties of the single action potential. Figure 4A illustrates two representative traces of a single action potential reaching the firing threshold. These traces were obtained by injecting 10 ms of somatic depolarizing current into either the control or the shNup358 neurons. To assess the detailed action potential kinetics in detail, we assembled phase plots from overlaid means of single action potentials.

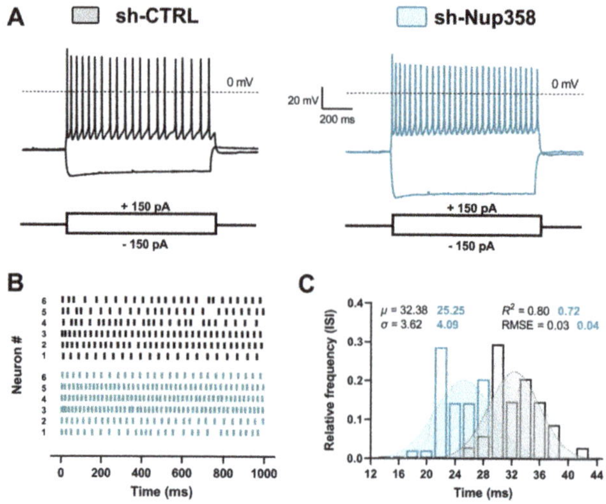

Figure 3. The downregulation of Nup358 increases the firing frequency of cortical neurons. (**A**) Representative recordings of membrane potentials in scrambled and shNup358-treated neurons in response to 1000 ms long whole-cell current injections. (**B**) Raster plot showing the temporal locations of action potential firing over current injection trials. Each bar in the raster indicates one action potential, and each row represents an independently recorded neuron. (**C**) Relative frequency distributions of the mean Inter Spike Interval (ISI) computed from the neurons assessed in (**B**). $n = 6 + 6$. μ, Gaussian fit mean; σ, Gaussian fit S.D.; RMSE, root-mean-square deviation.

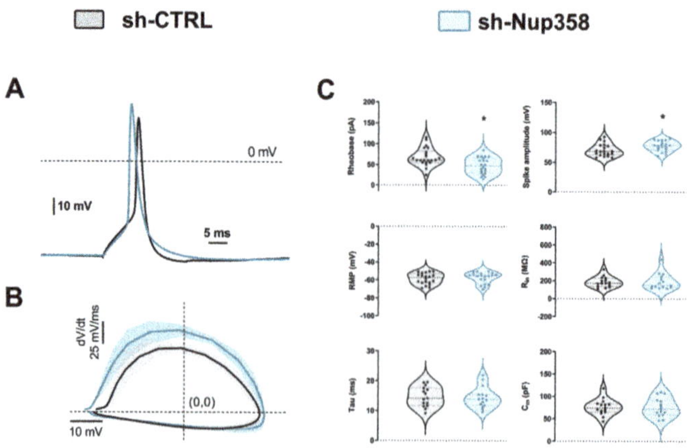

Figure 4. Downregulation of Nup358 alters the intrinsic excitability of neurons. (**A**) Representative recordings of single APs elicited with a 10-millisecond current pulse delivered to control or shNup358-silenced neurons. (**B**) Phase plane plots (dV/dt versus membrane voltage) from averaged single APs. The central line represents the mean, and the SEM is shown as a faded area. (**C**) Violin graph comparing the rheobase and spike amplitude (as well as passive membrane properties) in Nup358-downregulated neurons. Graphs of grouped data show individual cells and mean ± SEM from cortical neurons pooled from at least four independent brain dissections for each condition. * $p < 0.05$, Student's unpaired t-test. $n = 19$. RMP, resting membrane potential; R_{in}, input resistance; Tau, membrane time constant; C_m, membrane capacitance.

We examined changes in different parameters of individual spikes' waveforms (Figure 4C). The amplitude of the AP in shNup358-silenced neurons was higher and significantly different from control neurons, with a computed increase of 7.3 ± 1.90 mV. The minimum amount of current required to evoke an AP (rheobase) was 47.50 ± 4.24 mV smaller in Nup358-silenced neurons compared to control neurons, which had 68.00 ± 4.58 mV (Figure 4B). No such differences were observed between the mean of resting membrane potential, input resistance, capacitance, or membrane time constant (tau) membrane properties.

Overall, Nup35-silenced neurons exhibited a faster firing and larger amplitude of action potentials. These results thus suggest that Nup358 modulates the active membrane properties, that is, the input–output relationships and the repetitive AP firing of individual neurons.

Since the AIS—by defining a specific neuronal compartment with the lowest firing threshold—serves as the site for action potential (AP) initiation, we investigated the potential impact of ionic conductances involved in AP generation, considering that Nup358 interacts with the scaffold protein Ankyrin-G, which plays a crucial role in the assembling the axon initial segment (AIS) and the recruiting of other molecules like voltage-dependent sodium channels (Na_V) [22]. Hence, using the voltage-clamp configuration and clamping neurons V_m at −70 mV, we examined whether the voltage-gated total ionic currents underlie the increase in neuronal activity of shNup358-silenced cortical neurons.

The step stimulation elicited voltage-gated macroscopic currents composed of Na^+-associated inward and K^+-associated outward currents (Figure 5A). To record the ionic responses belonging to the AIS, we placed the patch pipette in a position distal to the somatic compartment and delivered the voltage pulses (Figure 5(A_i)). The analysis of the steady-state (outward) or peak (inward) total currents at the maximum amplitude revealed an increase of 39.05 ± 3.56% in the amplitude of the inward current in shNup358-treated cells as compared to control neurons (Figure 5(B,B_i)). We did not observe modifications between the control and Nup358-silenced neurons when the potassium-associated outward component was compared (Figure 5C).

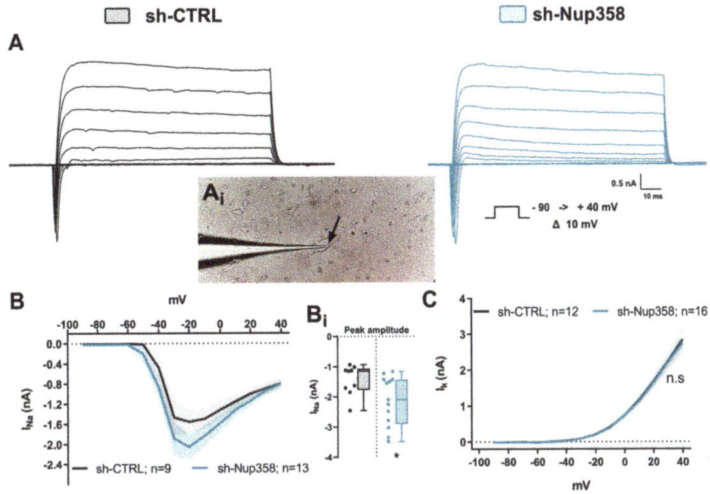

Figure 5. Inward total ionic current increase in shNup358-treated neurons. (**A**) Ionic current traces obtained from voltage-clamped cortical neurons. The neurons were held at −70 mV, and then voltage steps of 100 milliseconds from −90 to +40 mV were applied in 10 mV increments. (Inset $\mathbf{A_i}$) Exemplification of cultured cortical neurons from mice at 12DIV sealed with a glass micropipette (**left**). All the whole-cell patch clamp recordings in this work were performed in the optically identified AIS surrounding area (black arrow). (**B**) Current–voltage (I–V) relationships of the inward deflection (peak) of the traces are depicted in (**A**), left. (Inset $\mathbf{B_i}$) Analysis of the mean differences in maximum current amplitude at −20 mV. (**C**) Current–voltage relationships of the outward deflection (steady state) of the traces depicted in (**A**) right. The total number of assessed neurons is shown in each panel. * $p < 0.05$, Student's unpaired t-test; n.s., not significant.

Based on the previous finding, we sought to dissect sodium-associated inward currents using pharmacological and ionic substitution procedures (Figure 6A). The amplitude of the Na^+ currents was increased in shNup358-treated neurons at the maximum peak of the curve (−20 mV). We measured the current density using the ratio of peak current amplitude to the cell membrane capacitance (pA/pF). This formal analysis computed an increase of 56% ± 3.2 of the normalized current density obtained from silenced Nup358 neurons when compared to control neurons (Figure 6B). To measure the voltage dependence of sodium conductance, the IV curve data from Figure 6B were then employed to construct a conductance/voltage (GV) plot. Then, the normalized (G/G_{max}) GV relationships were fit using a single Boltzmann function (see methods for details). From this analysis, we obtained the potential of the half-maximal activation ($V_{1/2}$) and the slope value (K). In control neurons, the calculated values for $V_{1/2}$ and K were −24.37 ± 0.68 mV and 3.49 ± 0.61, respectively. Interestingly, in neurons treated with shNup358, the $V_{1/2}$ showed a slight hyperpolarization of −26.65 ± 0.35 mV, and the K value decreased to 1.96 ± 0.27 (Figure 6C). At this point, the results clearly indicate that Nup358 is a key modulator of sodium conductance in the AIS.

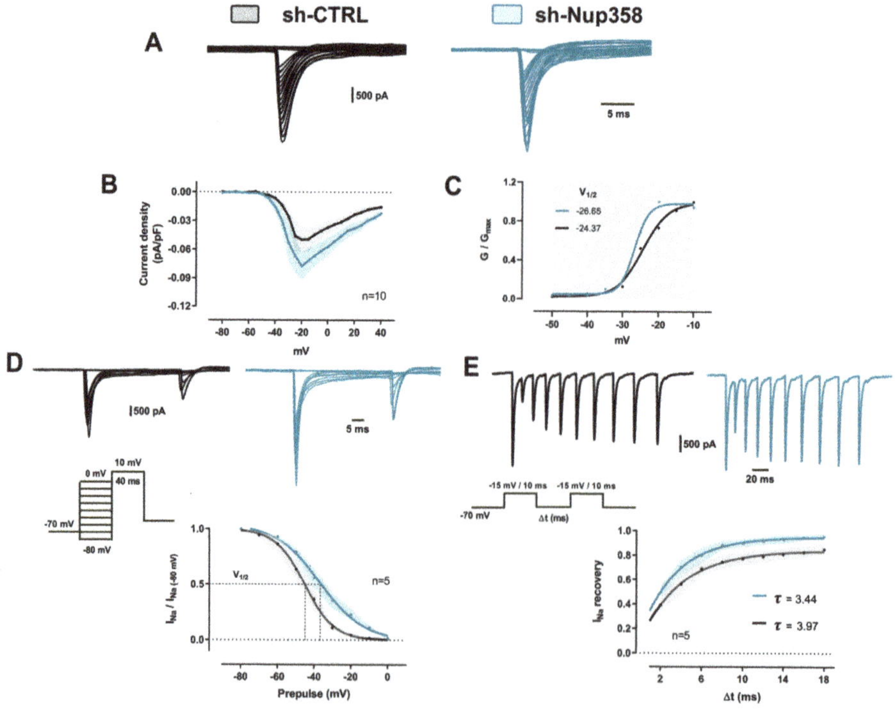

Figure 6. Nup358-dependent modifications of sodium currents' biophysical properties. (**A**) Cs-based isolation of voltage-gated sodium currents (I_{Na}) in response to voltage steps between −80 and +40 mV. (**B**) I-V plot of normalized peak I_{Na} currents plotted with current density as a function of test voltage. (**C**) Normalized mean conductance versus voltage plot of I_{Na} activation of the data in (**B**). The line represents the average Boltzmann fit to data. (**D**) Representative currents of I_{Na} steady-state inactivation (inset, voltage protocol). Currents were normalized to the current elicited from a holding potential of −80 mV and used to construct the steady-state inactivation. Plotted points represent the mean of the normalized currents and are fitted with a Boltzmann function. (**E**) Representative I_{Na} current traces were obtained using the protocol shown in the inset to study the recovery time course from fast inactivation. The plot represents the current recovery as a function of the interstimulus interval (Δt). Recoveries were fitted with one exponential.

To gain more insight into the biophysical mechanisms by which the Na^+ conductances could be altered in the absence of Nup358, we measured steady-steady fast inactivation of Na^+ currents elicited with voltage commands (pre-pulse) from −80 to 0 mV (10 mV step increase, 40 ms) in cortical neurons held at −70 mV. This allows the channels to undergo fast inactivation without permitting the channels to slowly inactivate and then to depolarize to a test pulse (10 mV) to measure the fraction of channels that did not fast inactivate. As Figure 6D shows, Nup358 silencing induced a positive shift in the I_{Na+} (voltage-gated sodium current) inactivation curve ($V_{1/2}$ = −36.97 ± 0.68 mV and *K* factor = −12.04 ± 1.15) compared with the control ones ($V_{1/2}$ = −45.19 ± 0.92 mV and K factor = −8.07 ± 0.85). Finally, we used a multiple-pulse protocol with recovery durations between 2 ms and 18 ms to further characterize the recovery from fast inactivation properties (inset, Figure 6D). Compared to the control, Na^+ channels from Nup358-silenced neurons recovered rapidly from inactivation when held at negative voltages between pulses. The computed recovery time constant in shNup358 was 3.44 ms, a value considerably lower than the 3.97 ms estimated from scrambled neurons (Figure 6E). Taken together, the data obtained in this work strongly suggest that Nup358 promotes a fine-tuned control of sodium conductance in the AIS.

4. Discussion

4.1. Nup358 Downregulation and the Neuronal Excitability Alteration

Our results provide the first evidence that the downregulation of Nup358 results in increased neuronal excitability associated with changes in the biophysics of voltage-gated sodium channels. Previously, the Macchi group highlighted the unique localization of Nup358 protein expression at the AIS of cultured neurons and its relevance to neuronal activity [14]. Here, we examine in depth the electrophysiological implications of reduced Nup358 expression in cortical neurons. It is well known that the AIS has a pivotal role in triggering APs and determining neuronal output. The biophysical properties of the AIS are primarily determined by the molecular assembly and the concentration of voltage-gated channels' proteins in this specialized region. The increased excitability observed with Nup358 silencing is consistent with the hypothesis that nucleoporin regulates the abundance, membrane distribution, or activity of some proteins, such as voltage-gated ion channels.

Previous studies have examined the role of individual Nups in the modulation of voltage-gated ion channels in cardiac electrophysiology. For instance, Nup50 induces an increase in the transcription and translation of the *Kcna4* gene, which encodes a K^+ voltage-gated channel (Kv1.4) of shaker-related subfamily member 4, which is essential for repolarization in cardiomyocytes [16]. Nup107 also regulates cardiac bioelectricity by controlling the nucleocytoplasmic trafficking of a sodium channel mRNA (*Scn5a*) [17,23]. As is well known, both sodium and potassium voltage-gated ion channels are fundamental to the initiation and the conduction of APs in excitable cells, respectively [24]. These studies thus underscore the role of individual Nups in regulating the activity of excitable cells by modulating the expression of voltage-gated channels. In a similar manner, Nup358 might modulate the expression of proteins either through direct/indirect transcriptional control of the genes encoding the ion channel subunits or by changing the expression of genes encoding enzymes altering ion channel activity through post-translational modifications.

An emerging number of reports indicate the role of nucleoporins in the homeostasis impairment of neuronal cells and thus having ramifications in neurodegenerative disease. For instance, the alteration of the Nup153 expression leads to the functional impairment of hippocampal neural stem cells in the AD mouse model [25]. Moreover, Ranbp2/Nup358 has been implicated in the regulation of the manifestations associated with Parkinson's disease [26]. The conditional ablation of Nup358 in mouse motor neurons causes disruptions in cellular functions and the development of amyotrophic lateral sclerosis (ALS)-like syndromes [12]. Furthermore, an in vitro/in vivo study reports that Nup358 contributes to neuronal activity and that its downregulation disrupts the functional properties of neurons regardless of the molecular domain of expression [27].

Using patch-clamp recordings, we observed a marked increase in evoked and spontaneous APs discharge upon Nup358 silencing in neurons. This hyperexcitability of neurons due to a reduction in Nup358 expression is the first demonstration linking a formerly recognized nuclear protein with an electrophysiological event. We found an increment in the amplitude of I_{Na} from Nup358-silenced neurons. Guan et al. [17] likewise found a rise in the amplitude of Nav1.5 channel currents in cardiac myocytes, but that increment was the cellular result of Nup107 overexpression.

We further explored the biophysical mechanisms of voltage-gated sodium currents. To note, we did not find changes in the voltage-gated potassium currents between control and Nup358-silenced cortical neurons, as reported for another nucleoporin family [16]. Since the activity-dependent alterations of Na^+ channels would be predicted to have functional effects on neuronal excitability [28], we explored the kinetic properties of the activation/inactivation of voltage-gated Na^+ channels in cortical neurons subjected to Nup358 downregulation. The kinetics of the voltage dependency of activation and the steady state from the fast inactivation of voltage-dependent Na^+ channels were modified in Nup358-silenced neurons. In accordance with this, others have reported a functional alteration of the voltage-dependent Na^+ channel kinetics in neurons from the central nervous system [29,30]. Collectively, these studies demonstrate that alterations induced by Nup358 in Na^+ channel functions may underlie cortical neuron hyperexcitability. Among the alterations, we report (1) the presence of increased inward current, (2) a shift in steady-state activation or inactivation, and (3) a modification of channel recovery from fast inactivation). Moreover, we have noted that Nup358 is a fast-reactive protein responding to neuronal insults [13]. The chemical depolarization of cortical neurons with KCl modified the subcellular distribution and protein expression of Nup358 [13].

Taken together, our data show that, besides its role in nucleocytoplasmic transport, Nup358 markedly contributes to neuronal activity by modulating sodium currents. However, we cannot exclude those additional molecular mechanisms, such as anchoring proteins like AnkG, which could contribute to neuronal plasticity and in turn exert a fine-tuned regulation of Na^+ channels kinetics. Moreover, according to our functional data, the involvement of other ionic channels/conductance (e.g., potassium, calcium, etc.) also in Nup-regulated neuronal cells, as reported for K^+ ionic channel in cardiomyocytes [16,17], cannot be excluded a priori and could be worth investigating. Nevertheless, the functional implication of Nup358 downregulation has been limited to single-cell excitability. Finally, the study of the role of endogenously expressed Nup358 in regulating or preventing cell hyperexcitability would be of great translational interest.

4.2. *Nup358 Downregulation: A Promising Window into Ion Channel Pathophysiology in Neurodegenerative Diseases?*

On the basis of our results, it would be interesting to examine the effect of the conditional depletion of Nup358 on the final output of a neuronal network to assess the potential contribution (i.e., the active role) to the pathogenesis and/or pathophysiology of NDDs.

Nowadays, it is certain that the intracellular machinery constituted by Nuclear-Pore-Complex–Nucleoporins–Nucleocytoplasmic Transport (NCP-Nup-NCT) is emerging as a key player in the cell alterations as well as in the pathophysiology of NDDs [2–4,31]. Although the cause–effect relationship between NCT defects and NDD pathogenesis has still not been fully assessed, several lines of evidence lead to the conclusion that they are undoubtedly associated with many neurodegenerative disorders [23,31]. The link between the alterations of the aforementioned machinery and the pathophysiology of NDDs has been demonstrated at molecular and cellular levels, e.g., in AD (Alzheimer's disease) [25,32]; ALS (amyotrophic lateral sclerosis) [33,34], encephalopathy [35], and, very recently, primary tauopathies [36].

Nevertheless, how the NPC-Nup-NCT machinery alterations influence neuronal excitability remains less understood. Accordingly, our electrophysiological results obtained in non-pathological cell models, to our knowledge, are the first ones in the field to demonstrate an active role of Nup358 in modulating the neuronal activity in native mouse cortical neurons, mainly through the gating of voltage-gated sodium channels.

Consequently, we are confident in hypothesizing and verifying that our results obtained in native/NUP downregulated cortical cells could be a proof of concept to verify if common biophysical and pathophysiological mechanisms related to the altered neuronal excitability via Nup358 downregulation could also occur in NDDs affected neuronal cell models.

5. Conclusions

Taken together, our data show that Nup expression tunes the cell membrane excitability in non-pathological cortical cells mainly acting on the gating of voltage-gated Na channels.

Although they were obtained on non-pathological cell models, we hope that our findings might constitute seminal results towards the functional study of Nup-related defects, which will need future experiments. Succeeding in this, the electrical correlate of altered neuron membrane excitability due to ion channel dysregulation, as already effectively proposed as a putative target/marker of the disease [18–20,37], could be extended at the NPC-Nup-NCT machinery level. Accordingly, our future goals will comprise the developing of ion channel-targeted neurotherapies based on the Nup pathogenetic cascade, according to modern pharmacogenetics and personalized pharmacology guidelines [38,39] established for neurological disorders.

Author Contributions: Conceptualization, V.A.M.-R., G.P., P.M. and C.M.; methodology, V.A.M.-R., G.P., P.M. and C.M.; software, V.A.M.-R., F.P. and I.R.-M.; validation, G.P., P.M. and C.M.; formal analysis, V.A.M.-R., I.R.-M. and C.M.; investigation, V.A.M.-R., F.P. and B.K.; resources, P.M., G.P. and C.M.; data curation, V.A.M.-R., F.P., P.M. and C.M.; writing—original draft preparation, V.A.M.-R. and C.M.; writing—review and editing, V.A.M.-R., P.M. and C.M.; visualization, V.A.M.-R., F.P., I.R.-M. and B.K.; supervision, P.M. and C.M.; project administration, P.M. and C.M.; funding acquisition, P.M. and C.M. All authors have read and agreed to the published version of the manuscript.

Funding: This research was partially funded by Fondazione CARITRO, Trento, Italy (to CM and VAMR). VAMR was granted by Young Post-Doc Call 2018, Fondazione CARITRO, Trento, Italy.

Institutional Review Board Statement: Animal care and experimental procedures were conducted in accordance with the University of Trento ethic committee and approved by the Italian Ministry of Health, following the European Union guidelines adopted by the Italian Ministry of Health (approval code: 162/2022-PR, approval date: 1 March 2022, based on Italian Law Art. 31, D.lgs. 26/2014).

Data Availability Statement: Relevant data are contained within the article.

Acknowledgments: We are grateful to Elena Gerola (IBF-CNR Trento, I) for her administrative support and fund management. We warmly thank Michael Whalen (IBF-CNR, Trento, I) for the accurate revision of the English and comments on the final version of the manuscript.

Conflicts of Interest: The authors declare no conflict of interest.

References

1. Beck, M.; Hurt, E. The nuclear pore complex: Understanding its function through structural insight. *Nat. Rev. Mol. Cell Biol.* **2016**, *18*, 73–89. [CrossRef] [PubMed]
2. Grossman, E.; Medalia, O.; Zwerger, M. Functional architecture of the nuclear pore complex. *Annu. Rev. Biophys.* **2012**, *41*, 557–584. [CrossRef] [PubMed]
3. Ferreira, P.A. The coming-of-age of nucleocytoplasmic transport in motor neuron disease and neurodegeneration. *Cell. Mol. Life Sci.* **2019**, *76*, 2247–2273. [CrossRef] [PubMed]
4. Coyne, A.N.; Rothstein, J.D. Nuclear pore complexes—A doorway to neural injury in neurodegeneration. *Nat. Rev. Neurol.* **2022**, *18*, 348–362. [CrossRef] [PubMed]
5. Bodoor, K.; Shaikh, S.; Salina, D.; Raharjo, W.H.; Bastos, R.; Lohka, M.; Burke, B. Sequential recruitment of npc proteins to the nuclear periphery at the end of mitosis. *J. Cell Sci.* **1999**, *112*, 2253–2264. [CrossRef] [PubMed]
6. Chatel, G.; Fahrenkrog, B. dynamics and diverse functions of nuclear pore complex proteins. *Nucleus* **2012**, *3*, 162–171. [CrossRef] [PubMed]

7. Guan, T.; Kehlenbach, R.H.; Schirmer, E.C.; Kehlenbach, A.; Fan, F.; Clurman, B.E.; Arnheim, N.; Gerace, L. Nup50, a nucleoplasmically oriented nucleoporin with a role in nuclear protein export. *Mol. Cell. Biol.* **2000**, *20*, 5619–5630. [CrossRef] [PubMed]
8. Rabut, G.; Doye, V.; Ellenberg, J. Mapping the dynamic organization of the nuclear pore complex inside single living cells. *Nat. Cell Biol.* **2004**, *6*, 1114–1121. [CrossRef]
9. Bernad, R.; van der Velde, H.; Fornerod, M.; Pickersgill, H. Nup358/RanBP2 attaches to the nuclear pore complex via association with Nup88 and Nup214/CAN and plays a supporting role in CRM1-mediated nuclear protein export. *Mol. Cell. Biol.* **2004**, *24*, 2373–2384. [CrossRef]
10. Goldberg, M.W. Nuclear pore complex tethers to the cytoskeleton. *Semin. Cell Dev. Biol.* **2017**, *68*, 52–58. [CrossRef]
11. Vyas, P.; Singh, A.; Murawala, P.; Joseph, J. Nup358 interacts with dishevelled and APKC to regulate neuronal polarity. *Biol. Open* **2013**, *2*, 1270–1278. [CrossRef] [PubMed]
12. Cho, K.I.; Yoon, D.; Qiu, S.; Danziger, Z.; Grill, W.M.; Wetsel, W.C.; Ferreira, P.A. Loss of Ranbp2 in motoneurons causes disruption of nucleocytoplasmic and chemokine signaling, proteostasis of HnRNPH3 and Mmp28, and development of amyotrophic lateral sclerosis-like syndromes. *Dis. Models Mech.* **2017**, *10*, 559–579. [CrossRef]
13. Cho, K.; Yoon, D.; Yu, M.; Peachey, N.S.; Ferreira, P.A. Microglial activation in an amyotrophic lateral sclerosis-like model caused by Ranbp2 loss and nucleocytoplasmic transport impairment in retinal ganglion neurons. *Cell. Mol. Life Sci.* **2019**, *76*, 3407–3432. [CrossRef] [PubMed]
14. Khalaf, B.; Roncador, A.; Pischedda, F.; Casini, A.; Thomas, S.; Piccoli, G.; Kiebler, M.; Macchi, P. Ankyrin-G induces nucleoporin Nup358 to associate with the axon initial segment of neurons. *J. Cell Sci.* **2019**, *132*, jcs222802. [CrossRef] [PubMed]
15. Hamdan, H.; Lim, B.C.; Torii, T.; Joshi, A.; Konning, M.; Smith, C.; Palmer, D.J.; Ng, P.; Leterrier, C.; Oses-Prieto, J.A.; et al. Mapping axon initial segment structure and function by multiplexed proximity biotinylation. *Nat. Comm.* **2020**, *11*, 100. [CrossRef] [PubMed]
16. Gao, X.; Yu, S.; Guan, Y.; Shen, Y.; Xu, L. Nucleoporin 50 mediates Kcna4 transcription to regulate cardiac electrical activity. *J. Cell Sci.* **2021**, *134*, jcs.256818. [CrossRef] [PubMed]
17. Guan, Y.; Gao, X.; Tang, Q.; Huang, L.; Gao, S.; Yu, S.; Huang, J.; Li, J.; Zhou, D.; Zhang, Y.; et al. Nucleoporin 107 facilitates the nuclear export of Scn5a mRNA to regulate cardiac bioelectricity. *J. Cell. Mol. Med.* **2019**, *23*, 1448–1457. [CrossRef] [PubMed]
18. Jimenez-Garduño, A.M.; Juarez-Hernandez, L.J.; Polanco, M.J.; Tosatto, L.; Michelatti, D.; Arosio, D.; Basso, M.; Pennuto, M.; Musio, C. Altered ionic currents and amelioration by IGF-1 and PACAP in motoneuron-derived cells modelling SBMA. *Biophys. Chem.* **2017**, *229*, 68–76. [CrossRef]
19. Martínez-Rojas, V.A.; Jiménez-Garduño, A.M.; Michelatti, D.; Tosatto, L.; Marchioretto, M.; Arosio, D.; Basso, M.; Pennuto, M.; Musio, C. ClC-2-like chloride current alterations in a cell model of spinal and bulbar muscular atrophy, a polyglutamine disease. *J. Mol. Neurosci.* **2021**, *71*, 662–674. [CrossRef]
20. Martínez-Rojas, V.A.; Arosio, D.; Pennuto, M.; Musio, C. Clenbuterol-sensitive delayed outward potassium currents in a cell model of spinal and bulbar muscular atrophy. *Pflügers Arch. Eur. J. Physiol.* **2021**, *473*, 1213–1227. [CrossRef]
21. Zona, C.; Pieri, M.; Carunchio, I. Voltage-dependent sodium channels in spinal cord motor neurons display rapid recovery from fast inactivation in a mouse model of amyotrophic lateral sclerosis. *J. Neurophysiol.* **2006**, *96*, 3314–3322. [CrossRef] [PubMed]
22. Zhong, G.; He, J.; Zhou, R.; Lorenzo, D.; Babcock, H.P.; Bennett, V.; Zhuang, X. Developmental mechanism of the periodic membrane skeleton in axons. *Elife* **2014**, *3*, e04581. [CrossRef]
23. Hutten, S.; Dormann, D. Nucleocytoplasmic transport defects in neurodegeneration—Cause or consequence? *Semin. Cell Dev. Biol.* **2020**, *99*, 151–162. [CrossRef] [PubMed]
24. Hille, B. *Ionic Channels of Excitable Membrane*, 3rd ed.; Oxford University Press: Oxford, UK, 2001; pp. 1–814.
25. Leone, L.; Colussi, C.; Gironi, K.; Longo, V.; Fusco, S.; Li Puma, D.D.; D'Ascenzo, M.; Grassi, C. Altered Nup153 expression impairs the function of cultured hippocampal neural stem cells isolated from a mouse model of alzheimer's disease. *Mol. Neurobiol.* **2019**, *56*, 5934–5949. [CrossRef] [PubMed]
26. Cho, K.I.; Searle, K.; Webb, M.; Yi, H.; Ferreira, P.A. Ranbp2 haploinsufficiency mediates distinct cellular and biochemical phenotypes in brain and retinal dopaminergic and glia cells elicited by the parkinsonian neurotoxin, 1-Methyl-4-Phenyl-1,2,3,6-Tetrahydropyridine (MPTP). *Cell. Mol. Life Sci.* **2012**, *69*, 3511–3527. [CrossRef] [PubMed]
27. Ogawa, Y.; Rasband, M.N. Endogenously expressed Ranbp2 is not at the axon initial segment. *J. Cell Sci.* **2021**, *134*, jcs256180. [CrossRef] [PubMed]
28. Grubb, M.S.; Burrone, J. Activity-dependent relocation of the axon initial segment fine-tunes neuronal excitability. *Nature* **2010**, *465*, 1070–1074. [CrossRef] [PubMed]
29. Katz, E.; Stoler, O.; Scheller, A.; Khrapunsky, Y.; Goebbels, S.; Kirchhoff, F.; Gutnick, M.J.; Wolf, F.; Fleidervish, I.A. Role of sodium channel subtype in action potential generation by neocortical pyramidal neurons. *Proc. Natl. Acad. Sci. USA* **2018**, *115*, E7184–E7192. [CrossRef]
30. Yang, J.; Xiao, Y.; Li, L.; He, Q.; Li, M.; Shu, Y. Biophysical properties of somatic and axonal voltage-gated sodium channels in midbrain dopaminergic neurons. *Front. Cell. Neurosci.* **2019**, *13*, 317. [CrossRef] [PubMed]
31. Liu, J.; Hetzer, M.W. Nuclear pore complex maintenance and implications for age-related diseases. *Trends Cell Biol.* **2022**, *32*, 216–227. [CrossRef]

32. Sheffield, L.G.; Miskiewicz, H.B.; Tannenbaum, L.B.; Mirra, S.S. Nuclear pore complex proteins in Alzheimer disease. *J. Neuropathol. Exp. Neurol.* **2006**, *65*, 45–54. [CrossRef] [PubMed]
33. Fallini, C.; Khalil, B.; Smith, C.L.; Rossoll, W. Traffic jam at the nuclear pore: All roads lead to nucleocytoplasmic transport defects in ALS/FTD. *Neurobiol. Dis.* **2020**, *140*, 104835. [CrossRef] [PubMed]
34. Chandra, S.; Lusk, C.P. Emerging connections between nuclear pore complex homeostasis and ALS. *Int. J. Mol. Sci.* **2022**, *23*, 1329. [CrossRef] [PubMed]
35. Fichtman, B.; Harel, T.; Biran, N.; Zagairy, F.; Applegate, C.D.; Salzberg, Y.; Gilboa, T.; Salah, S.; Shaag, A.; Simanovsky, N.; et al. Pathogenic variants in NUP214 cause "plugged" nuclear pore channels and acute febrile encephalopathy. *Am. J. Hum. Genet.* **2019**, *105*, 48–64. [CrossRef] [PubMed]
36. Dickson, J.R.; Frosch, M.P.; Hyman, B.T. Altered localization of nucleoporin 98 in primary tauopathies. *Brain Commun.* **2022**, *5*, fcac334. [CrossRef] [PubMed]
37. Huang, H.; Shakkottai, V.G. Targeting Ion channels and Purkinje neuron intrinsic membrane excitability as a therapeutic strategy for cerebellar ataxia. *Life* **2023**, *13*, 1350. [CrossRef] [PubMed]
38. Cacabelos, R. New paradigms in pharmaceutical development. *Life* **2022**, *12*, 1433. [CrossRef] [PubMed]
39. Cacabelos, R.; Naidoo, V.; Martínez-Iglesias, O.; Corzo, L.; Cacabelos, N.; Pego, R.; Carril, J.C. Personalized management and treatment of Alzheimer's disease. *Life* **2022**, *12*, 460. [CrossRef]

MDPI AG
Grosspeteranlage 5
4052 Basel
Switzerland
Tel.: +41 61 683 77 34

Life Editorial Office
E-mail: life@mdpi.com
www.mdpi.com/journal/life

www.ingramcontent.com/pod-product-compliance
Lightning Source LLC
Chambersburg PA
CBHW061500210926
51850CB00008B/1774